VOLUME SIX HUNDRED AND FIFTY

METHODS IN ENZYMOLOGY

Rare-Earth Element Biochemistry: Methanol Dehydrogenases and Lanthanide Biology

METHODS IN ENZYMOLOGY

VOLUME SIX HUNDRED AND FIFTY

METHODS IN ENZYMOLOGY

Rare-Earth Element Biochemistry: Methanol Dehydrogenases and Lanthanide Biology

Edited by

JOSEPH A. COTRUVO, Jr.
Department of Chemistry,
The Pennsylvania State University,
University Park, PA, United States

ELSEVIER

ACADEMIC PRESS
An imprint of Elsevier

Academic Press is an imprint of Elsevier
50 Hampshire Street, 5th Floor, Cambridge, MA 02139, United States
525 B Street, Suite 1650, San Diego, CA 92101, United States
The Boulevard, Langford Lane, Kidlington, Oxford OX5 1GB, United Kingdom
125 London Wall, London, EC2Y 5AS, United Kingdom

First edition 2021

Notices
Knowledge and best practice in this field are constantly changing. As new research and experience broaden our understanding, changes in research methods, professional practices, or medical treatment may become necessary.

Practitioners and researchers must always rely on their own experience and knowledge in evaluating and using any information, methods, compounds, or experiments described herein. In using such information or methods they should be mindful of their own safety and the safety of others, including parties for whom they have a professional responsibility.

To the fullest extent of the law, neither the Publisher nor the authors, contributors, or editors, assume any liability for any injury and/or damage to persons or property as a matter of products liability, negligence or otherwise, or from any use or operation of any methods, products, instructions, or ideas contained in the material herein.

ISBN: 978-0-12-823856-1
ISSN: 0076-6879

For information on all Academic Press publications
visit our website at https://www.elsevier.com/books-and-journals

Publisher: Zoe Kruze
Developmental Editor: Federico Paulo S. Mendoza
Production Project Manager: James Selvam
Cover Designer: Alan Studholme

Typeset by SPi Global, India

Contents

Contributors

Pavlo Bohutskyi
Pacific Northwest National Laboratory, Richland, WA, United States

Ludmila Chistoserdova
Department of Chemical Engineering, University of Washington, Seattle, WA, United States

Joseph A. Cotruvo Jr.
Department of Chemistry, The Pennsylvania State University, University Park, PA, United States

Ralph Valentine Crisostomo
Department of Biological Sciences, San José State University, San José, CA, United States

Lena J. Daumann
Department of Chemistry, Ludwig-Maximilians-Universität München, München, Germany

Emily R. Featherston
Department of Chemistry, The Pennsylvania State University, University Park, PA, United States

Nathan M. Good
Department of Plant and Microbial Biology, University of California, Berkeley, CA, United States

Joseph D. Groom
Department of Chemical Engineering, University of Washington, Seattle, WA, United States

Sophie M. Gutenthaler
Department of Chemistry, Ludwig-Maximilians-Universität München, München, Germany

Bodo Hattendorf
Laboratory of Inorganic Chemistry, Department of Chemistry and Applied Biosciences, ETH Zurich, Zurich, Switzerland

Lucas Hemmerle
Institute of Microbiology, Department of Biology, ETH Zurich, Zurich, Switzerland

Jing Huang
Department of Parasitology, Xiangya School of Medicine, Central South University, Changsha, Hunan, China

Zachary J. Johnson
Department of Biology, San Diego State University, San Diego, CA, United States

Marina G. Kalyuzhnaya
Department of Biology, San Diego State University, San Diego, CA, United States

Alison L. Knasin
P. Roy and Diana T. Vagelos Laboratories, Department of Chemistry, University of Pennsylvania, Philadelphia, PA, United States

Dennis D. Krutkin
Department of Biology, San Diego State University, San Diego, CA, United States

Mary E. Lidstrom
Department of Chemical Engineering; Department of Microbiology, University of Washington, Seattle, WA, United States

N. Cecilia Martinez-Gomez
Department of Plant and Microbial Biology, University of California, Berkeley, CA, United States

Joseph A. Mattocks
Department of Chemistry, The Pennsylvania State University, University Park, PA, United States

Ryoji Mitsui
Department of Biochemistry, Faculty of Science, Okayama University of Science, Okayama, Japan

Tomoyuki Nakagawa
Faculty of Applied Biological Sciences, Gifu University, Gifu, Japan

Andrea M. Ochsner
Institute of Microbiology, Department of Biology, ETH Zurich, Zurich, Switzerland

Manh Tri Phi
Department of Chemistry, Ludwig-Maximilians-Universität München, München, Germany

Eric J. Schelter
P. Roy and Diana T. Vagelos Laboratories, Department of Chemistry, University of Pennsylvania, Philadelphia, PA, United States

Helena Singer
Department of Chemistry, Ludwig-Maximilians-Universität München, München, Germany

Elizabeth Skovran
Department of Biological Sciences, San José State University, San José, CA, United States

Gabriel A. Subuyuj
Department of Biological Sciences, San José State University, San José, CA, United States

Akio Tani
Institute of Plant Science and Resources, Okayama University, Okayama, Japan

Jonathan L. Tirsch
Department of Chemistry, The Pennsylvania State University, University Park, PA, United States

Thomas Vonderach
Laboratory of Inorganic Chemistry, Department of Chemistry and Applied Biosciences, ETH Zurich, Zurich, Switzerland

Julia A. Vorholt
Institute of Microbiology, Department of Biology, ETH Zurich, Zurich, Switzerland

Huong N. Vu
Department of Biological Sciences, San José State University, San José, CA, United States

Zheng Yu
Department of Microbiology, Xiangya School of Medicine, Central South University, Changsha, Hunan, China

Yue Zheng
CAS Key Laboratory of Urban Pollutant Conversion, Institute of Urban Environment, Chinese Academy of Sciences, Xiamen, China

Preface

Until recently, "rare-earth element biochemistry" largely constituted (1) utilization of the 15 lanthanides (lanthanum to lutetium), plus yttrium and scandium, as nonnative, spectroscopic probes for binding sites of other metals, and (2) small-molecule metal complexes used as biological imaging agents (e.g., MRI contrast or luminescence-based probes). While these studies continue to be important, the discovery in 2011 that bacteria that utilize one-carbon compounds like methane and methanol—methylotrophs—specifically incorporate certain lanthanides in a methanol dehydrogenase enzyme of previously cryptic function has ushered in a new era of lanthanides in biology and biochemistry.

In the last decade, this field has blossomed into several interconnected branches. First, on a *systems* level, the catalog of methylotrophic and nonmethylotrophic organisms encoding lanthanide-dependent processes is rapidly expanding. Identifying and studying lanthanide utilization in these organisms has important implications for ecology, geochemistry, understanding of microbial communities and their interactions with other organisms (e.g., plants), and more. Second, a field called "*lanthanomics*" has emerged, aimed at identifying the molecular players involved in lanthanide utilization in these organisms (the "lanthanome") and how they fit together. This area encompasses identifying and characterizing lanthanoenzymes and other lanthanoproteins, understanding lanthanide-dependent regulation of methylotrophy and other processes, and probing lanthanide homeostasis mechanisms (acquisition, trafficking, utilization, storage). Third, while microbial carbon metabolism is critical to the global carbon cycle, agriculture, and much more, *f*-elements themselves have many important *technological applications*. Characterization of the systems involved in biological lanthanide utilization has great potential to revolutionize these fields: from developing new environmental and chemical biological detection tools, to more sustainable extraction and separations of rare earths and actinides, to medical imaging and therapeutic agents, to new catalysts, and more. The discoveries of lanthanide biochemistry have already begun to have an impact for these applications.

The goal of these two volumes of *Methods in Enzymology*, Vol. 650 and 651, is to provide detailed methods related to interactions of rare earths with biological systems, spanning the research directions mentioned above, from fundamental biology of methylotrophs to emerging, far-reaching applications of this exciting field. Broadly, the first volume is focused on understanding the lanthanide-dependent processes of methylotrophs—genetic methods, protein expression and purification, activity assays of alcohol dehydrogenases, and model chemistry. The second volume is focused on the biochemical and biophysical characterization of natural and non-natural lanthanide-binding biomolecules, as well as emerging ways in which rare earths are being applied to medical imaging, therapeutics, screening, and sensing.

In this first volume, we begin with a contribution from some of the researchers who made the initial discoveries that lanthanides are used specifically by methylotrophic bacteria. Nakagawa et al. give a historical overview of that discovery, along with providing methods for isolating lanthanide-dependent bacteria and studying their lanthanide-dependent methanol dehydrogenases (Chapter 1).

In Chapter 2, Knasin and Schelter describe the synthesis and characterization of the first model complex for a lanthanide-dependent methanol dehydrogenase. Many questions remain about the enzymatic mechanism of alcohol oxidation, and studies of this complex and others will provide important insights into this chemistry, along with similarities and differences versus the calcium-dependent homologs.

In Chapter 3, Daumann et al. update the classic Day and Anthony *Methods in Enzymology* article from 1990 describing both dye-linked and *c*-type cytochrome-linked activity assays of methanol dehydrogenases.

In Chapter 4, Chistoserdova et al. describe protocols for expressing and characterizing various alcohol dehydrogenases and the associated *c*-type cytochrome, XoxG, in addition to reviewing and tabulating the catalytic properties of many of these enzymes.

In Chapter 5, Good and Martinez-Gomez discuss the expression of lanthanide-dependent enzymes in *Methylorubrum extorquens* AM1, a model methylotrophic bacterium.

In Chapter 6, our laboratory discusses protocols, including two procedures presented for the first time, for *E. coli*-based expression, purification, and characterization of several proteins involved in lanthanide homeostasis and utilization, including lanmodulin, XoxG, and others.

In Chapter 7, Skovran et al. describe the use of transposon mutagenesis to screen for genes involved in lanthanide-dependent physiology, a strategy that can be extended to elucidation of genes involved in similar processes in other organisms.

In Chapter 8, Kalyuzhnaya et al. describe the design and analysis of RNA sequencing studies to understand conditional gene expression profiles, as applied to lanthanide-utilizing methylotrophs.

Many methylotrophs are associated with plants (the phyllosphere), and the long-recognized positive effect of lanthanides on plant growth may be attributable, at least in part, to lanthanide utilization by the associated bacteria. In Chapter 9, Vorholt et al. present mass spectrometry-based proteomics of phyllosphere bacteria, using *Arabidopsis thaliana* as a model plant system, as well as inductively coupled plasma mass spectrometry (ICP-MS) methods for quantifying rare earths.

Finally, in Chapter 10, Groom and Lidstrom present protocols for culture and metal uptake studies in the methanotrophic bacterium, *Methylotuvimicrobium buryatense* 5GB1C, which, unlike many of the other bacteria discussed in this volume, grows optimally at alkaline pH and high salt concentrations.

I would like to thank the editors-in-chief of *Methods in Enzymology*, David Christianson and Anna Pyle, for inviting me to edit these volumes, as well as the editorial staff, especially Zoe Kruze, Shellie Bryant, and Paulo Mendoza, for keeping the authors and me on track. I am also indebted to all of the contributors for their hard work on their chapters. Any volume like this is necessarily incomplete; unfortunately, a few important players in the field were unable to contribute due to the difficult circumstances of the past year. Nevertheless, I am delighted that these volumes highlight not only well-established investigators but also many young investigators, all of whom are propelling this new field forward.

I hope that the dissemination of the methods included in these volumes will bring new researchers into this exciting field, accelerate the pace of discovery, and lead to new research directions. While these volumes may constitute the first compendium of methods to study the native biology, biochemistry, and bio-applications of rare earth elements, I am sure that it will not be the last. I look forward to the next decade of rare-earth element biochemistry!

Joseph A. Cotruvo, Jr.
Department of Chemistry
The Pennsylvania State University
University Park, PA, United States

CHAPTER ONE

Discovery of lanthanide-dependent methylotrophy and screening methods for lanthanide-dependent methylotrophs

Akio Tani[a,†], Ryoji Mitsui[b,†], and Tomoyuki Nakagawa[c,*,†]
[a]Institute of Plant Science and Resources, Okayama University, Okayama, Japan
[b]Department of Biochemistry, Faculty of Science, Okayama University of Science, Okayama, Japan
[c]Faculty of Applied Biological Sciences, Gifu University, Gifu, Japan
[*]Corresponding author: e-mail address: t_nakaga@gifu-u.ac.jp

Contents

[†] All authors contributed equally to this work.

Methods in Enzymology, Volume 650
ISSN 0076-6879
https://doi.org/10.1016/bs.mie.2021.01.031

Abstract

The lanthanide elements (Lns) affect the physiology and growth of certain microorganisms known as "Ln-responsive microorganisms." Among them, in 2011, it was first reported that strains of *Methylobacterium* exhibited high methanol dehydrogenase (MDH) activity when grown in the presence of Lns; the purified Ln-inducible MDH was identified as XoxF-type MDH, whose catalytic function had previously been unknown. XoxF was the first enzyme to be identified as Ln-dependent, and its function in methylotrophy is more fundamental and important than that of the corresponding Ca^{2+}-dependent MDH MxaFI. XoxF is encoded in the genomes of methylotrophic as well as non-methylotrophic bacteria. Thus, Lns are among the most fascinating and important growth factors for bacteria that potentially utilize methanol. Bacteria that require Lns for methanol growth are called "Ln-dependent methylotrophs." Recent findings indicate that these microorganisms comprise an "Ln-dependent ecosystem" that we have not been able to reconstruct under laboratory conditions without Lns. In this chapter, we summarize methods for (1) screening of Ln-responsive microorganisms, (2) purification of native XoxFs from Ln-dependent methylotrophs, and (3) screening of Ln-dependent methylotrophs from natural environments, while providing a history of the discovery of the Ln-dependent methylotrophs.

1. Introduction

The lanthanides (Lns) consist of 15 distinct but chemically similar elements. Together with Sc^{3+} and Y^{3+}, they are also called the rare earth elements. Lns are abundantly distributed in soils, though Lns with relatively low atomic weights, namely, La^{3+}, Ce^{3+}, Pr^{3+}, and Nd^{3+}, are generally more abundant than the heavier Lns (Aide, 2019). Among the light-Lns (*L*-Ln), Ce^{3+} is the most abundant in the soil, with a concentration of over 60 mg/kg of soil (Aide, 2019). These elements are known to be important for modern high-tech industries, but their roles in biological systems were long not typically considered, since they were not believed to be essential to any biological activities.

A decade ago it was reported that Ln-responsive microorganisms, including *Methylobacterium*, *Bradyrhizobium*, *Streptomyces*, and other genera, exhibit specific phenotypes in the presence of Lns (Fitriyanto, Nakamura, et al., 2011; Hibi, Okuda, Sakuma, Iwama, & Kawai, 2011; Kamijo, Suzuki, Kawai, Fujii, & Murase, 1999). In *Streptomyces*, Lns induce the production of secondary metabolites such as antibiotics (Kawai, Wang, Okamoto, & Ochi, 2007); in *Bradyrhizobium*, Lns induce the production of *exo*-polysaccharides (Fitriyanto, Nakamura, et al., 2011). *Methylobacterium* strains also form large colonies in the presence of Lns, even on low-nutrient medium (Hibi, Okuda, et al., 2011).

In *Methylobacterium* and *Bradyrhizobium* strains, Lns were found to be an essential cofactor for XoxF-type methanol dehydrogenase (MDH), whose function had previously been a mystery (Fitriyanto, Fushimi, et al., 2011; Hibi, Asai, et al., 2011; Nakagawa et al., 2012). Furthermore, it was proved that XoxF acts as a key enzyme in Ln-dependent methylotrophy [in this chapter we use "methylotrophy" as the ability to grow on methanol in a narrow meaning, whereas it is originally defined as the ability to grow on reduced carbon compounds containing one or more carbon atoms but containing no carbon-carbon bonds (Anthony, 1982)]. Based on these discoveries, the microorganisms requiring Lns for their methylotrophy were defined as "Ln-dependent methylotrophs."

Since typical microbial media do not contain Lns, the Lns have been an overlooked but important factor, inducing cryptic microbial functions that are not evident in typical laboratory environments. Therefore, it is now expected that there are novel types of microorganisms using Lns to promote their activities in natural environments. To contribute to the development of "*Ln biochemistry*" as an emerging field, we have developed methods of screening for novel Ln-dependent methylotrophs. Using these methods, we have discovered new genera and species that require the presence of various Lns to grow on methanol.

In this chapter, we summarize methods of screening for Ln-responsive microorganisms and Ln-dependent methylotrophs and our methods of purifying native XoxFs from Ln-dependent methylotrophs, while presenting a history of the discovery of Ln-dependent methylotrophs.

2. Screening for Ln-responsive microorganisms

The research group of Professor Keiichi Kawai (Gifu University, Japan) aimed to clarify the effects of Lns on the physiology of microorganisms by identifying microorganisms with unique growth behaviors on a low-nutrient plate medium containing Lns. Through their screening experiments, they discovered microbial strains that exhibited unique phenotypes only in the presence of Lns, such as *Streptomyces* sp. YB-1 (Kamijo et al., 1999), *Methylobacterium* sp. MAFF 211642 (Hibi, Okuda, et al., 2011) and *Bradyrhizobium* sp. Ce-3 (= MAFF 211645) (Fitriyanto, Nakamura, et al., 2011). In this section, we summarize the screening methods for, and important properties of, the isolated Ln-responsive microorganisms.

2.1 Media and cultivation conditions for the screening of Ln-responsive microorganisms

In screening for Ln-responsive microorganisms, a 100-fold diluted nutrient medium (1/100 N medium) was used as the screening medium, mimicking an oligotrophic natural environment. The 1/100 N medium consisted of 0.005% meat extract, 0.01% peptone and 0.005% NaCl. This medium was supplemented with 30 μM (final concentration) of each of several Lns. Ln compounds are readily available from chemical companies as chlorides ($LnCl_3$) and oxides (Ln_2O_3). As Ln_2O_3 is almost insoluble in water, $LnCl_3$ is more suitable for use as an Ln source, although oxides are also reported to work as Ln sources for methanotrophic bacteria, which are able to utilize methane as a carbon source (Kato, Takashino, Igarashi, & Kitagawa, 2020). Bacteria were cultivated at 30 °C for 5–7 days using solid media.

Fig. 1 depicts the Kawai group's strategy for screening for Ln-responsive microorganisms. Soil, water, or plant samples are suspended in sterile water. For soil and plant samples, the addition of 0.1% Silwet L-77 (PhytoTechnology Laboratories, Shawnee Mission, KS, USA) may facilitate bacterial cell removal from the materials. The sample suspensions should be diluted 10^2 to 10^5 times with distilled water and spread onto the 1/100 N/Ln plates. Because nutrients, cofactors, and growth inhibitors may be exchanged among bacterial colonies, dilution is important for isolation. The plates are incubated at 30 °C for 5–7 days. The colonies are then picked up using 10P chips, loops, or sterile toothpicks, and streaked onto new 1/100 N/Ln plates for at least two to three rounds to confirm their purity. As is often the case with isolation of bacteria, special care is needed to verify colony purity, as microcolonies that are visible only through microscopy may be associated with the main colonies. Growth of the isolates is then tested on both 1/100 N and 1/100 N/Ln plates, and the isolates that exhibit different phenotypes on the two plate types are selected.

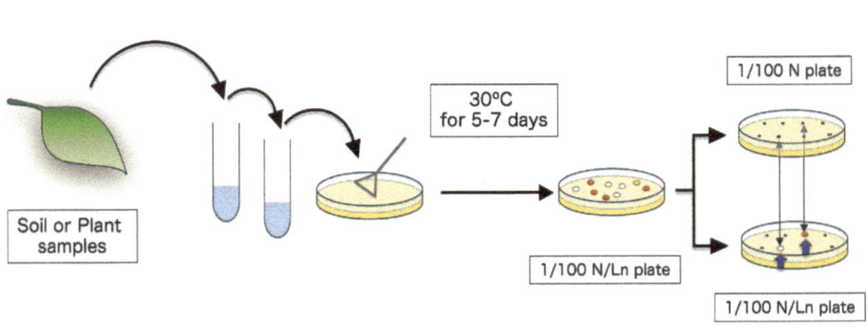

Fig. 1 Screening strategy for isolation of Ln-responsive microorganisms from the natural environment.

2.2 Ln-responsive actinomycetes

Professor Kawai and his research group set out to identify oligotrophic microorganisms capable of decreasing the concentration of 30 μM Yb^{3+} in the 1/100 N medium (Kamijo et al., 1999). They successfully isolated five strains which they identified as Yb^{3+}-decreasing oligotrophic microorganisms. Among them, *Streptomyces* sp. YB-1, isolated from soil, produced a reddish-purple pigment in an Yb^{3+}-dependent manner. This pigment was identified as a kind of naphthoquinone, which has a structure similar to that of nanaomycin (Kamijo et al., 1999). Subsequent to this discovery, several strains in genus *Streptomyces* have been reported to overproduce antibiotics in the presence of Ln; for example, actinorhodin is produced by *S. coelicolor*, actinomycin by *S. antibioticus*, and streptomycin by *S. griseus* (Kawai et al., 2007). Lns also activate expression of the secondary metabolite-biosynthetic gene clusters in *S. coelicolor* (Tanaka, Hosaka, & Ochi, 2010).

2.3 Ln-responsive methylotrophs

Three Pink-Pigmented Facultative Methylotroph (PPFM) strains, which formed larger colonies on the 1/100 N/Eu^{3+} plate, were also isolated as Ln-responsive microorganisms. These strains were identified as *Methylobacterium* sp. (Hibi, Okuda, et al., 2011). Among them, strain MAFF 211642 showed a 1.6-fold larger colony diameter when grown in the presence of Sm^{3+} versus in the absence of any Ln (Hibi, Okuda, et al., 2011). Fig. 2A shows an example of another PPFM strain that we isolated. This strain, identified as *Methylorubrum* sp. strain GM77, produces larger colonies on the 1/100 N/Sm^{3+} plate. This phenotype is not common among all *Methylobacterium* and *Methylorubrum* species; however, *Methylorubrum extorquens* strain AM1, for example, did not show this phenotype (unpublished data).

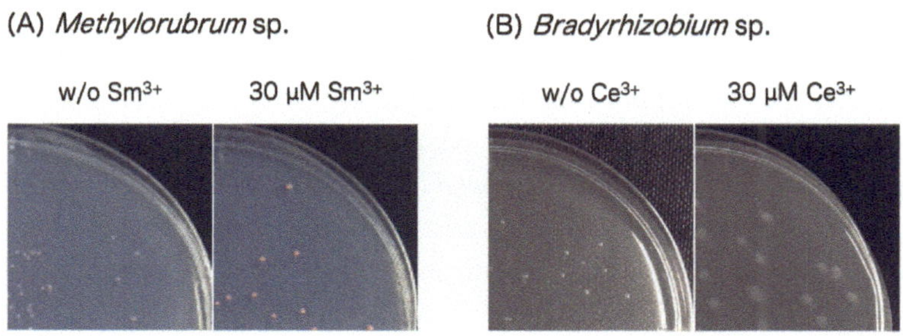

Fig. 2 Colony morphology of selected Ln-responsive microorganisms. (A) *Methylorubrum* sp. strain GM77. (B) *Bradyrhizobium* sp. strain Ce-3.

2.4 Ln-responsive rhizobia

In another experiment, slime production was induced in an isolate from soil in Ise City, Mie Prefecture, Japan, on 1/100 N/Ce^{3+} medium (Fig. 2B). The strain was identified as *Bradyrhizobium* sp. Ce-3 (= MAFF 211645), which belongs to *Bradyrhizobium* clade II (Fitriyanto, Nakamura, et al., 2011, Pastawan, Suganuma, et al., 2020). The slime produced by strain Ce-3 was identified as an exopolysaccharide (EPS) known as rhamnan, more than 95% of which is composed of L-rhamnose (Fitriyanto, Nakamura, et al., 2011). Several other strains that exhibit unique phenotypes in response to Lns have been isolated from natural environments by similar means.

3. Discovery of Ln-dependent MDH and Ln-dependent methylotrophy

The studies described above identified *Methylobacterium* and *Bradyrhizobium* species as Ln-responsive microorganisms. As the *Methylobacterium* species are well known to be methylotrophs, Professor Kawai thought that Lns might enhance the methanol metabolism of *Methylobacterium* isolates. Indeed, MDH activity was induced by Lns with methanol in *Methylobacterium* sp. strain MAFF 211642 (Hibi, Okuda, et al., 2011) and in *Methylobacterium radiotolerans* NBRC 15690^T. The MDH from *Methylobacterium radiotolerans* NBRC 15690^T cells grown in the presence of La^{3+} was purified to homogeneity, and the protein was identified as XoxF-type MDH based on its N-terminal amino acid sequence. Meanwhile, the enzyme purified from the strain grown without La^{3+} was MxaFI-type MDH (Hibi, Asai, et al., 2011). This was the very first report to suggest that XoxF is an Ln-responsive enzyme. In a follow-up study, we demonstrated that XoxF from *Methylorubrum extorquens* AM1 is an Ln-dependent MDH that contains Ln in the molecule. More importantly, we also showed that an *mxaF* mutant of the strain showed growth on methanol only in the presence of La^{3+} (Nakagawa et al., 2012). These results indicated that XoxF is an active Ln-containing MDH and that it supports methylotrophic growth depending on XoxF and Ln, even in the absence of MxaF. Later, the crystal structure of XoxF-type MDH from *Methylacidiphilum fumariolicum* SolV, an acidophilic methanotroph that grows only in water from the mudpot that is its original habitat or else with Ln supplementation, was identified, and the protein was shown to contain Ce^{3+} at its catalytic site (Pol et al., 2014).

As MDHs are highly expressed in methylotrophic bacteria, their purification is relatively straightforward. A simple purification method for MxaF-type MDH has been demonstrated already (Liu, Kirchhoff, Faehnle, Viola, & Hudson, 2006). In this section, we summarize the purification method for native Ln-dependent MDHs.

3.1 Purification of native Ln-inducible MDHs

Several native Ln-dependent MDHs have been purified from *Methylobacterium radiotolerans* NBRC 15690^T, *Methylorubrum extorquens* AM1, *Bradyrhizobium* sp. Ce-3, *B. diazoefficiens* USDA110, and *Methylacidiphilum fumariolicum* SolV (Cotruvo, Featherston, Mattocks, Ho, & Laremore, 2018; Fitriyanto, Fushimi, et al., 2011; Hibi, Asai, et al., 2011; Nakagawa et al., 2012; Pol et al., 2014; Wang et al., 2020, 2019). Here, we introduce an effective method for the purification of native Ln-dependent MDHs as an alternative to the methods already described.

The genome of *Methylorubrum extorquens* AM1 encodes Ca^{2+}-dependent MxaF and Ln-dependent XoxF1 as well as the XoxF1 paralogue XoxF2 and the Ln-dependent ethanol dehydrogenase ExaF (Good et al., 2016). For the purification of Ln-inducible MDH, Δ*xoxF2* is the optimal strain because, although most of the product is XoxF1, Δ*xoxF2* often permits the co-purification of XoxF2 as well (unpublished data). The concentration of Ln that was added to the medium was 30 μM. This concentration may have been too high compared to that in natural environments (~1 μM even in acidic environments; Pol et al., 2014); indeed, 0.1–1 μM Ln was later found to be sufficient to induce XoxF-type MDHs (Vu et al., 2016). In *Methylomonas* sp. LW13, which contains both types of MDH, 30–60 μM La^{3+} induces the expression of XoxF (Zheng, Huang, Zhao, & Chistoserdova, 2018). Because Ln^{3+} readily forms insoluble salts with phosphate, the effective Ln concentration depends on the medium composition and the solubilizing ability of the strain.

Approximately 10 g of wet cells of strain AM1 Δ*xoxF2* grown on 6 L of Hypho medium (Delaney et al., 2013) containing 0.5% methanol and 30 μM La^{3+} or Ce^{3+} were suspended in 100 mL 50 mM acetate buffer (pH 5.0) and disrupted using a BSP-1107900 Bead-Beater (Biospec Products, Bartlesville, OK, USA) or an Ultrasonic Disruptor UD-201 (Tomy Seiko Co., Ltd., Tokyo, Japan). The cell debris was removed by centrifugation (9000 × *g* at 4 °C for 15 min).

The cell-free extracts were fractionated by the addition of solid ammonium sulfate ($AmSO_4$, 70% saturation [472 g/L]) with stirring at 4 °C. MDH activity was recovered in the supernatant fraction, and the precipitated protein was removed by centrifugation (9000 × *g* at 4 °C for 15 min). MDH activity was measured by the dye-linked assay, and one unit of the activity was defined as the amount of protein that reduce 1 μmol DCIP (dichloroindophenol) per min, using its molar extinction coefficient at 600 nm (ε_{600}, 22 mM^{-1} cm^{-1}) (Wang et al., 2019). However, the dye-linked assay includes some known problems; (i) auto-bleaching (reduction of DCIP without addition of the substrate) of the reaction mixture (Anthony, 1982), and (ii) phenazine methosulfate (PMS) is used as artificial electron acceptor instead of *c*-type cytochrome as the physiological electron acceptor (Featherston et al., 2019; Versantvoort et al., 2019; Zheng et al., 2018). Therefore, it is difficult to determine the correct MDH activity by the method. Recently, a modified method for dye-linked MDH activity was reported (Jahn et al., 2020), which we believe the readers should follow.

To apply the active MDH sample to hydrophobic interaction chromatography, 1.2 mol of $AmSO_4$ per L was added to the supernatant, and the sample was loaded onto a Butyl TOYOPEARL 650S column (φ26 mm × 110 mm, Tosoh, Tokyo, Japan) pre-equilibrated with 25 mM acetate buffer containing 1.2 M $AmSO_4$ (pH 5.0). The protein was eluted by a stepwise decrease in $AmSO_4$ concentration from 1.0 to 0.5 M in 6 steps (flow rate, 2 mL/min). Active MDH fractions were eluted at 0.8–0.7 M $AmSO_4$. The fractions were collected and desalted by dialysis against 25 mM acetate buffer (pH 5.0), and the protein solution was loaded onto a HiTrap SP HP (1 mL) (GE Healthcare, Little Chalfont, UK) pre-equilibrated with 25 mM acetate buffer (pH 5.0). The protein was eluted by a linear increase in NaCl concentration from 0 to 500 mM (flow rate, 1 mL/min). Active MDH fractions were eluted at ca. 300 mM NaCl. The fractions were combined, desalted, and concentrated using an Amicon Ultra-50 K (EMD Millipore, Billerica, MA, USA). This method can also be used for the purification of Ln-inducible MDHs from *Bradyrhizobium* strains (Wang et al., 2019).

The Ln-inducible MDHs are dimers of the α-subunit, which is encoded by *xoxF1* or *xoxF* homologs. The purified MDHs showed high levels of MDH activity (Fitriyanto, Fushimi, et al., 2011; Hibi, Asai, et al., 2011) and contained Lns as cofactors (Nakagawa et al., 2012; Wang et al., 2019, 2020). The enzymatic properties of these native MDHs containing

La^{3+} or Ce^{3+} were similar to each other; for example, the K_m values (for methanol) of XoxFs from strains AM1 and Ce-3 were 67.7 and 67.0 μM, respectively, and the V_{max} values were 29.3 and 12.9 U/mg, respectively (Wang et al., 2020, 2019).

Recombinant Ln-inducible MDHs have also been purified from several microorganisms, and their properties are summarized in other chapters.

3.2 Discovery of Ln-dependent methylotrophy

Several research groups have sought to reveal the functions of XoxF in *Rhodobacter sphaeroides* (Wilson, Gleisten, & Donohue, 2008), *Paracoccus denitrificans* (Harms et al., 1996) and *B. diazoefficiens* (Sudtachat et al., 2009) as well as *Methylobacterium* species. In strain AM1, an *xoxF* insertion mutant was able to grow on methanol/Ca^{2+}, albeit with a lower growth rate (Chistoserdova & Lidstrom, 1997) and less competitive in-planta outcomes compared to the wild type (Schmidt, Christen, Kiefer, & Vorholt, 2010). The prerequisite of this experiment was the discovery that XoxF-type MDH protein as well as MxaF-type MDH is highly abundant in the phyllospheric microbial communities that include *Methylobacterium* species as major bacteria, suggesting that XoxF is active in natural environments (Delmotte et al., 2009). However, several years went by before the catalytic function of XoxFs as Ln-dependent MDH was confirmed.

Upon the discovery of Ln-dependent MDH, we decided to verify the metabolic role of the Ln-dependent MDH in strain AM1. We showed that Δ*mxaF* did not show any methanol growth without Ln^{3+}, but the strain could grow on methanol/Ln medium (based on Hypho medium) (Nakagawa et al., 2012). This result indicated that MxaFI is not essential, as XoxF1 can serve as the main MDH for methylotrophy in the presence of Ln. This was the first report to identify Ln-dependent methylotrophy.

Bradyrhizobium species were also known to contain methylotrophy genes including *xoxF* (also called *mxaF'*), but their methanol growth had not been demonstrated in laboratory conditions (Sudtachat et al., 2009). High activity levels of MDH were first found in Ce^{3+}-containing culture of *Bradyrhizobium* sp. Ce-3, and the purified enzyme was identified as XoxF-type MDH. Because a complex medium was used in the study (Fitriyanto, Fushimi, et al., 2011), however, it did not prove that *Bradyrhizobium* species are capable of utilizing methanol. Wang et al. (2019) first reported the growth of *B. diazoefficiens* USDA110 on methanol, following up with another report on *Bradyrhizobium* sp. Ce-3 (Pastawan, Suganuma, et al., 2020). The *xox* gene cluster, which is composed of four genes (*xoxF*, *xoxG*, *fldA*, and *gfaA*), is distributed widely in the genus *Bradyrhizobium* (Pastawan, Fitriyanto, & Nakagawa, 2020; Pastawan, Suganuma, et al., 2020; Sudtachat et al., 2009). Thus, *Bradyrhizobium* species are regarded as Ln-dependent methylotrophs. Since bacterial C_1 assimilation pathways (i.e., the serine pathway and the ribulose monophosphate pathway) are not found in *Bradyrhizobium* species, it is presumed that RubisCO-dependent CO_2 fixation occurs (Keltjens, Pol, Reimann, & Op den Camp, 2014). Therefore, in a strict sense, *Bradyrhizobium* species are Ln-dependent methanol-utilizing autotrophs, as they use methanol as an energy source and CO_2 as a carbon source.

4. Screening for Ln-dependent methylotrophs

The screening method for Ln-responsive microorganisms described in the previous section may not be sufficient to identify Ln-dependent methylotrophs, since the medium used in this method contains other potential carbon sources. Given that XoxF is an Ln-dependent MDH, identifying Ln-dependent methylotrophs requires a synthetic medium that contains methanol or methane as a sole carbon source and the application of selective pressure. The fact that *xoxF* is more widely distributed than *mxaF* across bacterial genomes implies the existence of methylotrophs that require Ln to grow on C_1 compounds. One example of this is *Methylacidiphilum fumariolicum* SolV, a novel *Verrucomicrobial* bacterium whose growth on methane is dependent on Ln-XoxF (Pol et al., 2014).

In the final section, we introduce (1) a screening strategy for Ln-dependent methylotrophs, (2) the Ln-dependent methylotrophs we have identified using MALDI-TOF/MS, and (3) the distribution of these Ln-dependent methylotrophs in the natural environment.

4.1 Media and cultivation conditions for the screening of Ln-dependent methylotrophs

Hypho medium (Delaney et al., 2013) supplemented with Lns can be used for the screening of Ln-dependent methylotrophs. Other common minimal media for methylotrophs such as NMS medium (Kelly & Wood, 2010), MM medium (Alamgir, Masuda, Fujitani, Fukuda, & Tani, 2015), and MP medium (Delaney et al., 2013) are also used instead of the Hypho medium. The effective concentrations of Lns may vary depending on the medium chosen. Initially, 30 μM Ln was used for *Methylobacterium* and *Bradyrhizobium* species (Fitriyanto, Fushimi, et al., 2011; Hibi, Okuda, et al., 2011), as Ln concentrations in the soil are reported to range from 10 to 50 μM (Peverill, Meehan, Buckingham, & Maheswaran, 1997). XoxFs are known to be induced mainly by four L-Ln, namely, La^{3+}, Ce^{3+}, Pr^{3+}, and Nd^{3+}, with methanol (Pastawan, Suganuma, et al., 2020; Vu et al., 2016; Wang et al., 2020). Moreover, Ca^{2+} does not repress the induction of XoxF1 in the presence of Ln species, and XoxF1 becomes a dominant MDH in strain AM1 grown on Hypho medium containing 0.5% methanol and 30 μM of $CaCl_2$ with 10 μM of $LaCl_3$ (data not shown).

From these facts, we can conclude that minimal media containing 30 μM of $CeCl_3$ or $LaCl_3$ may be suitable for screening for Ln-dependent methylotrophs. However, researchers should be very careful to avoid Ln contamination in any media prepared in glassware. Ln species are absorbed by glass surfaces and released by chelators produced by microorganisms. The Ln concentrations in a solution prepared in glassware fall almost below the detection limit upon preparation, yet any glassware that has been used with Ln (even if it has been washed with laboratory detergents) can support the growth of Ln-dependent mutants, such as an *mxaF* mutant of *Methylobacterium* species (unpublished data). When glassware is needed, it should be washed extensively with 0.1–1 M HCl. Plasticware is strongly recommended to avoid Ln contamination.

4.2 Screening strategy for Ln-dependent methylotrophs

Methylotrophs that contain the MxaF-type MDH as well as XoxF-type MDH (such as *Methylobacterium* species) grow irrespective of the presence of Ln. We refer to them here as "facultative Ln-dependent methylotrophs." Methylotrophs that do not possess MxaF-type MDH and do not show any growth on methanol in the absence of Ln are called "obligate Ln-dependent methylotrophs."

Our screening strategy for Ln-dependent methylotrophs mimics the classic screening strategy for methylotrophs but occurs in the presence of Ln (Fig. 3). Enrichment cultivation using methane or methanol as a sole carbon source is a good strategy. Usually cell growth can be recognized by the turbidity of the media. When methane is used, it is usually paired with NMS medium, which contains nitrate as a nitrogen source because ammonium ions inhibit methane monooxygenase. Direct spreading of suspension samples onto solid mineral medium containing Ln and methanol, or one

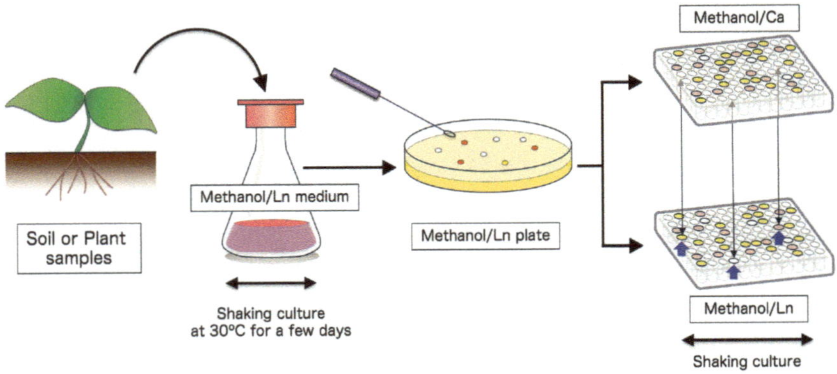

Fig. 3 Screening strategy for isolation of Ln-dependent methylotrophs from the natural environment.

containing Ln without a carbon source but under an atmosphere containing methane (10%–20% *v*/v) prepared in a desiccator or gas-impermeable plastic bags, will support the growth of methanol- and methane-utilizing bacteria, respectively. In this case, supplementation with cycloheximide (50 mg/L) is recommended to suppress fungal growth. Incubation at 30 °C will allow the growth of mesophilic bacteria. The colonies must be transferred onto new plate media until purity is achieved.

The growth of isolates on methane or methanol as a sole carbon source should be tested in liquid media (150–200 μL) prepared in 96-well plates in the presence and absence of Ln. The plate should be rotary-shaken (~300 rpm) to facilitate aerobic growth. The use of an air-permeable plate cover may enable better growth and avoid cross-contamination of volatile carbon sources (as metabolic intermediates) between the wells. The isolates that show growth only in the presence of Ln can be regarded as obligate Ln-dependent methylotrophs.

Identification of bacteria should follow standard methods including 16S rRNA gene sequencing (Lane, 1991). The presence of *xoxF* and *mxaF* may be verified by PCR using degenerate primers (McDonald & Murrell, 1997; Ramachandran & Walsh, 2015). Genome sequencing will give more precise information on the presence of MDH genes and the number and types of *xoxF*, as one bacterial strain may harbor several copies of different types of *xoxF*.

4.3 Identification of Ln-dependent methylotrophs using MALDI-TOF/MS

During screening, one may wish to identify a bacterial species without sequencing its 16S rRNA gene. Although 16S rRNA gene sequencing is a promising identification method, it is tedious and costly when one has hundreds of isolates. Matrix-assisted laser desorption and ionization time-of-flight mass spectrometry (MALDI-TOF/MS) is an ideal tool for identifying or classifying microorganisms without sequencing.

To prepare a sample for identification, a loopful of bacterial colonies is smeared onto the MALDI target plate, and a matrix solution (saturated solution of sinapinic acid in 50% acetonitrile and 2.5% trifluoroacetic acid) is applied onto the smear. Mass spectrometric data are obtained for each sample; most of the prominent peaks can be attributed to ribosomal proteins, which are small ($m/z < 20{,}000$) and positively charged. The data are further processed by clustering to classify the isolates, or by database matching to identify the isolates. Classification enables grouping of the isolates, reducing the number of isolates that require further characterization (such as 16S rRNA gene sequencing or phenotyping). The identification process necessitates a good database of mass spectrometric data of known strains. The commercially available databases usually contain many clinical and pathogenic isolates but not environmental bacteria. Therefore, identification will not give a good match-but at the same time it means that your isolates may be novel. Both analyses can be performed by software programs (such as Bruker BioTyper or Shimadzu Strain solution). For precise identification, 16S rRNA gene sequencing is still necessary. A successful application of this technique for *Methylobacterium* species has been reported (Tani et al., 2012).

4.4 Ln-dependent methylotrophs isolated from natural environments

Using various plant and soil samples as isolation sources, enrichment cultivation was performed with 0.5% methanol or 20% (vol/vol) methane as a sole carbon source in the presence of 30 μM $LaCl_3$ or $HoCl_3$.

PPFMs can be isolated as facultative Ln-dependent methylotrophs using methanol as a sole carbon source. On the other hand, strain SM30 was isolated from the rice rhizosphere using methane-enrichment culture, although this strain is not a methanotroph but rather a facultative Ln-dependent methylotroph. Its $\Delta mxaF$ mutant showed growth on methanol only in the presence of light Lns (Fig. 4) (Lv & Tani, 2018). The strain was identified as a new genus, *Oharaeibacter diazotrophicus* SM30, which is related to *Pleomorphomonas* species (Lv, Sahin, & Tani, 2018).

Furthermore, the isolates growing weakly on NMS solid medium under methane as a carbon source were tested for Ln-dependent methylotrophy. Five Ln-dependent methanol-utilizing bacteria were isolated, and all were

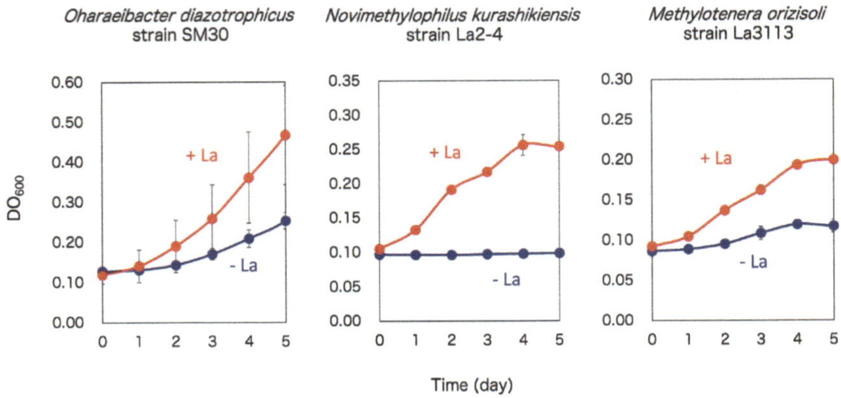

Fig. 4 Methanol growth of newly isolated methylotrophs, such as *Oharaeibacter diazotrophicus* strain SM30, *Novimethylophilus kurashikiensis* strain La2–4, and *Methylotenera orizisoli* strain La3113 on methanol/Ca^{2+} (blue, −La) and methanol/La^{3+} plates (red, +La) (based on NMS medium).

shown to be obligate Ln-dependent methylotrophs, since none of them grew at all on methanol without Ln supplementation. They were identified as *Burkholderia ambifaria* sp. La20-1, *Cupriavidus necator* sp. Ho1-7, *Dyadobacter endophyticus* sp. Ho17-2, *Methylotenera mobilis* sp. La3113, and La2-4, a strain whose 16S rRNA gene sequence shows only 93.4% identity to that of *Methylophilus luteus* Mim^T. Methanol growth of the strain La2-4 occurs only with supplementation with four L-Ln (La^{3+}, Ce^{3+}, Pr^{3+}, and Nd^{3+}) (Lv et al., 2018). Based on its genomic and phenotypic characteristics, the isolate La2-4 was determined to belong to a new genus and was named *Novimethylophilus kurashikiensis* (Lv et al., 2018). The genome of La2-4 encodes five XoxF-type MDHs (of the XoxF3 and XoxF4 family) but not the MxaF-type MDH. It also encodes the tetrahydromethanopterin (H_4MPT) pathway and the ribulose monophosphate (RuMP) cycle for formaldehyde metabolism. Strain La3113 was also identified as a new species, *Methylotenera oryzisoli,* whose genome encodes three XoxF-type MDHs (of the XoxF4 and XoxF5 family) but no MxaF (Lv, Sahin, & Tani, 2020).

5. Summary

Since the discovery of Ln-responsive microorganisms, it has been widely known that they actively use Lns to enable unique activities. We have successfully identified novel types of Ln-responsive microorganisms and expect to identify others in the future using the screening method described here. The observable effect of Ln on microbial growth may be small in strains that do not require Ln for growth, so researchers must observe the colony morphology very carefully. The main difference between our screening method and ordinary microbial screening is the addition of Ln to the medium. When the possible carbon sources are limited to methane and methanol, Ln-dependent methylotrophs will appear as growing colonies. In the case of heterotrophic conditions, colony morphology should be observed. In the experiments described in this chapter, we used 1/100 N medium, but other suitable media may be tested. Among the Ln-responsive microorganisms, *Bradyrhizobium* species were found to be Ln-dependent methylotrophs. Their methylotrophy is dependent on Ln-dependent MDH, and Ln is required to complete the metabolic pathway. Further screening for Ln-dependent methylotrophs revealed additional new genera and species which would not have been isolated if Ln had not been added to the medium. These results suggest that there are more unknown microorganisms whose growth is dependent on Ln as well as more known microorganisms that exhibit as-yet-undiscovered Ln-dependent methylotrophy.

The screening method for Ln-dependent methylotrophs targets the functionality of XoxF as an Ln-dependent MDH. Another strategy targeting a different microbial functionality may be necessary to find additional microorganisms whose growth is dependent on Lns, if any; to date, however, there has been no report on Ln-dependent enzymes other than XoxF-type MDH and ExaF-type ADH (Good et al., 2016), both of which are bacterial, PQQ-dependent periplasmic alcohol dehydrogenases. *Methylobacterium* species have long been studied as a model for methylotrophy, and they are also found to contain Ln- and XoxF-dependent methylotrophic pathways. This finding has introduced a new insight into our knowledge of bacterial methylotrophy. Since these species contain MxaF, the effect of Ln is not drastic, in contrast to the obligate Ln-dependent methylotrophs. Due to the high activity levels of XoxF, however, the addition of Ln results in obvious growth promotion in many cases. *Methylobacterium* species switch between expressing MxaF and XoxF depending on Ln availability. XoxF functions not only as an Ln-dependent MDH but also as an essential factor for MxaF expression (Skovran, Raghuraman, & Martinez-Gomez, 2019). The mechanism and biological relevance of this Ln-switch is one of the interesting new areas of research that will shed further light on the meaning of Ln for bacteria.

Because of these findings, the existence of Ln-dependent methylotrophy is now common knowledge. It should be noted, however, that the

Ca^{2+}-dependent methylotrophy that we have been studying in our laboratory conditions is an exception, as methylotrophy appears to have originated as an Ln-dependent microbial metabolic process. Therefore, it is considered that Ln-dependent methylotrophs may be one of the most important contributors to the C_1 global carbon cycle in ecosystems as well as to "*Ln biochemistry*."

Acknowledgment

We thank Professor Keiichi Kawai, whose contributions to this field are unparalleled, for his collaboration with us.

References

Aide, M. (2019). Lanthanide soil chemistry and its importance in understanding soil pathways: Mobility, plant uptake, and soil health. In N. S. Awwad, & A. T. Mubarak (Eds.), *Lanthanides*. London, UK: IntechOpen. https://doi.org/10.5772/intechopen.79238.

Alamgir, K. M., Masuda, S., Fujitani, Y., Fukuda, F., & Tani, A. (2015). Production of ergothioneine by *Methylobacterium* species. *Frontiers in Microbiology*, *6*, 1185.

Anthony, C. (1982). *The biochemistry of methylotrophs*. New York, NY: Academic Press.

Chistoserdova, L., & Lidstrom, M. E. (1997). Molecular and mutational analysis of a DNA region separating two methylotrophy gene clusters in *Methylobacterium extorquens* AM1. *Microbiology*, *143*, 1729–1736.

Cotruvo, J. A., Jr., Featherston, E. R., Mattocks, J. A., Ho, J. V., & Laremore, T. N. (2018). Lanmodulin: A highly selective lanthanide-binding protein from a lanthanide-utilizing bacterium. *Journal of the American Chemical Society*, *140*, 15056–15061.

Delaney, N. F., Kaczmarek, M. E., Ward, L. M., Swanson, P. K., Lee, M.-C., & Marx, C. J. (2013). Development of an optimized medium, strain and high-throughput culturing methods for *Methylobacterium extorquens*. *PLoS One*, *8*, e62957.

Delmotte, N., Knief, C., Chaffron, S., Innerebner, G., Roschitzki, B., Schlapbach, R., et al. (2009). Community proteogenomics reveals insights into the physiology of phyllosphere bacteria. *Proceedings of the National Academy of Sciences of the United States of America*, *106*, 16428–16433.

Featherston, E. R., Rose, H. R., McBride, M. J., Taylor, E. M., Boal, A. K., & Cotruvo, J. A., Jr. (2019). Biochemicaland structural characterization of XoxG and XoxJ and their roles in lanthanide-dependent methanol dehydrogenase activity. *Chembiochem*, *20*, 2360–2372.

Fitriyanto, N. A., Fushimi, M., Matsunaga, M., Pertiwiningrum, A., Iwama, T., & Kawai, K. (2011). Molecular structure and gene analysis of Ce^{3+}-induced methanol dehydrogenase of *Bradyrhizobium* sp. MAFF211645. *Journal of Bioscience and Bioengineering*, *111*, 613–617.

Fitriyanto, N. A., Nakamura, M., Muto, S., Kato, K., Yabe, T., Iwama, T., et al. (2011). Ce^{3+}-induced exopolysaccharide production by *Bradyrhizobium* sp. MAFF211645. *Journal of Bioscience and Bioengineering*, *111*, 146–152.

Good, N. M., Vu, H. N., Suriano, C. J., Subuyuj, G. A., Skovran, E., & Martinez-Gomez, N. C. (2016). Pyrroloquinoline quinone ethanol dehydrogenase in *Methylobacterium extorquens* AM1 extends lanthanide-dependent metabolism to multicarbon substrates. *Journal of Bacteriology*, *198*, 3109–3118.

Harms, N., Ras, J., Koning, S., Reijnders, W. N. M., Stouthamer, A. H., & van Spanning, R. J. M. (1996). Genetics of C_1 metabolism regulation in *Paracoccus denitrificans*. In M. E. Lidstrom, & F. R. Tabita (Eds.), *Microbial growth on C_1 compounds* (pp. 126–132). Dordrecht, Netherlands: Kluwer Academic Publishers.

Hibi, Y., Asai, K., Arafuka, H., Hamajima, M., Iwama, T., & Kawai, K. (2011). Molecular structure of La^{3+}-induced methanol dehydrogenase-like protein in *Methylobacterium radiotolerans*. *Journal of Bioscience and Bioengineering, 111*, 547–549.

Hibi, Y., Okuda, M., Sakuma, R., Iwama, T., & Kawai, K. (2011). Effects of Eu and Sm on *Methylobacterium* sp. –growth behaviors and expression of methanol dehydrogenase. *Kango Gijutsu, 40*, 108–114 (in Japanese).

Jahn, B., Jonasson, N. S. W., Hu, H., Singer, H., Pol, A., Good, N. M., et al. (2020). Understanding the chemistry of the artificial electron acceptors PES, PMS, DCPIP and Wurster's Blue in methanol dehydrogenase assays. *Journal of Biological Inorganic Chemistry, 25*, 199–212.

Kamijo, M., Suzuki, T., Kawai, K., Fujii, T., & Murase, H. (1999). Ytterbium-decreasing *Streptomyces* sp. and its naphthoquinone-pigment production in the presence of rare-earth elements. *Journal of Bioscience and Bioengineering, 87*, 340–343.

Kato, S., Takashino, M., Igarashi, K., & Kitagawa, W. (2020). Isolation and genomic characterization of a proteobacterial methanotroph requiring lanthanides. *Microbes and Environments, 35*, 1–7.

Kawai, K., Wang, G., Okamoto, S., & Ochi, K. (2007). The rare earth, scandium, causes antibiotic overproduction in *Streptomyces* spp. *FEMS Microbiology Letters, 274*, 311–315.

Kelly, D. P., & Wood, A. P. (2010). Isolation and characterization of methanotrophs and methylotrophs: Diversity of methylotrophic organisms and of one-carbon substrates. In K. N. Timmis (Ed.), *Handbook of hydrocarbon and lipid microbiology* (pp. 3827–3845). Berlin, Heidelberg: Springer.

Keltjens, J. T., Pol, A., Reimann, J., & Op den Camp, H. J. M. (2014). PQQ-dependent methanol dehydrogenases: Rare-earth elements make a difference. *Applied Microbiology and Biotechnology, 98*, 6163–6183.

Lane, D. J. (1991). 16S/23S rRNA sequencing. In E. Stackebrandt, & M. Goodfellow (Eds.), *Nucleic acid techniques in bacterial systematics* (pp. 115–175). New York: John Wiley and Sons.

Liu, Q., Kirchhoff, J. R., Faehnle, C. R., Viola, R. E., & Hudson, R. A. (2006). A rapid method for the purification of methanol dehydrogenase from *Methylobacterium extorquens*. *Protein Expression and Purification, 46*, 316–320.

Lv, H., Sahin, N., & Tani, A. (2018). Isolation and genomic characterization of *Novimethylophilus kurashikiensis* gen. nov. sp. nov., a new lanthanide-dependent methylotrophic species of *Methylophilaceae*. *Environmental Microbiology, 20*, 1204–1223.

Lv, H., Sahin, N., & Tani, A. (2020). *Methylotenera oryzisoli* sp. nov., a lanthanide-dependent methylotrophic bacteria isolated from rice field soil. *International Journal of Systematic and Evolutionary Microbiology, 70*, 2713–2718.

Lv, H., & Tani, A. (2018). Genomic characterization of methylotrophy of *Oharaeibacter diazotrophicus* strain $SM30^T$. *Journal of Bioscience and Bioengineering, 126*, 667–675.

McDonald, R. I., & Murrell, J. C. (1997). The methanol dehydrogenase structural gene *mxaF* and its use as a functional gene probe for methanotrophs and methylotrophs. *Applied and Environmental Microbiology, 63*, 3218–3224.

Nakagawa, T., Mitsui, R., Tani, A., Sasa, K., Tashiro, S., Iwama, T., et al. (2012). A catalytic role of XoxF1 as La^{3+}-dependent methanol dehydrogenase in *Methylobacterium extorquens* strain AM1. *PLoS One, 7*, e50480.

Pastawan, V., Fitriyanto, N. A., & Nakagawa, T. (2020). Biological function of lanthanide in plant-symbiotic bacteria: Lanthanide-dependent methanol oxidation system. *Reviews in Agricultural Science, 8*, 186–198.

Pastawan, V., Suganuma, S., Mizuno, K., Wang, L., Tani, A., Mitsui, R., et al. (2020). Regulation of lanthanide-dependent methanol oxidation pathway in the legume symbiotic nitrogen-fixing bacterium *Bradyrhizobium* sp. strain Ce-3. *Journal of Bioscience and Bioengineering, 130*, 582–587. https://doi.org/10.1016/j.jbiosc.2020.07.012.

Peverill, K., Meehan, B., Buckingham, S., & Maheswaran, J. (1997). The role of applications of rare earth elements in enhancement of crop and pasture production. *Rare Earths, 31*, 69–85.

Pol, A., Barends, T. R. M., Dietl, A., Khadem, A. F., Eygensteyn, J., Jetten, M. S. M., et al. (2014). Rare earth metals are essential for methanotrophic life in volcanic mudpots. *Environmental Microbiology, 16*, 255–264.

Ramachandran, A., & Walsh, A. D. (2015). Investigation of XoxF methanol dehydrogenases reveals new methylotrophic bacteria in pelagic marine and freshwater ecosystems. *FEMS Microbiology Ecology, 91*, fiv105.

Schmidt, S., Christen, P., Kiefer, P., & Vorholt, J. A. (2010). Functional investigation of methanol dehydrogenase-like protein XoxF in *Methylobacterium extorquens* AM1. *Microbiology, 156*, 2575–2586.

Skovran, E., Raghuraman, C., & Martinez-Gomez, N. C. (2019). Lanthanides in methylotrophy. *Current Issues in Molecular Biology, 33*, 101–116.

Sudtachat, N., Ito, N., Itakura, M., Masuda, S., Eda, S., Mitsui, H., et al. (2009). Aerobic vanillate degradation and C_1 compound metabolism in *Bradyrhizobium japonicum*. *Applied and Environmental Microbiology, 75*, 5012–5017.

Tanaka, Y., Hosaka, T., & Ochi, K. (2010). Rare earth elements activate the secondary metabolite-biosynthetic gene clusters in *Streptomyces coelicolor* A3(2). *Journal of Antibiotics (Tokyo), 63*, 477–481.

Tani, A., Sahin, N., Matsuyama, Y., Enomoto, T., Nishimura, N., Yokota, A., et al. (2012). High-throughput identification and screening of novel *Methylobacterium* species using whole-cell MALDI-TOF/MS analysis. *PLoS One*, 7, e40784.

Versantvoort, W., Pol, A., Daumann, L. J., Larrabee, J. A., Strayer, A. H., Jetten, M. S. M., et al. (2019). Characterization of a novel cytochrome c_{GJ} as the electron acceptor of XoxF-MDH in the thermoacidophilic methano- troph *Methylacidiphilum fumariolicum* SolV. *Biochimica et Biophysica Acta, Proteins and Proteomics, 1867*, 595–603.

Vu, H. N., Subuyuj, G. A., Vijayakumar, S., Good, N. M., Martinez-Gomez, N. C., & Skovran, E. (2016). Lanthanide-dependent regulation of methanol oxidation systems in *Methylobacterium extorquens* AM1 and their contribution to methanol growth. *Journal of Bacteriology, 198*, 1250–1259.

Wang, L., Hibino, A., Suganuma, S., Ebihara, A., Iwamoto, S., Mitsui, R., et al. (2020). Preference for particular lanthanide species and thermal stability of XoxFs in *Methylorubrum extorquens* strain AM1. *Enzyme and Microbial Technology, 136*, 109518.

Wang, L., Suganuma, S., Hibino, A., Mitsui, R., Tani, A., Matsumoto, T., et al. (2019). Lanthanide-dependent methanol dehydrogenase from the legume symbiotic nitrogen-fixing bacterium *Bradyrhizobium diazoefficiens* strain USDA 110. *Enzyme and Microbial Technology, 130*, 109371.

Wilson, S. M., Gleisten, M. P., & Donohue, T. J. (2008). Identification of proteins involved in formaldehyde metabolism by *Rhodobacter sphaeroides*. *Microbiology, 154*, 296–305.

Zheng, Y., Huang, J., Zhao, F., & Chistoserdova, L. (2018). Physiological effect of XoxG(4) on lanthanide-dependent methanotrophy. *MBio, 9*, e02430-17.

CHAPTER TWO

Synthetic modeling of the structure and function of the rare-earth dependent methanol dehydrogenase cofactor

Alison L. Knasin and Eric J. Schelter*
P. Roy and Diana T. Vagelos Laboratories, Department of Chemistry, University of Pennsylvania, Philadelphia, PA, United States
*Corresponding author: e-mail address: schelter@sas.upenn.edu

Contents

Abstract

Historically, rare-earth ions have been considered irrelevant to biology. Recently, the active sites of certain methanol dehydrogenase (MDH) enzymes have been shown to contain a redox-inactive, rare-earth (RE) cation coordinated by the redox-active pyrroloquinoline quinone (PQQ) cofactor. Importantly, it was demonstrated that rare earths were essential for the growth of certain methylotrophs that incorporated the XoxF-MDH. In this chapter, we summarize the optimized synthesis of a previously published rare-earth complex that serves as a model of the active site of this RE-containing MDH enzyme. The structure and reactivity of the metalated complex, $[La(L_{QQ})(NO_3)_3]$ are also discussed. $[La(L_{QQ})(NO_3)_3]$ catalytically oxidizes the test alcohol substrate,

Methods in Enzymology, Volume 650
ISSN 0076-6879
https://doi.org/10.1016/bs.mie.2021.01.037

p-methylbenzyl alcohol, ^{4Me}BnOH, to *p*-methylbenzaldehyde, ^{4Me}PhCHO, in the presence of a base (2,6-lutidine) and a terminal oxidant (ferrocenium hexafluorophosphate) with ~17 turnovers. By studying this synthetic model, we have developed a body of evidence about both the reactivity and the mechanism of dehydrogenation of alcohols as a molecular analogue to a native, rare-earth dependent enzyme.

Abbreviations

AE addition-elimination
DFT density functional theory
HT hydride transfer
MDH methanol dehydrogenase
PQQ pyrroloquinoline quinone
RE rare earth

1. Introduction

The 17 rare-earth (RE) elements comprise the group III elements Sc, Y, and the lanthanide series (La-Lu). Although they are indicated as "rare," these elements have natural abundances similar to that of d-block metals, including copper, cobalt, and zinc (Haxel, Hedrick, & Orris, 2002; Rudnick & Gao, 2003). The RE elements display similar chemical properties under ambient conditions, with the 3+ oxidation state being the most chemically stable. The RE elements are electropositive, kinetically labile, and Lewis acidic. Along the lanthanide series, there is a decrease in the size of the elements' atomic and ionic radii, due to population of the 4f-electron shell, an effect referred to as the lanthanide contraction. The 4f valence orbitals exhibit almost entirely ionic bonding, which leads to complexes of the rare earths exhibiting only small crystal field effects with typical coordination numbers between 8 and 12 (Cheisson & Schelter, 2019). The contraction also results in an increase in the Lewis acidity of these elements across the series; La^{3+} is the least Lewis acidic cation and Lu^{3+} is the most Lewis acidic of the lanthanide series. The RE elements have many uses in everyday life. Their electronic characteristics are applied in electronic displays, magnets, and wind turbines, among other applications. Previously, one of the only applications for these elements within biology was use as contrast agents within medical imaging tools (Cheisson & Schelter, 2019). However, it has now been shown that these elements have a role in biology.

Microbiologists have demonstrated that the rare-earth elements are essential for some methanotrophic life (Pol et al., 2014). Through comparison of the amounts of depleted rare earths to methane concentration associated with the *Deepwater Horizon* oil spill, methanotrophs were found to uptake and use REs during metabolism of methane. These results contributed to evidence that implicates REs as biologically significant metal ions (Huang et al., 2019; Shiller, Chan, Joung, Redmond, & Kessler, 2017). Recently, Cotruvo and coworkers found that a newly discovered rare-earth sequestering protein, lanmodulin, exhibits a $\sim 10^8$ higher affinity toward rare-earth ions (RE^{3+}) compared to calcium ions (Ca^{2+}), indicating that Nature has evolved a distinct transport mechanism for rare-earth cations (Cotruvo, Featherson, Mattocks, Ho, & Laremore, 2018). Collectively, these results show that biological systems affect rare-earth dependent reactivity, as well as potential rare-earth sequestration. Recently, others have found prokaryotes that can uptake and use rare-earth elements, which indicates previously unrecognized biological significance (Bonificio & Clarke, 2016; Cotruvo, 2019; Daumann, 2019; Park, Brewer, Reed, Lammers, & Jiao, 2017). Whereas the biology of RE-containing cofactors has been the focus of increased investigation, far less research has focused on using molecular chemistry to model RE-based biochemistry.

Methanol dehydrogenase (MDH) enzymes contain a redox-active pyrroloquinoline quinone (PQQ) cofactor bound to a metal cation. They oxidize methanol to formaldehyde, generating carbon sources used as energy and as biomass. In 2011, researchers found some methanotrophs to contain certain MDH enzymes (termed XoxF-type MDHs) which could uptake and use rare-earth cations (RE^{3+}), rather than typically observed calcium cations (Ca^{2+}), to catalyze alcohol oxidation (Hibi et al., 2011; Nakagawa et al., 2012) (Fig. 1). The following year, Nakagawa and coworkers implicated a catalytic role for the La^{3+} cation within XoxF-type MDH (Nakagawa et al., 2012). Subsequently, in 2014, certain acidophilic, methanotrophic bacteria were shown to contain an XoxF-type MDH. This enzyme shows optimal activity with RE^{3+}, whereas MxaF-MDH displays optimal activity with Ca^{2+} ions (Huang, Yu, & Chistoserdova, 2018; Keltjens, Pol, Reimann, & Op den Camp, 2014; Pol et al., 2014; Shiller et al., 2017). The XoxF- and MxaF-type MDHs contain about a 50% amino acid sequence homology (Deng, Ro, & Rosenzweig, 2018). The inner-coordination sphere of these MDHs differs only by one aspartate residue that binds to the RE^{3+} ions in the XoxF-type homodimers, presumably due to differences in both size and charge of the redox-inactive metal center (Fig. 2).

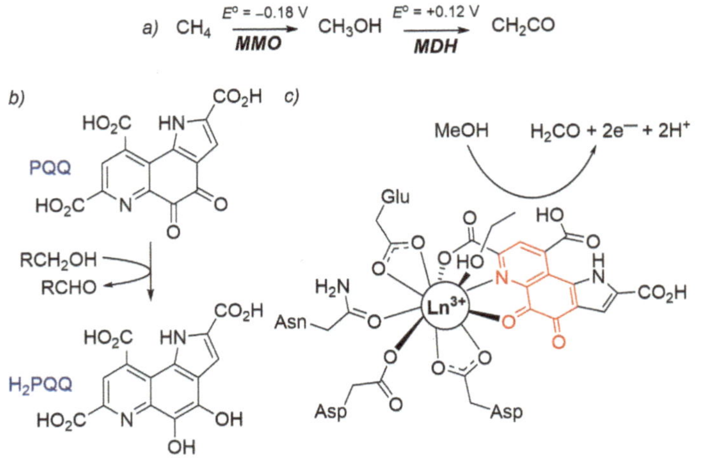

Fig. 1 (A) Enzymatic reactions for the oxidation of methane to formaldehyde, (B) Oxidation of substrates by PQQ, (C) Active site of a rare-earth dependent MDH with ethanol as substrate. A portion of PQQ is highlighted to show the conserved motif within the biomimetic ligand described in this work. *Adapted with permission from the American Chemical Society: McSkimming, A., Cheisson, T., Carroll, P. J., & Schelter, E. J. (2018). Functional synthetic model for the lanthanide-dependent quinoid alcohol dehydrogenase active site.* Journal of the American Chemical Society, 140*(4), 1223–1226. https:/doi.org/10.1021/jacs.7b12318. Copyright 2018 American Chemical Society.*

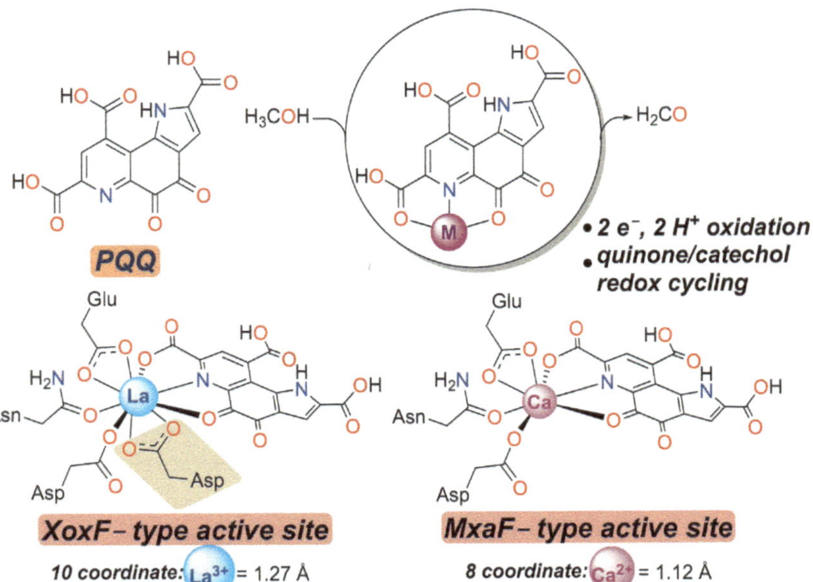

Fig. 2 Overview of MDH reactivity (top) and comparison of active sites of XoxF- and MxaF-type MDHs with ionic radii of La^{3+} and Ca^{2+} (Shannon, 1976). These two depictions of complexes show an open coordination site reserved for binding of either a solvent molecule or substrate (bottom).

Molecular model compounds have contributed to understanding systems such as methane monooxygenases (both particulate and soluble methane monooxygenases, pMMO and sMMO, respectively) and the oxygen-evolving complex of Photosystem II (Cady, Crabtree, & Brudvig, 2008; Culpepper & Rosenzweig, 2012; Kanady, Tsui, Day, & Agapie, 2011; Tinberg & Lippard, 2011). These synthetic models give researchers the opportunity to investigate the mechanistic details of enzymatic active sites without requiring the specific conditions that biological systems often require. Thus, it would be advantageous to use molecular chemistry to study the active sites and the reactivity of RE-containing enzymes.

Only a few reports of PQQ model complexes have been published; these reports do not necessarily include structural characterization data. Additionally, there have been few reports of synthetic rare-earth complexes of PQQ or its analogues (Itoh et al., 1995; Itoh, Kawakami, & Fukuzumi, 1997, 1998; Mitome, Ishizuka, Shiota, Yoshizawa, & Kojima, 2013, 2015; Nakamura, Kohzuma, Kuma, & Suzuki, 1994; Suzuki, Sakurai, Itoh, & Ohshiro, 1988; Wanner, Sixt, Klinkhammer, & Kaim, 1999). The challenges associated with modeling the MDH active site include possible non-natural binding motifs, such as binding through the two quinone oxygen atoms, or the possibility of multiple quinoline quinone moieties binding to a single metal center (Chen et al., 2012; Liu et al., 2009; Wei et al., 2015). Recently, Daumann and coworkers isolated and crystallized Eu-PQQ outside of a protein environment for the first time (Lumpe et al., 2020). They found that the complex crystallizes as a dimer, Eu_2PQQ_2, with Eu^{3+} binding to the biologically relevant sites of PQQ.

Our group published the synthesis of a ligand framework, L_{QQ}, to serve as the model complex of PQQ (Fig. 3). The metalated complex with lanthanum, $[La(L_{QQ})(NO_3)_3]$, was the first RE-containing model system. This system can be used to study the mechanism of RE-dependent, catalytic alcohol oxidation. In this chapter, we discuss our route to synthesize the ligand L_{QQ}, and the optimization of its synthesis and purification. Additionally, we discuss the metalation of L_{QQ} with lanthanum and the reactivity of this complex. We also describe the computational work performed using this complex to study the mechanism of dehydrogenation of *p*-methylbenzyl alcohol (^{4Me}BnOH) to *p*-methylbenzaldehyde (^{4Me}PhCO). The development of $[La(L_{QQ})(NO_3)_3]$ allowed us to investigate reactivity and mechanism, and to draw a connection with the native biological enzyme.

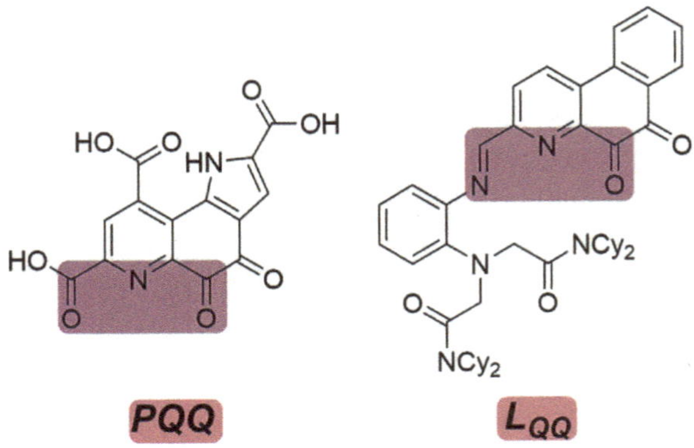

Fig. 3 Structural comparison of PQQ and L_{QQ}.

2. Ligand design and realization

Interest in XoxF-type MDHs was sparked when the X-ray structure of this enzyme, isolated from *Methylacidiphilum fumariolicum* SolV (PDB: 4MAE), was shown to contain a rare-earth cation within its active site that was directly bound to the redox-active PQQ cofactor (Bogart, Lewis, & Schelter, 2015; Huang et al., 2018; Keltjens et al., 2014; Pol et al., 2014; Shiller et al., 2017). The RE-dependent MDH showed faster methanol oxidation rates (Keltjens et al., 2014; Picone & Op den Camp, 2019). The Michaelis constants (K_M) for XoxF-type Ce-MDH from *M. fumariolicum* SolV and MxaF-type Ca-MDH from *Pseudomonas* W1 were reported as 0.8 and 15 μM, respectively (Keltjens et al., 2014). The k_{cat}/K_M values, also known as the specificity constant, for the oxidation of methanol to formaldehyde for these two strains of MDHs were reported at 11,600 and 240 s^{-1} mM^{-1}, respectively (Keltjens et al., 2014). This RE-dependent MDH was shown to also oxidize formaldehyde to formic acid, which was not observed for the Ca-containing analogue. These observations contributed to earlier reports of rare-earth cations having a biological role and a competitive advantage over Ca-dependent MDH. It was hypothesized that rare-earth cations were more effective than calcium ions in their role of substrate activation, due to increased Lewis acidity. Density functional theory (DFT) was used to evaluate and compare the electronic structure of the active sites of XoxF- and MxaF-type MDHs (Bogart et al., 2015).

Notably, two different mechanistic pathways have been suggested for methanol dehydrogenation in Ca-MDHs, a hydride transfer (HT) or addition-elimination (AE) pathway. The rate-limiting step of the hydride transfer mechanism is a concomitant transfer of a hydride from the methyl group to the electrophilic quinone carbon and the proton of the alcohol to the carboxylates from pendant amino acid residues (Fig. 4). Alternatively, for the addition-elimination step, the rate-limiting step is a nucleophilic addition of the alcohol to the electrophilic carbon and concomitant proton transfer to the acetal oxygen atom (Fig. 5). To rationalize the biological mechanism of oxidizing methanol, a synthetic model can be used to investigate these different mechanisms.

The reported solution electrochemistry of PQQ (and its requisite salts) has been challenging to measure unambiguously, due to a strong pH dependence. Our group began to study the electrochemical properties of the quinone cofactor PQQ, by synthesizing a series of PQQ surrogates **3a–c** (Scheme 1) (Dorfner, Carroll, & Schelter, 2015). This approach allowed for a predictive method of determining the first and second reduction potentials, with the goal of predicting these properties of other PQQ analogues not yet synthesized. To achieve this predictive power, the computed ΔG values of various *o*- and *p*-quinones were plotted against their experimental reduction potentials, which allowed for an experimental vs computational correlation to be established.

We expected to use this predictive method to study and synthetically modify the electronic properties of quinone cofactors. These species would then be coordinated with rare-earth cations and examined to gain information about XoxF-type MDH. However, once coordinated to rare earths, the resulting complexes were found to be reactive with solvent and the coordination chemistry was difficult to control and study. Thus, we were motivated to synthesize a different ligand that featured steric motifs enforced by XoxF-type MDH, to ultimately control the binding motif of the rare-earth cation.

The ligand L_{QQ} (Scheme 2) was synthesized to probe the mechanism for a model complex of RE-dependent MDH (McSkimming, Cheisson, Carroll, & Schelter, 2018). A 14-step synthesis was developed and optimized (Scheme 3), with the goal to isolate large quantities of this ligand to allow for exploratory synthetic and mechanistic experimentation. The ligand L_{QQ} is sterically saturated to replicate the structure and bonding of native MDH. Additionally, the nitrate counterions serve as a "proton shuttle," similarly to the carboxylates in amino acids within the natural system.

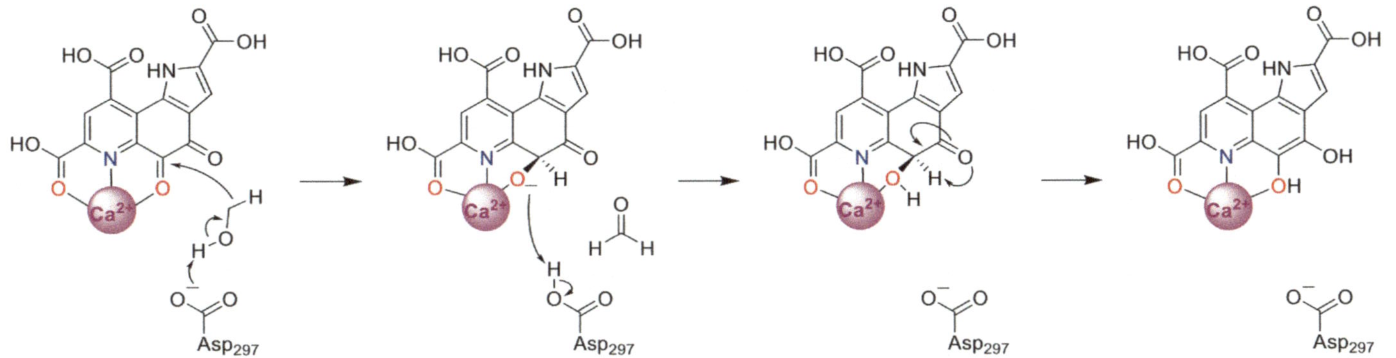

Fig. 4 Dehydrogenation of methanol to formaldehyde by Ca-MDH through the hydride transfer mechanism.

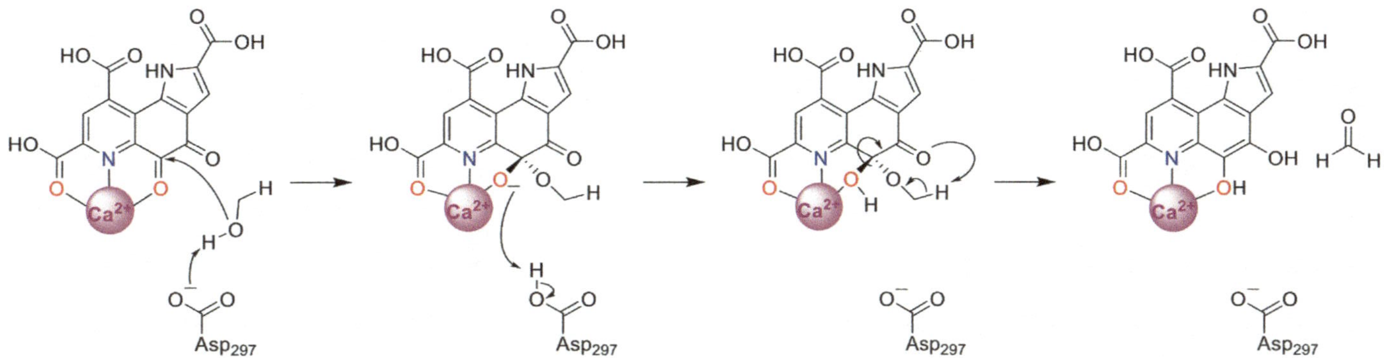

Fig. 5 Dehydrogenation of methanol to formaldehyde by Ca-MDH through the addition-elimination mechanism.

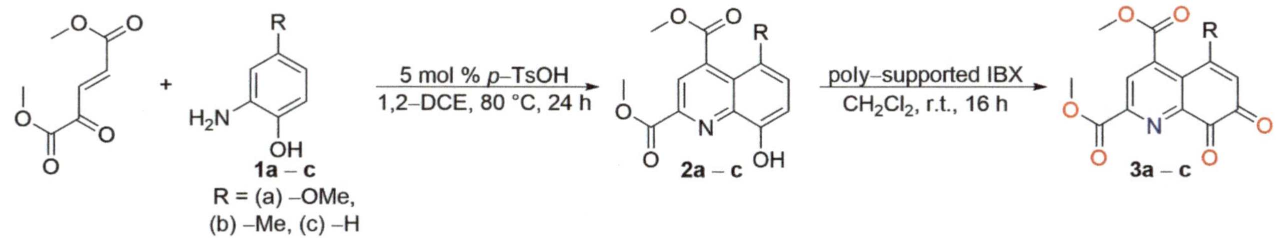

Scheme 1 Initial attempt to synthesize a series of PQQ analogues.

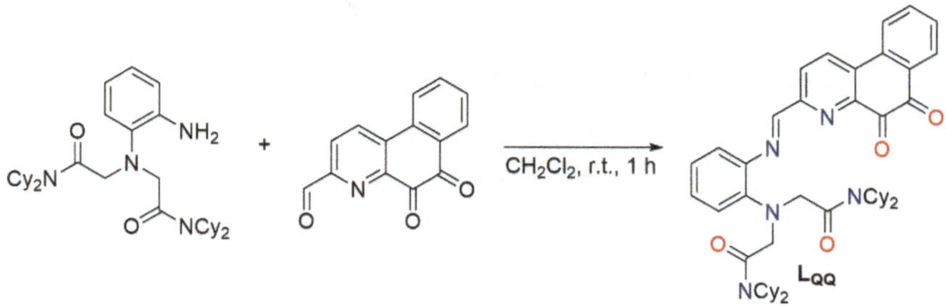

Scheme 2 Synthesis of L_{QQ}.

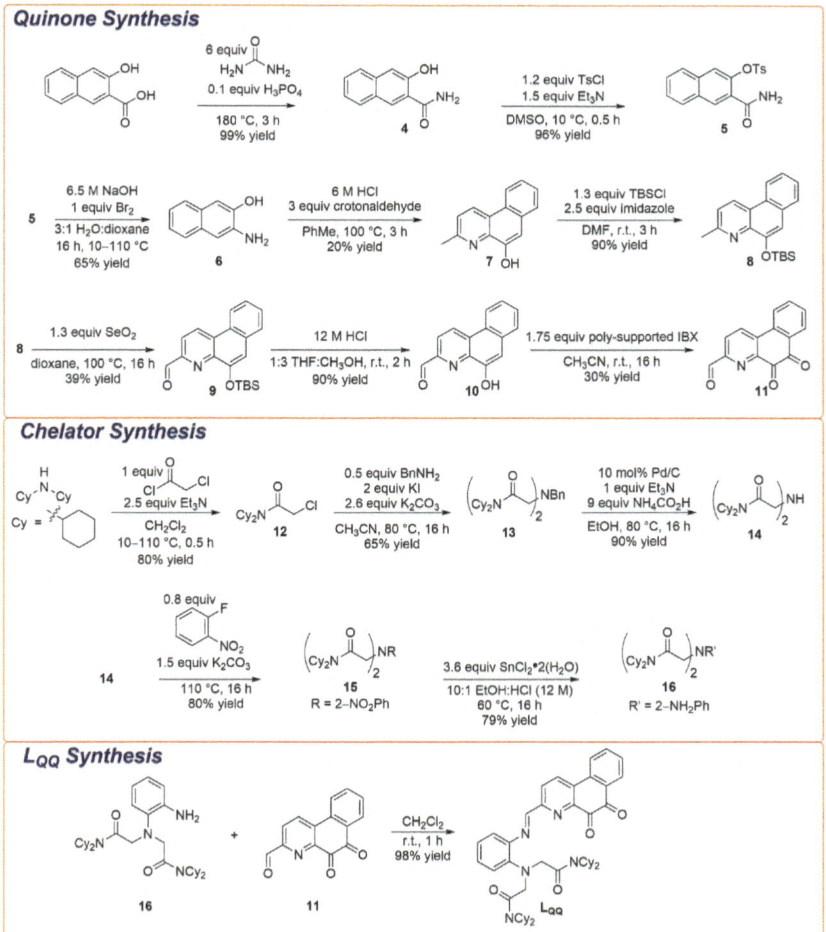

Scheme 3 Modified, optimized synthesis of L_{QQ}.

3. L_{QQ}—Synthetic pathway and purification

The reaction conditions for three of the steps of the synthesis of L_{QQ} were improved from the previously published synthetic pathway to reproducibly provide higher yields and purities (McSkimming et al., 2018). This convergent synthesis consists of two parallel pathways, one comprising quinone **11** and one comprising aniline **16**, followed by a Schiff base condensation coupling as the terminal step (Scheme 2). It should be noted that the synthesis described was scalable to a 1.5 mol scale for the amidation reaction to yield compound **4**. In performing the synthesis of compounds **4–7**, a heating mantle and a mechanical stirrer should be used in place of magnetic stir bar and oil bath conditions. At each step, purity of isolated material was confirmed through use of ^{1}H NMR spectroscopy.

The quinone synthesis begins with an amidation of 3-hydroxy-2-naphthoic acid using urea and phosphoric acid, giving primary amide **4** (Scheme 4), a procedure that was adapted from a reported synthesis (Georg, 1953). For this reaction, 6 equiv. urea were used as solvent and reagent. Once the urea was molten, 3-hydroxy-2-naphthoic acid was added as a solid and 0.1 equiv. phosphoric acid were added dropwise to the reaction mixture at 150 °C. Note: the addition of acid to the reaction mixture is an exothermic process, which could lead to effervescence. This operation should be performed slowly and closely monitored over the course of the addition of acid. After the addition, the reaction was stirred at 180 °C for 3 h. Decomposition occurs when the temperature exceeds 180 °C, evidenced by the mixture changing from an orange to a red color. Orange solid was precipitated from solution by addition of 6 L water to the reaction mixture, which was subsequently cooled to room temperature. The reaction mixture was filtered and dried on the filter, giving **4** as an orange solid.

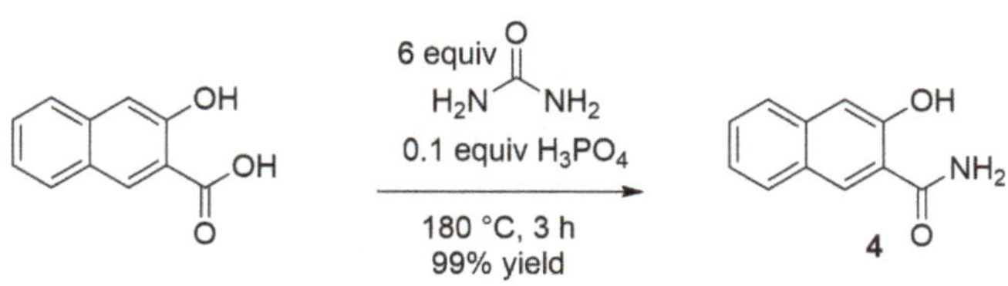

Scheme 4 Amidation of 3-hydroxy-2-naphthoic acid with phosphoric acid and urea to yield primary amide **4**.

In the next step, the naphthol of amide **4** was protected using tosyl chloride, to give tosylated **5** (Scheme 5). Once **4** was dissolved in dimethyl sulfoxide (DMSO) and 1.5 equiv. triethyl amine (Et_3N) were added, the reaction mixture was cooled until nearly frozen, as the addition of tosyl chloride was exothermic. 1.2 equiv. tosyl chloride should be added slowly and the reaction mixture stirred for 0.5 h while still cold. A yellow solid precipitated from solution by adding water (three times the volume of DMSO originally added). The yellow solid was filtered and washed with a copious amount of water. This solid was resuspended in isopropanol, filtered and dried on the filter, giving amide **5** as an off-white solid.

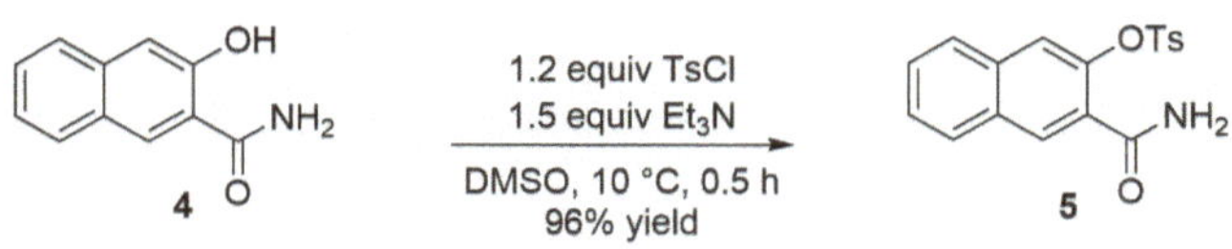

Scheme 5 Protection of compound **4** with tosyl chloride to yield compound **5**.

Compound **6** was obtained through a Hofmann rearrangement and deprotection (Scheme 6). A solution of 6.5 M sodium hydroxide (NaOH) and 1,4-dioxane was cooled to 10 °C. While cold, 1 equiv. bromine was added dropwise using an addition funnel. Compound **5** was added to solution slowly and the reaction mixture was stirred for 0.5 h while still cold. The mixture was then refluxed at 100 °C for 16 h. After 1 h, the light brown suspension changed to a dark brown solution. The reaction mixture was cooled to room temperature and acidified to pH 0 with 12 M hydrochloric acid (HCl). As a safety measure, ice should be added into the flask, along with the HCl. The aqueous phase was separated using diethyl ether (Et_2O). This aqueous layer was subsequently filtered. The solution was adjusted to pH 3 with 3 M NaOH and then to pH 8 with sodium carbonate (Na_2CO_3). The light brown solid that precipitated (**6**) was filtered and dried on a Büchner funnel.

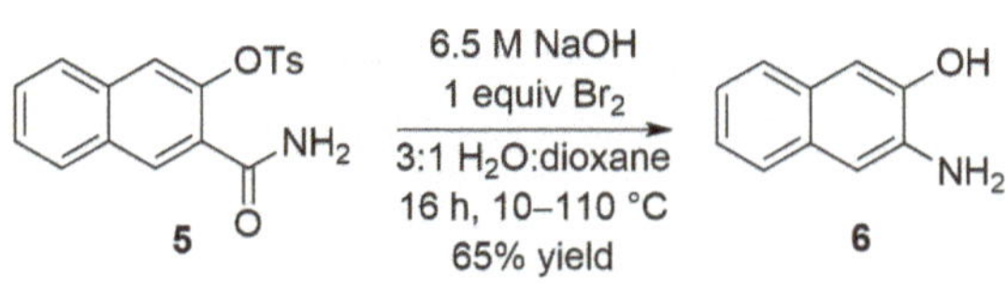

Scheme 6 Hofmann rearrangement and deprotection of compound **5**, leading to compound **6**.

The following step is a Skraup reaction to produce **7** (Scheme 7). Aniline **6** was dissolved in a 3:1 solution of 6 M HCl and toluene and heated to 100 °C. 3 equiv. crotonaldehyde were added using an addition funnel in a 1:1 solution in toluene. This reaction mixture was stirred at 100 °C for 3 h. Once cooled to room temperature, the organic layer was separated using additional toluene and 6 M HCl. The aqueous portion was adjusted to pH 3 with 3 M NaOH, and then to pH 8 with sodium bicarbonate ($NaHCO_3$). The organic layer was separated with ethyl acetate, washed with brine, and dried over sodium sulfate (Na_2SO_4). The reaction mixture was filtered and concentrated under reduced pressure to yield a brown oil. This crude product was purified using flash chromatography (95:5 hexanes: ethyl acetate eluent), as indicated by work from the Williams group, and recrystallized twice from methanol to obtain **7** as a white solid (Bakewell et al., 2012).

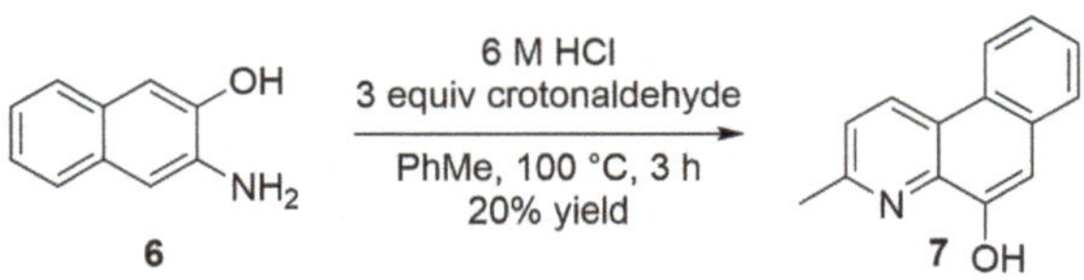

Scheme 7 Skraup reaction of compound **6** to yield compound **7**.

The ensuing step consists of a *tert*-butyldimethylsilyl (TBS) protection of **7** to produce **8** as a white solid (Scheme 8). Compound **7** was dissolved in *N*,*N*′-dimethylformamide (DMF) and 2.5 equiv. imidazole dissolved in DMF were added. 1.3 equiv. *tert*-butyldimethylsilyl chloride (TBSCl) were added portion-wise and the reaction mixture was stirred at room temperature for 3 h. A copious amount of distilled water was added slowly (200 mL over 10 min) to precipitate **8** from solution. This solid was filtered and dried on a medium porosity fritted filter.

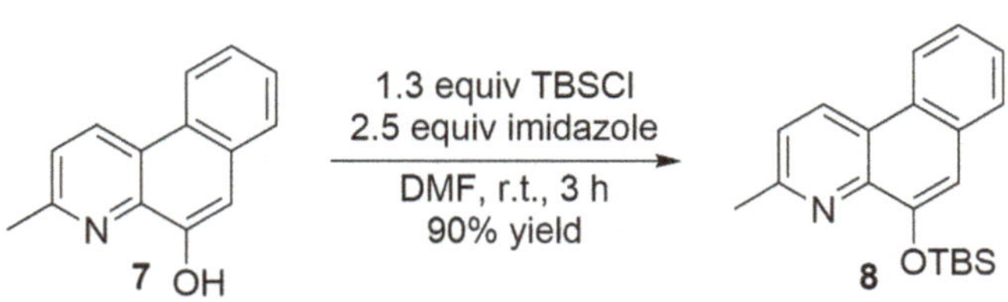

Scheme 8 *tert*-Butyldimethylsilyl protection of compound **7** to obtain compound **8**.

To obtain aldehyde **9**, 1.3 equiv. selenium dioxide (SeO_2) were dissolved in dioxane (Scheme 9). Compound **8** was dissolved in 1,4-dioxane and the two solutions were mixed. This mixture was refluxed at 110 °C for 16 h. Once cooled to room temperature, 250 mL hexanes were added and the crude mixture was passed through a pad of silica (~5 cm in diameter and ~10 cm in height) using Et_2O as the eluent. The filtrate was concentrated under reduced pressure and then 20 mL of a 1:1 solution of dichloromethane (CH_2Cl_2) and hexanes was added. This solution was passed through a second pad of silica (~5 cm in diameter and ~10 cm in height) using a 1:10 solution of Et_2O and hexanes as the eluent. The filtrate was once again concentrated under reduced pressure to yield **9** as a yellow solid. No further purification was performed on the compound.

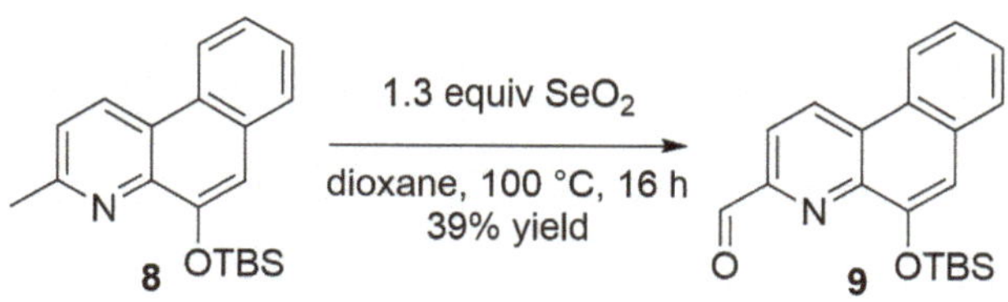

Scheme 9 Selenium dioxide oxidation of compound **8** to compound **9.**

For the next step, a TBS deprotection of **9**, led to aldehyde **10**. Protected aldehyde (**8**) was dissolved in a 1:3 mixture of tetrahydrofuran (THF) and methanol (Scheme 10). Enough 12 M HCl was added dropwise to solubilize **8**, and the solution was stirred at room temperature for 2 h. A solid was precipitated from solution once 100 mL saturated aqueous solution of $NaHCO_3$ were added. This yellow solid (**10**) was filtered, washed with hexanes, and dried on a medium porosity fritted filter.

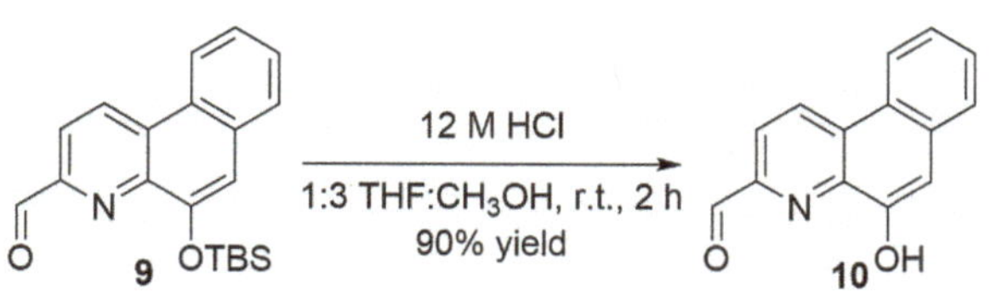

Scheme 10 Deprotection of compound **9** with hydrochloric acid to yield compound **10.**

In the final step of the quinone synthesis, the oxidation of phenol **10** to quinone **11** was originally reported to employ IBX (*o*-iodoxybenzoic acid); however, it was found that this gave quinone **11** together with *o*-iodobenzoic acid as an impurity, which was difficult to remove. Changing the oxidant to polymer-supported IBX yielded quinone **11** without the *o*-iodobenzoic acid impurity (Scheme 11) (Dorfner et al., 2015). For this oxidation, compound **10** was dissolved in CH_3CN. 1.75 equiv. polymer-supported IBX were added as a solid and the reaction mixture was stirred at r.t. for 16 h. The suspension was filtered and the solid was washed with CH_3CN until the filtrate was colorless. The filtrate was passed through a pad of Celite (~5 cm in diameter and ~10 cm in height) using a 1:1 solution of Et_2O and CH_2Cl_2 as the eluent. The filtrate was concentrated under reduced pressure and compound **11** was precipitated out of solution with addition of 1,2-dichloroethane. This orange solid was filtered and dried on a medium porosity fritted filter. No further purification was required.

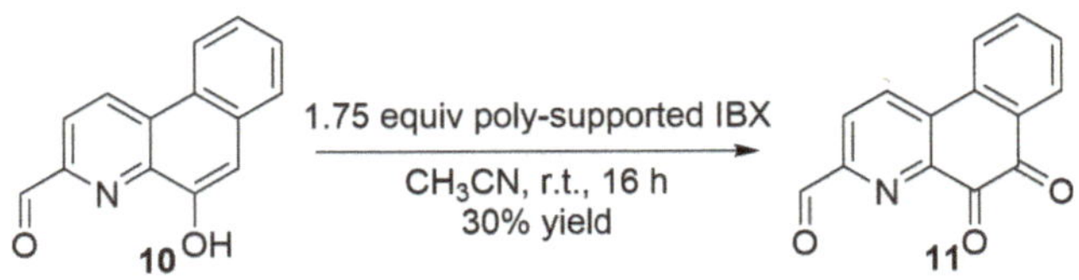

Scheme 11 Oxidation of compound **10** with polymer-supported *o*-iodoxybenzoic acid to yield compound **11**.

The chelator synthesis begins with a previously published peptide coupling to give **12** as a brown solid (Scheme 12) (Joshi et al., 2013). 1 equiv. dicyclohexylamine was dissolved in CH_2Cl_2 and cooled to 10 °C. To this solution, 1.1 equiv. chloroacetyl chloride were added as a solid and 2.5 equiv. Et_3N were added dropwise with stirring. The reaction mixture was stirred for 0.5 h at 10 °C, allowed to warm to room temperature, and washed with 100 mL saturated $NaHCO_3$ and 100 mL of brine. Once dried over Na_2SO_4 and filtered, the filtrate was concentrated under reduced pressure to yield a brown solid. This crude solid required trituration with Et_2O to remove an insoluble impurity. To do so, the crude solid was stirred in 100 mL Et_2O and filtered. The filtrate contained the product, and the filtered solid contained the insoluble impurity. The filtrate was once again concentrated under reduced pressure to obtain **12** as a brown solid.

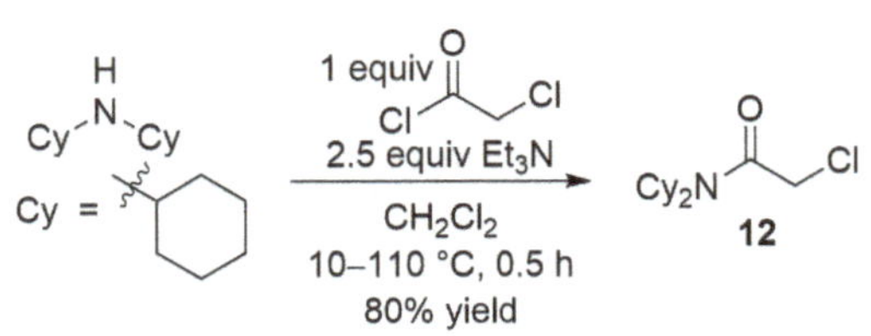

Scheme 12 Peptide coupling between dicyclohexylamine and chloroacetyl chloride to yield compound **12**.

The next step involved sequential S_N2 reactions with benzylamine (Scheme 13). Compound **12** was dissolved in acetonitrile (CH_3CN) with 0.5 equiv. benzylamine, 2 equiv. KI, and 2.6 equiv. K_2CO_3. This reaction mixture was stirred for 16 h at 80 °C. Once allowed to cool to room temperature, 300 mL distilled water were added, and the reaction mixture was extracted three times with 50 mL of a 3:1 solution of hexanes:ethyl acetate. It should be noted that copious amounts of water were required to avoid a three-phase mixture. Additional water increases the miscibility between the acetonitrile and water layers, which ensured a biphasic work-up with aqueous and organic layers. After extraction, the organic layer was washed with brine, dried over Na_2SO_4, and filtered. The filtrate was concentrated under reduced pressure and the crude solid was recrystallized from concentrated ethanol to obtain **13** as a white solid.

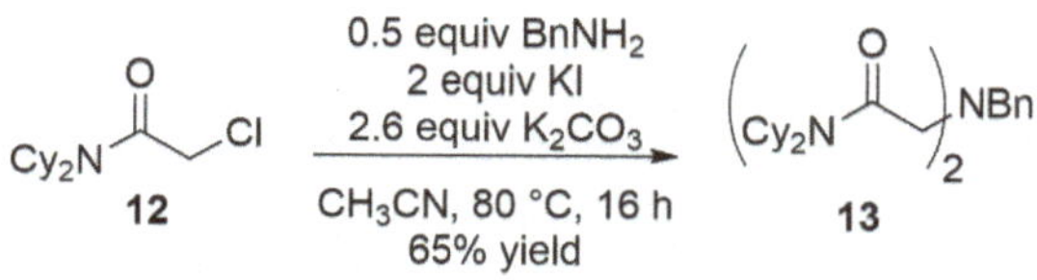

Scheme 13 S_N2 reaction of compound **12** with benzylamine to give compound **13.**

Benzylamine **13** was isolated as an ammonium salt from the above reaction as evidenced by a tertiary ammonium N-H stretch observed in the IR spectrum (2663 cm^{-1}). Attempted hydrogenations with the ammonium salt to produce **14** were unsuccessful, but the hydrogenation was successfully performed in the presence of an exogenous base (Et_3N) (Scheme 14). Once compound **13** was dissolved in 150 mL ethanol, 1 equiv. Et_3N was added and the solution was stirred at room temperature for 15 min. 9 equiv. ammonium formate and 10 mol% activated Pd/C were added as solids and the reaction mixture was refluxed at 80 °C for 16 h. Once allowed to cool to room temperature, the crude mixture was filtered over Celite and washed with ethanol until the eluent was colorless. The filtrate was concentrated under reduced pressure to yield **14** as a white solid.

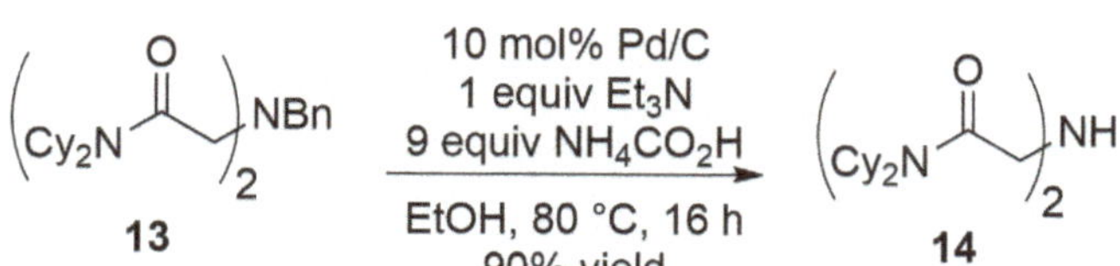

Scheme 14 Hydrogenation of compound **13** with palladium over carbon in the presence of an exogenous base to yield compound **14**.

The next step involves a nucleophilic aromatic substitution reaction (Scheme 15). A neat reaction of **14**, 0.8 equiv. 1-fluoro-2-nitrobenzene, and 1.5 equiv. K_2CO_3 were stirred at 110 °C for 16 h. The reaction was cooled to room temperature and the orange solid was dissolved in 250 mL Et_2O and filtered over Celite. To precipitate unreacted starting material, 150 mL of 2 M HCl were added and the resulting slurry was filtered over Celite. The organic layer was separated, washed with brine, dried over Na_2SO_4, and filtered. The filtrate was concentrated under reduced pressure and the crude material was triturated with hexanes.

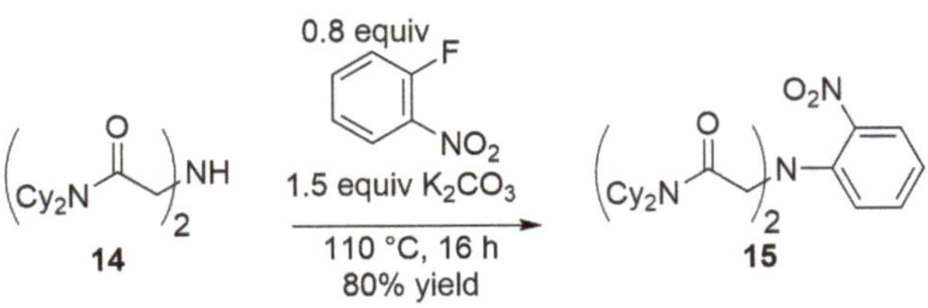

Scheme 15 Nucleophilic aromatic substitution reaction between compound **14** and 1-fluoro-2-nitrobenzene to yield compound **15**.

The final step of the chelator synthesis, reduction of **15** to **16** was reported using $NiCl_2{\cdot}2(H_2O)$ and $NaBH_4$ (McSkimming et al., 2018). However, we recently noted that $SnCl_2{\cdot}2(H_2O)$ in HCl gave clean conversion to the aniline (Scheme 16). Compound **15** was dissolved in 20 mL ethanol with minimal 12 M HCl and 3.6 equiv. $SnCl_2{\cdot}2(H_2O)$. The reaction mixture was stirred at 60 °C for 32 h. Once allowed to cool to room temperature, 75 mL water were added, along with 3 M NaOH until the solution was pH 14, which led to the formation of a white precipitate. A saturated solution of $NaHCO_3$ was added and the reaction mixture was stirred at r. t. for 15 min. The slurry was filtered over Celite. The eluent was extracted three times with 50 mL of CH_2Cl_2, washed with 50 mL of brine, dried over Na_2SO_4, and filtered. The filtrate was concentrated under reduced pressure, which afforded compound **16** as a brown solid.

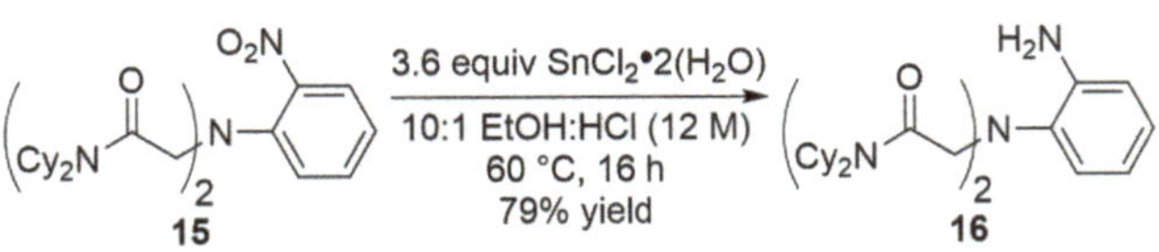

Scheme 16 Tin chloride reduction of compound **15** to give compound **16**.

The terminal step of the convergent synthesis of L_{QQ} was a Schiff base condensation of compounds **11** and **16** (Scheme 17). 1 equiv. **11** and 1.4 equiv. **16** were dissolved at r.t. in CH_2Cl_2 and stirred for 1 h. The reaction mixture was concentrated to about 5 mL and hexanes was added to precipitate L_{QQ} as a red powder. This solid was collected by filtration, washed three times with 25 mL hexanes, and dried on a medium porosity fritted filter. Prior to transfer into a glovebox, the ligand L_{QQ} was dried under dynamic vacuum overnight.

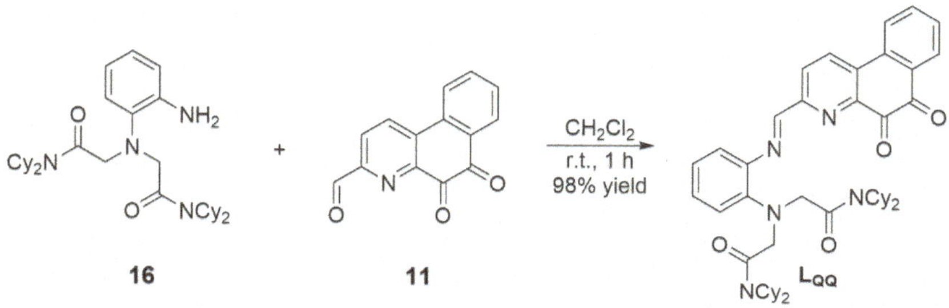

Scheme 17 Schiff base condensation between compounds **11** and **16** to give L_{QQ}.

4. Structure and reactivity of [La(L_{QQ})(NO_3)$_3$]

[La(L_{QQ})(NO_3)$_3$] was isolated from metalation of [La(NO_3)$_3$(THF)$_4$] with L_{QQ} as a moisture-sensitive solid. In a nitrogen-filled glovebox, 1 equiv. L_{QQ} was suspended in CH_3CN and 1.1 equiv. [La(NO_3)$_3$(THF)$_4$] were dissolved in CH_3CN. The [La(NO_3)$_3$(THF)$_4$] solution was added to the solution of L_{QQ} and the reaction mixture was stirred for 1 h. An orange solid precipitated from the solution which was shown to be the complex: [La(L_{QQ})(NO_3)$_3$]. The product was isolated by filtration, washed with CH_3CN, and dried under reduced pressure (Fig. 6). Crystals of [La(L_{QQ})(NO_3)$_3$] were obtained through slow diffusion of hexanes into dichloromethane. Analysis of the structure by X-ray crystallography showed a binding motif with strong similarities to the reported coordination environment of the XoxF-MDH enzymatic active site (Fig. 7) (McSkimming et al., 2018; Pol et al., 2014). Notably, the nitrate counteranions are bound in a κ^2-mode, analogous to the carboxylate groups of the amino acids within the active site in the natural system. Confirmation of the quinone moiety was determined by analysis of the bond length of C(2)-C(3). A typical C-C bond length for quinone, semiquinone, and catechol species are 1.530(5), 1.42(1)–1.448(9), and 1.348 (10)–1.415(6) Å, respectively (Pierpont & Buchanan, 1981). This C(2)-C(3) bond length within [La(L_{QQ})(NO_3)$_3$], as determined through analysis of the crystal structure, was 1.542(5) Å, which confirmed the quinone assignment.

Solution electrochemical experiments, using cyclic voltammetry, performed on L_{QQ} in CH_2Cl_2 with 0.1 M [^{n}Pr$_4$N][BAr$^F_4$] showed a reversible couple at $E_{1/2} = -0.95$ V vs Fc^+/Fc, which was assigned as the

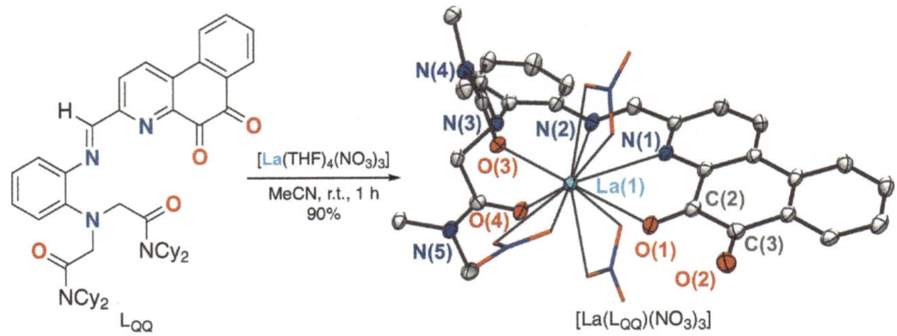

Fig. 6 Metalation of L_{QQ} with $[La(NO_3)_3(THF)_4]$ to give $[La(L_{QQ})(NO_3)_3]$. Ellipsoids are shown at the 30% probability level. Hydrogen atoms and cyclohexyl groups have been removed for clarity. *Adapted with permission from the American Chemical Society: McSkimming, A., Cheisson, T., Carroll, P. J., & Schelter, E. J. (2018). Functional synthetic model for the lanthanide-dependent quinoid alcohol dehydrogenase active site.* Journal of the American Chemical Society, 140*(4), 1223–1226. https://doi.org/10.1021/jacs.7b12318. Copyright 2018 American Chemical Society.*

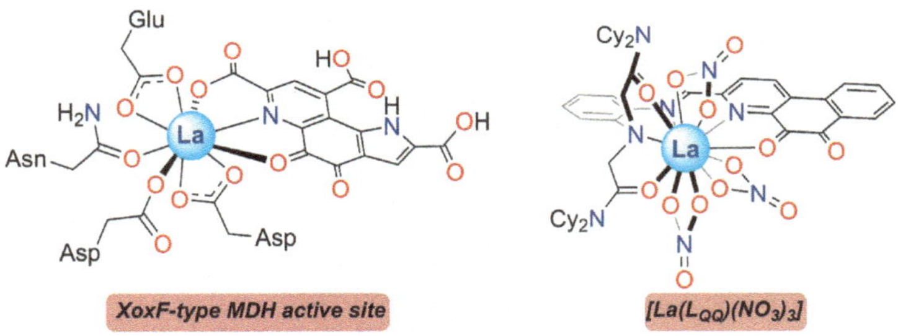

Fig. 7 Comparison of XoxF-type MDH active site with $[La(L_{QQ})(NO_3)_3]$.

quinone-semiquinone couple. There was also an irreversible process at $E_{pc} = -1.76$ V vs Fc^+/Fc, due to the formation of the catecholate dianion. The CV experiments performed on $[La(L_{QQ})(NO_3)_3]$ under the same conditions showed an anodic shift of the reversible quinone-semiquinone couple to -0.34 V vs Fc^+/Fc (Fig. 8), indicating an increased stability of this reduction when the lanthanum cation was coordinated. These electrochemical data were complicated and two quasi-reversible reduction events at -0.68 and -0.77 V vs Fc^+/Fc were not definitively assigned. They are possibly assignable to the loss of nitrate ligand(s) during the scan, giving another species upon reduction. However, the anodic shifting of the quinone-semiquinone couple by ~600 mV from free ligand to $[La(L_{QQ})(NO_3)_3]$ provided direct evidence that the semiquinone species was stabilized by the bound metal ion. Thus, it was hypothesized that $[La(L_{QQ})(NO_3)_3]$ would promote oxidation of alcohol, compared to free ligand, L_{QQ}.

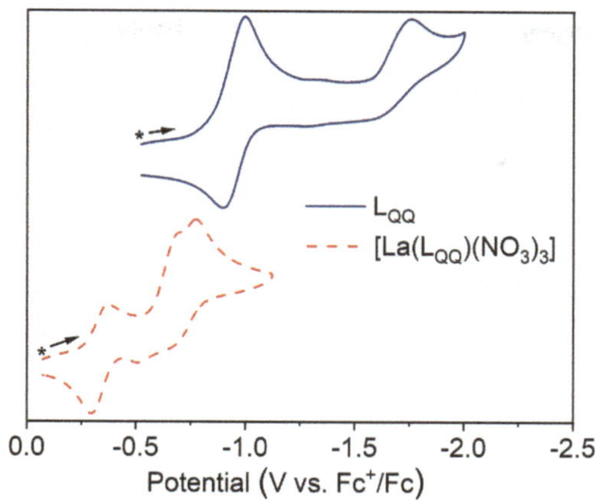

Fig. 8 Cyclic voltammograms of $[La(L_{QQ})(NO_3)_3]$ (solid line) and L_{QQ} (dashed line) in a 0.1 M solution of tetrapropylammonium tetrakis[3,5-bis(trifluoromethyl)phenyl]borate ($[^nPr_4N][BAr^F_4]$) in CH_2Cl_2 at a 100 mV s^{-1} scan rate. *Adapted with permission from the American Chemical Society: McSkimming, A., Cheisson, T., Carroll, P. J., & Schelter, E. J. (2018). Functional synthetic model for the lanthanide-dependent quinoid alcohol dehydrogenase active site.* Journal of the American Chemical Society, 140*(4), 1223–1226. https://doi.org/10.1021/jacs.7b12318. Copyright 2018 American Chemical Society.*

Subjecting free L_{QQ} in a CD_2Cl_2 solution to fourfold excess of ^{4Me}BnOH showed no reaction under anaerobic conditions, as determined by the absence of detectable aldehyde product, ^{4Me}PhCHO, using ^{1}H NMR spectroscopy, after 72 h (Gnanaprakasam, Zhang, & Milstein, 2010; Kawahara, Fujita, & Yamaguchi, 2012). Itoh and Fukuzumi demonstrated that addition of base promoted the dehydrogenation of methanol to formaldehyde (Itoh et al., 1997, 1998). However, when 2.2 equiv. of the base DBU (DBU= 1,8-diazabicyclo(5.4.0)undec-7-ene) was added to solution with L_{QQ} and ^{4Me}BnOH, only trace ^{4Me}PhCHO was formed: <3% yield after 30 min, 13% yield after 4 h by ^{1}H NMR spectroscopy. Conversely, $[La(L_{QQ})(NO_3)_3]$ in a CD_2Cl_2 solution demonstrated the ability to produce ^{4Me}PhCHO both with and without a base present, albeit more efficiently in the presence of a base (Table 1).

When performed stoichiometrically, the dehydrogenation of ^{4Me}BnOH using $[La(L_{QQ})(NO_3)_3]$ gave the oxidized product, ^{4Me}PhCHO, which was characterized by ^{1}H NMR spectroscopy, as well as $[La(L_{QQ}{}^{2-})(NO_3)]_2$ (Fig. 9). This dimer, where L_{QQ} was in the catechol redox state, was isolated independently by performing a chemical reduction with addition of 2 equiv. $[Cp_2Co]$ and characterized by both X-ray crystallography and infrared spectroscopy. The reduction of a quinone to a catechol is a two-electron redox process, as is the oxidation of an alcohol to an aldehyde. The product of

Table 1 Yields of *p*-methylbenzaldehyde from catalytic oxidation of *p*-methylbenzyl alcohol.

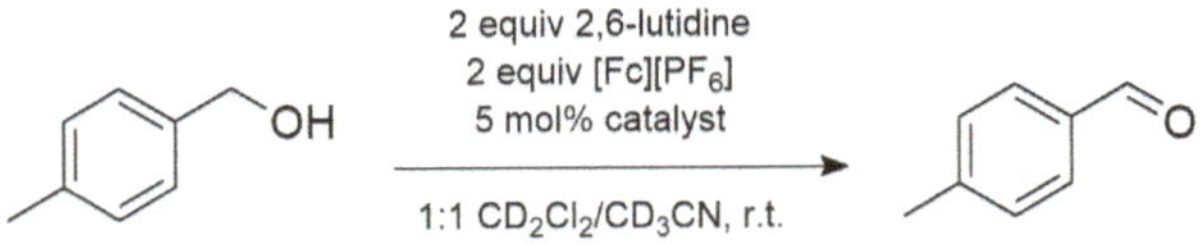

Catalyst	Time (h)	Yield (%)
None	24	<1
L_{QQ}	21	14
$[La(L_{QQ})(NO_3)_3]$	21	84
	44	96
$[La(NO_3)_3(THF)_4]$	21	5
$[La(L_{QQ})(NO_3)_3]$[a]	24	30
	72	67
$[La(L_{QQ})(NO_3)_3]$[b]	<10	67

[a]2.5 equiv. alcohol in CD_2Cl_2.
[b]Same conditions + 2.2 equiv. DBU.

this stoichiometric reaction was a dimer in the catechol redox state, which is reminiscent of the enzymatic cofactor that cycles between quinone and catechol states.

It was hypothesized that addition of a terminal oxidant and base had the potential to catalytically dehydrogenate ^{4Me}BnOH using $[La(L_{QQ})(NO_3)_3]$ (Table 1). To perform the catalytic oxidation, 1 equiv. ^{4Me}BnOH, 2 equiv. 2,6-lutidine, and about 0.006 mmol $(SiMe_3)_2O$ (as an internal standard) were dissolved in a 1:1 mixture of CD_2Cl_2 and CD_3CN (0.4 mL total). The 2,6-lutidine was added last, due to its volatility. It is important to note that the pK_a of 2,6-lutidine is ~7, which is closer to physiological pH (pH ~7.4) when compared to the pK_a of DBU, at ~12 (Jia, Ramstad, & Zhong, 2001; Kaupmees, Trumall, & Leito, 2014). The weaker base was chosen for these studies due to its more physiologically relevant pK_a. Once these three materials were dissolved, the reaction mixture was added to a J-Young tube, and an initial ^{1}H NMR spectrum was collected as a blank. After collecting the blank, the contents of the tube were transferred back into a vial and 5 mol % catalyst in minimal CD_2Cl_2 and 2 equiv. $[Fc][PF_6]$ were added. Note that $[Fc][PF_6]$ is poorly soluble. The solution within the vial was mixed for 21 and 44 h, at which points, ^{1}H NMR spectra were taken (by dissolving

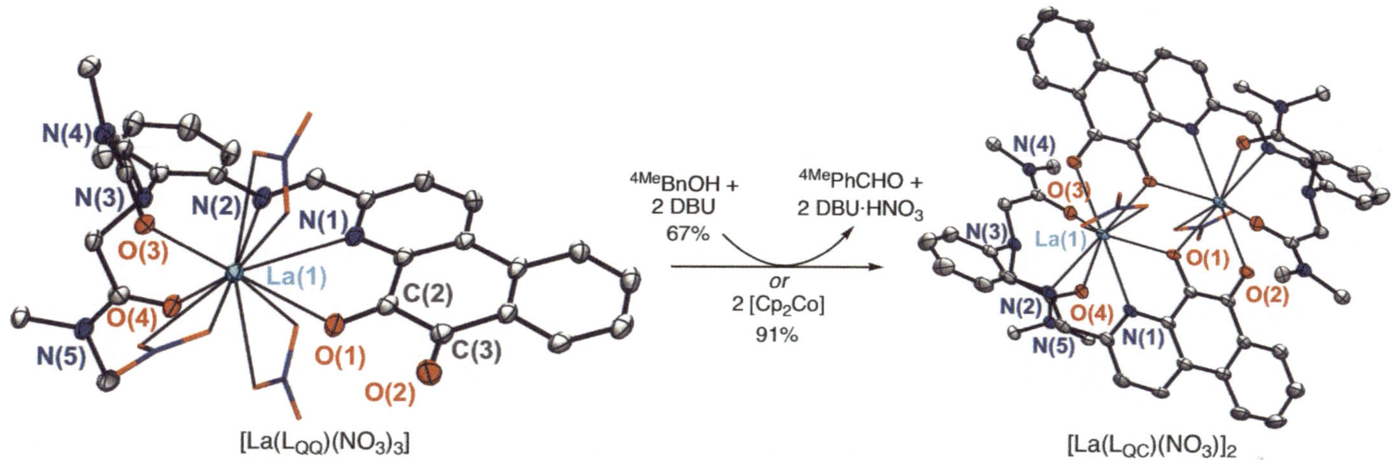

Fig. 9 Production of [La(L$_{QQ}^{2-}$)(NO$_3$)]$_2$ both stoichiometrically with DBU and ^{4Me}BnOH and independently using a chemical reduction with [Cp$_2$Co]. *Adapted with permission from the American Chemical Society: McSkimming, A., Cheisson, T., Carroll, P. J., & Schelter, E. J. (2018). Functional synthetic model for the lanthanide-dependent quinoid alcohol dehydrogenase active site.* Journal of the American Chemical Society, 140*(4), 1223–1226. https://doi.org/10.1021/jacs.7b12318. Copyright 2018 American Chemical Society.*

an aliquot of the vial in minimal deuterated solvent). Yields were determined through a comparison to the internal standard added.

Having established that [La(L_{QQ})(NO_3)$_3$] was a competent catalyst for alcohol dehydrogenation, this biomimetic system allowed the possible mechanistic pathways of the natural system to be explored. However, it is important to note the differences between the biomimetic system and the natural enzyme. For example, in [La(L_{QQ})(NO_3)$_3$], nitrates are present as the counteranion, analogous to the carboxylate groups of amino acids within the natural system. For our system, ^{4Me}BnOH was used as a test substrate for the dehydrogenation, as compared to methanol as the substrate in Nature. Finally, DBU was used in our experiments, but this base is not biologically relevant, as was previously described.

Despite the limitations, to better understand our system and gain information about the natural system, a computational study was performed in which reaction coordinates were determined for two possible mechanistic pathways (addition-elimination and hydride transfer) both in the presence and absence of a base (Fig. 10). These results indicated that, in the absence of a base, the hydride transfer mechanism was the most energetically feasible, albeit the reaction barrier is such that reactivity is slow at room temperature. In the presence of 2,6-lutidine, the hydride transfer was also favored. These

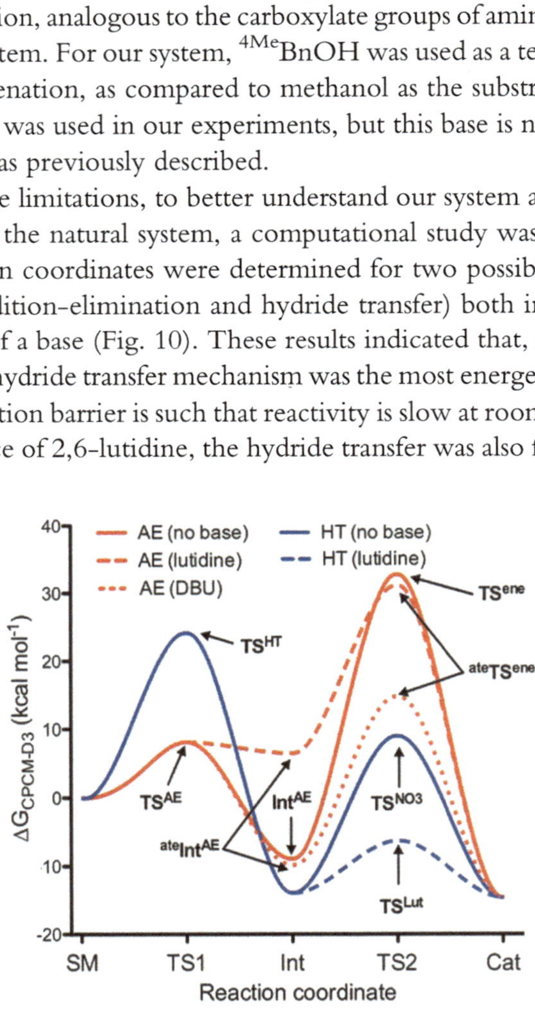

Fig. 10 Reaction coordinates for the dehydrogenation of ^{4Me}BnOH by [La(L_{QQ})(NO_3)$_3$], as calculated by DFT in CH_2Cl_2. Comparison of addition-elimination and hydride transfer mechanisms in the presence and absence of bases. *Adapted with permission from the American Chemical Society: McSkimming, A., Cheisson, T., Carroll, P. J., & Schelter, E. J. (2018). Functional synthetic model for the lanthanide-dependent quinoid alcohol dehydrogenase active site.* Journal of the American Chemical Society, *140(4), 1223–1226. https://doi.org/10.1021/jacs.7b12318. Copyright 2018 American Chemical Society.*

results were in agreement with the generally accepted mechanism for biological MDH enzymes (Idupulapati & Mainardi, 2010; Kay, Mennenga, Görisch, & Bittl, 2006; Zhang, Reddy, & Bruice, 2007; Zheng, Xia, Chen, Mathews, & Bruice, 2001).

5. Summary

In this chapter, we have detailed the synthesis and subsequent reactivity study of a synthetic model complex of a rare-earth containing enzyme. The ligand L_{QQ} is synthesized through a 14-step pathway. Once metalated with lanthanum, this system acts as a biomimetic surrogate for XoxF-type MDH, a rare-earth dependent enzyme that catalyzes the dehydrogenation of methanol to formaldehyde. The experimental findings for $[La(L_{QQ})(NO_3)_3]$ indicated that both stoichiometric and catalytic oxidization of ^{4Me}BnOH were achieved, showing that this complex serves as a biomimetic system. DFT computational work performed using the model suggested a hydride transfer mechanism as the lowest energy route to dehydrogenate ^{4Me}BnOH, which supports the hypothesis for reactivity for the biological system.

6. Experimental section

This section details alterations made to the previously reported synthesis of L_{QQ} (McSkimming et al., 2018). These changes, where noted, provided improvements in yields and/or reproducibility of the individual steps. Steps not explicitly included within this section were completed as described in the original work (McSkimming et al., 2018). All large-scale reactions were performed in a three-neck 12L round bottom flask. These reactions were stirred with an overhead stirrer, heated with a heating mantle, and cooled by immersion of the flask in a 5L bucket. A three-neck flask was chosen so reaction mixtures could be stirred with the overhead stirrer while reagents and solvents were added at the same time.

6.1 Preparation of compound 4

Note—This synthesis was adapted from conditions reported in a patent (Georg, 1953).

a. 600g of urea were heated to 150°C until melted.

b. 300g of 3-hydroxy-2-naphthoic acid were added to yield a brown mixture.

c. While still heating, 9mL phosphoric acid were added dropwise, and the heat was increased to 180°C. Safety note—addition of phosphoric acid leads to bubbling within the reaction. This addition is exothermic and can effervesce if added too quickly.

d. The solution was stirred while heating at 180 °C for 2 h, after which the reaction mixture was cooled to 150 °C.

e. Orange solid was precipitated by addition of 6 L H_2O to the reaction mixture. This solution was allowed to cool to room temperature, filtered and dried on a Büchner funnel.

f. The orange solid was dissolved in 1.5 L DMSO while heating to 110 °C.

g. Once fully dissolved, the solution was removed from heat and 4 L H_2O were added to precipitate **4** as a yellow-orange solid, which was filtered and dried on a Büchner funnel to give the compound in 90% yield.

^{1}H NMR (400 MHz, d_6-DMSO): δ 12.45 (s, 1H), 8.60 (s, 1H), 8.53 (d, J=3.8 Hz, 1H), 8.02 (s, 1H), 7.81 (dd, J=8.6, 4.0 Hz, 1H), 7.73 (dd, J=8.6, 4.1 Hz, 1H), 7.50 (m, 1H), 7.34 (m, 1H), 7.48 (d, J=4.0 Hz, 1H) ppm (Fig. 11, Scheme 18).

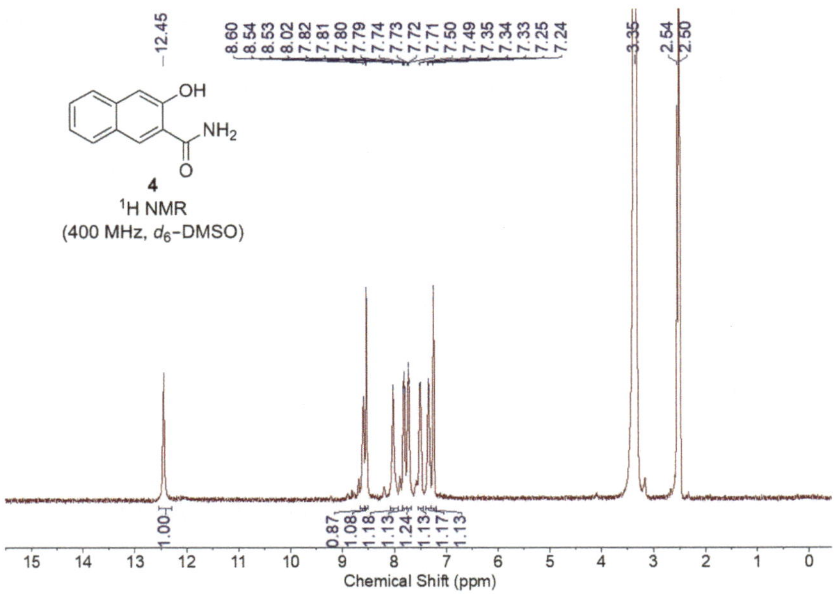

Fig. 11 ^{1}H NMR of **4** recorded at 400 MHz in d_6-DMSO. Peaks at δ 3.35 and 2.50 ppm are residual H_2O and DMSO, respectively.

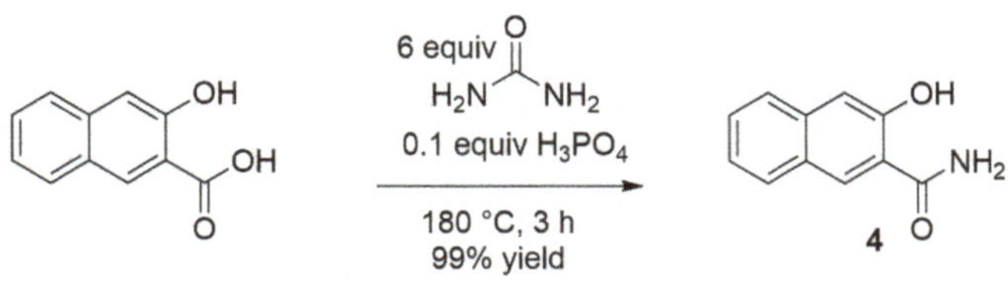

Scheme 18 Preparation of compound **4**.

6.2 Preparation of compound 5

a. 30 g of **4** were dissolved in 300 mL DMSO.

b. 33.8 mL Et_3N were added to the reaction mixture.

c. The reaction mixture was cooled in an ice bath (~10 °C) until nearly frozen.

d. 37.01 g of TsCl were added slowly and the reaction mixture was stirred while in the ice bath for 0.5 h.

e. The yellow-white solid was precipitated from solution by addition of 2 L H_2O, filtered and dried on a Büchner funnel.

f. The solid was resuspended in 2 L isopropanol, filtered, and dried on a Büchner funnel to obtain **5** as a white solid in a 96% yield.

^{1}H NMR (300 MHz, d_6-DMSO): δ 8.14 (s, 1H), 8.02 (m, 1H), 7.94 (m, 1H), 7.77 (d, J = 8.3 Hz, 2H), 7.76 (s, 1H), 7.67 (s, 1H), 7.60 (m, 2H), 7.50 (s, 1H), 7.45 (d, J = 8.0 Hz, 2H), 2.42 (s, 3H) ppm (Fig. 12, Scheme 19).

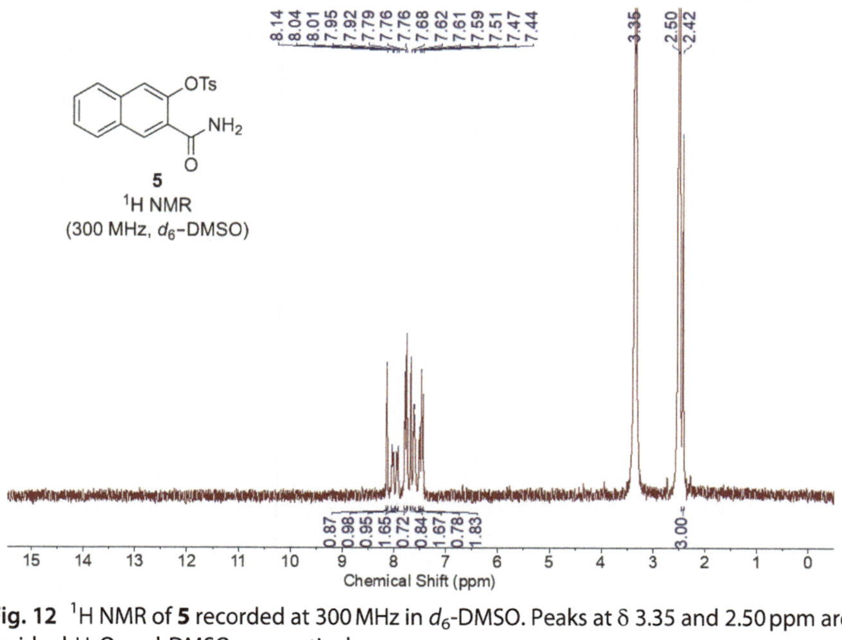

Fig. 12 ^{1}H NMR of **5** recorded at 300 MHz in d_6-DMSO. Peaks at δ 3.35 and 2.50 ppm are residual H_2O and DMSO, respectively.

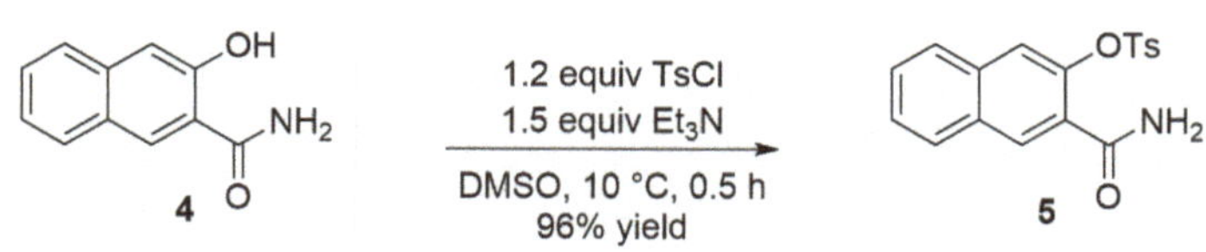

Scheme 19 Preparation of compound **5**.

6.3 Preparation of compound 6

a. 1 L of 6.5 M NaOH and 250 mL dioxane were cooled to 10 °C in an ice bath.

b. 11.5 mL Br_2 were added dropwise using an addition funnel, which produced a yellow solution.

c. **5** was added and the reaction mixture was stirred while in an ice bath for 0.5 h.

d. The reaction mixture was transferred to an oil bath and refluxed (with the use of a condenser) at 100 °C for 16 h, which produced a dark brown solution.

e. The reaction mixture was allowed to cool to room temperature and acidified to pH 0 with 12 M HCl.

f. 1 L Et_2O was added and the aqueous layer was separated and filtered.

g. The filtrate was treated with base to reach pH 3 using 3 M NaOH and then to pH 8 using a saturated solution of Na_2CO_3.

h. The resulting precipitate was filtered and dried on a Büchner funnel to obtain **6** as a light brown solid in a 65% yield.

^{1}H NMR (400 MHz, d_6-DMSO): δ 9.73 (s, 1H), 7.45 (d, J = 7.9 Hz, 1H), 7.41 (d, J = 7.8 Hz, 1H), 7.09 (t, J = 6.3 Hz, 1H), 7.03 (t, J = 6.6 Hz, 1H), 6.99 (s, 1H), 6.85 (s, 1H), 5.00 (s, 2H) ppm (Fig. 13, Scheme 20).

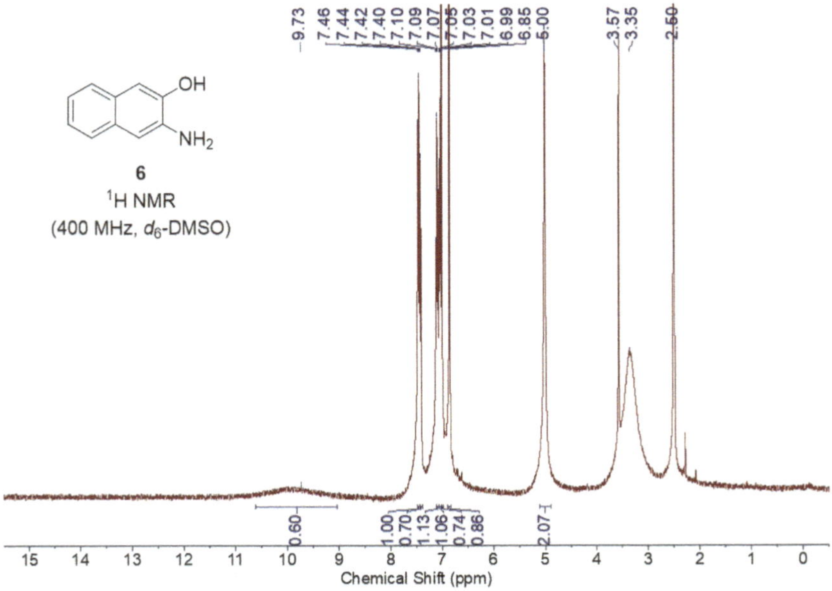

Fig. 13 ^{1}H NMR of **6** recorded at 400 MHz in d_6-DMSO. Peaks at δ 3.57, 3.35 and 2.50 ppm are 1,4-dioxane, residual H_2O, and DMSO, respectively.

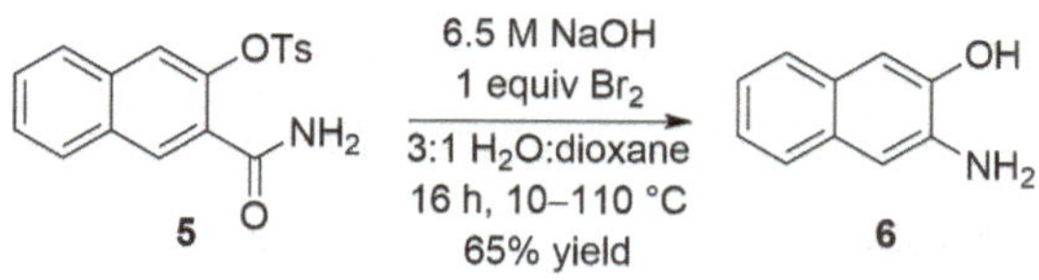

Scheme 20 Preparation of compound **6**.

6.4 Preparation of compound 11

a. 0.5 g of **10** were dissolved in 100 mL CH_3CN.
b. 3.2 g of polymer-supported IBX were added as a solid.
c. The reaction mixture was stirred at room temperature for 16 h.
d. The yellow-orange suspension was filtered on a medium porosity frit and the solid was washed with 3 × 15 mL CH_3CN until the filtrate was colorless.
e. The filtrate was passed through a pad of Celite (~5 cm in diameter and ~10 cm in height) using about 50 mL of a 1:1 solution of Et_2O and CH_2Cl_2 as eluent.
f. The combined filtrate was concentrated under reduced pressure.
g. Orange solid was precipitated out of solution through addition of 15 mL 1,2-dichloroethane.
h. This solid was filtered and dried on a medium porosity fritted filter to afford **11** as an orange solid in a 30% yield.

^1H NMR data were consistent with the reported spectrum for this compound (Fig. 14) (McSkimming et al., 2018).

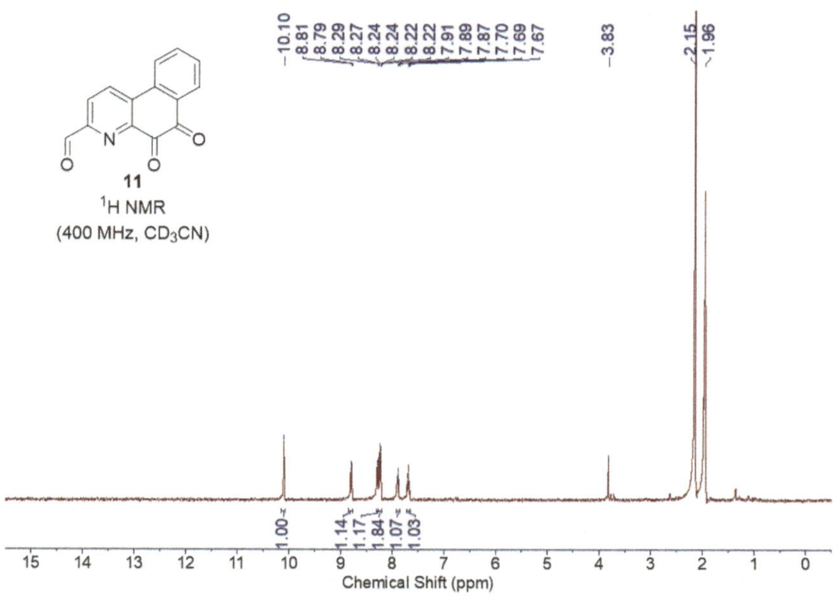

Fig. 14 ^1H NMR of **11** recorded at 400 MHz in CD_3CN. Peaks at δ 3.83, 2.15, and 1.96 ppm are 1,2-dichloroethane, residual H_2O, and acetonitrile, respectively.

^{1}H NMR (400 MHz, CD_3CN): δ 10.09 (s, 1H), 8.80 (d, J=8.4 Hz, 1H), 8.28 (d, J=7.9 Hz, 1H), 8.23 (d, J=8.5 Hz, 1H), 8.22 (d, J=7.8), 7.89 (t, J=7.7 Hz, 1H), 7.69 (t, J=7.5 Hz, 1H) ppm (Fig. 14, Scheme 21).

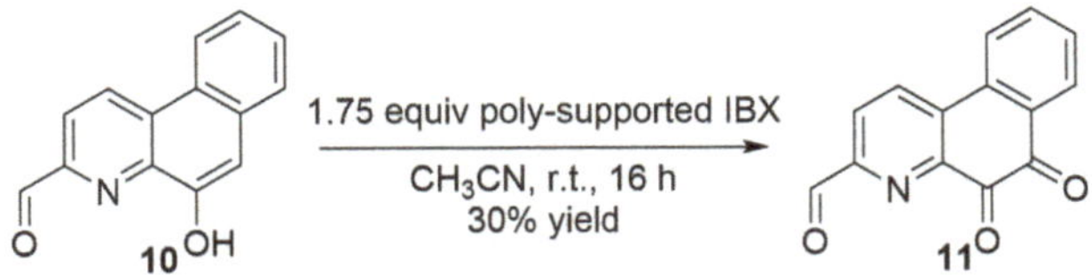

Scheme 21 Preparation of compound **11**.

6.5 Preparation of compound 13

a. 31.8 g of **12**, 6.6 g of $BnNH_2$, 57.3 g of KI, 44.3 g of K_2CO_3, and 650 mL CH_3CN were added to a flask and refluxed at 80 °C for 16 h.

b. The reaction mixture was cooled to room temperature and 400 mL H_2O were added in one portion to produce a brown, sticky precipitate.

c. The mixture was extracted twice with a solution of 3:1 hexanes:EtOAc, washed with brine, and dried over Na_2SO_4.

d. The mixture was filtered on a medium porosity frit and the filtrate was concentrated under reduced pressure to afford a sticky solid, which was recrystallized in ethanol to give **13** as a white solid in a 65% yield.

^{1}H NMR (400 MHz, $CDCl_3$): δ 7.41 (d, J=8.6 Hz, 2H), 7.29 (m, 3H), 3.95 (s, 2H), 3.46 (s, 4H), 2.48 (s, 4H), 1.74–1.02 (m, 40H) ppm (Fig. 15, Scheme 22).

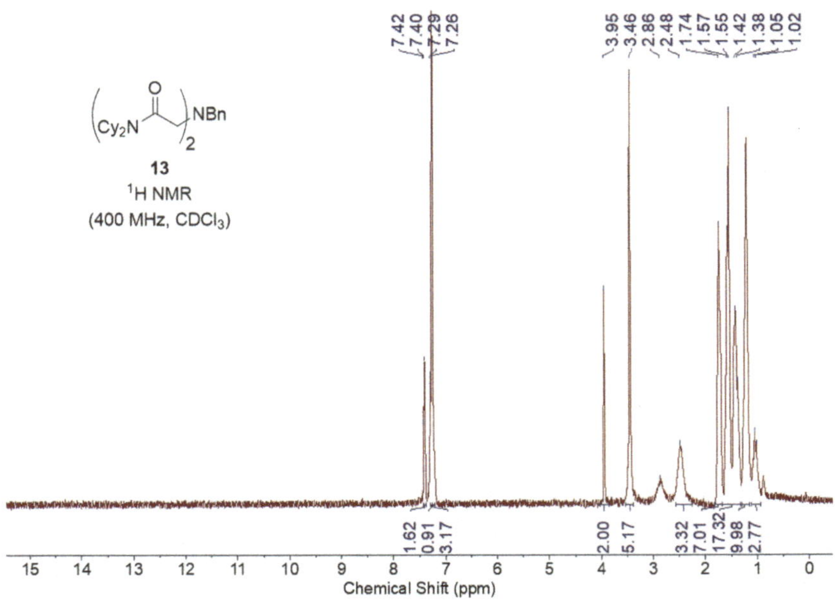

Fig. 15 ^{1}H NMR of **13** recorded at 400 MHz in $CDCl_3$. Peaks at δ 7.26 and 1.56 ppm are chloroform and residual H_2O, respectively.

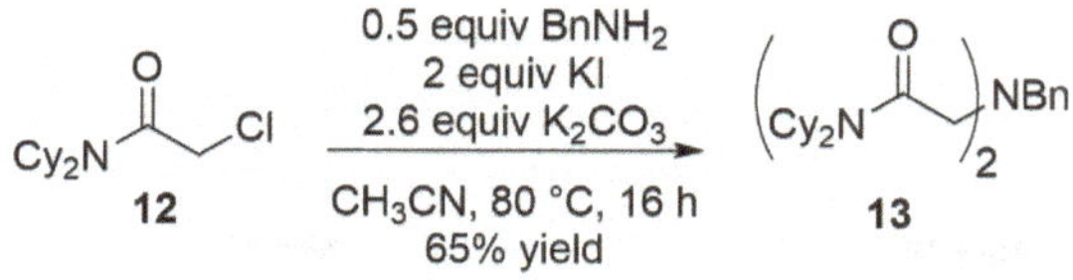

Scheme 22 Preparation of compound **13**.

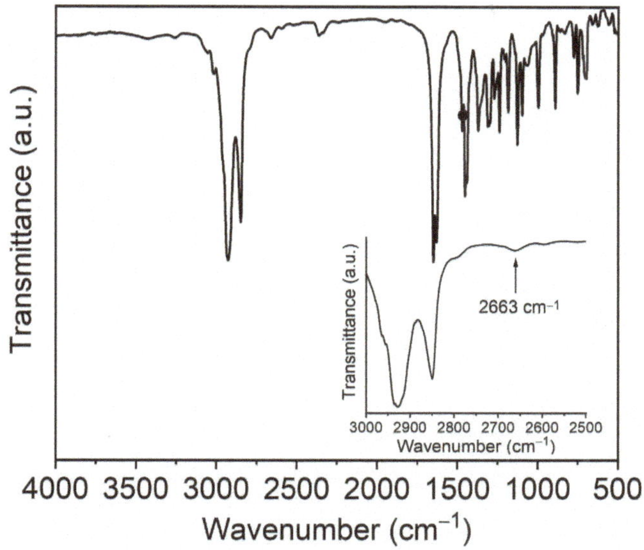

Fig. 16 IR spectrum of ammonium salt of compound **14**. Inset shows peak for tertiary ammonium stretch.

We determined that compound **13** was isolated as a secondary ammonium salt from the previous reaction conditions, as supported by a diagnostic N-H stretch in the IR spectrum (Fig. 16). Although we are not able to identify a discrete anion, the addition of 1 equiv. of base allowed the reaction to proceed. The modified synthesis, which allowed isolation of the compound, is included below.

6.6 Preparation of compound 14

a. 13.5 g of **13** was dissolved in 150 mL ethanol.

b. 2.8 mL Et_3N were added and the reaction mixture was stirred at room temperature.

c. After 15 min of stirring, 13.9 g of NH_4CO_2H and 2.2 g of 10% Pd/C were added and the reaction mixture was stirred at 80 °C for 16 h.

d. Once cooled to room temperature, the crude mixture was filtered over Celite and washed with ethanol until the eluent was colorless.

e. The filtrate was concentrated under reduced pressure to afford **14** as a white solid in a 90% yield.

^{1}H NMR data were consistent with the reported spectrum for compound **14** (Fig. 17, Scheme 23) (McSkimming et al., 2018).

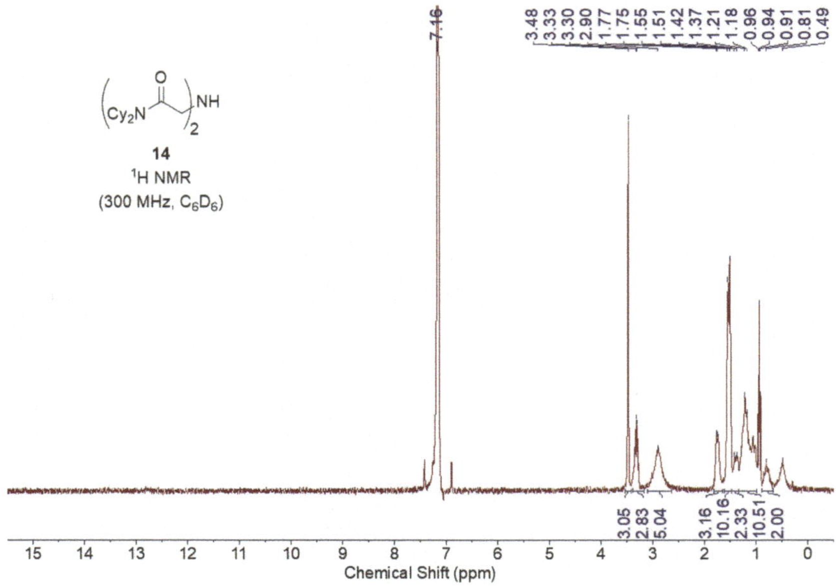

Fig. 17 ^{1}H NMR of **14** recorded at 300 MHz in C_6D_6. The peak at δ 7.16 ppm is due to benzene, and the peaks at δ 3.30, 0.96, 0.50 ppm are due to ethanol.

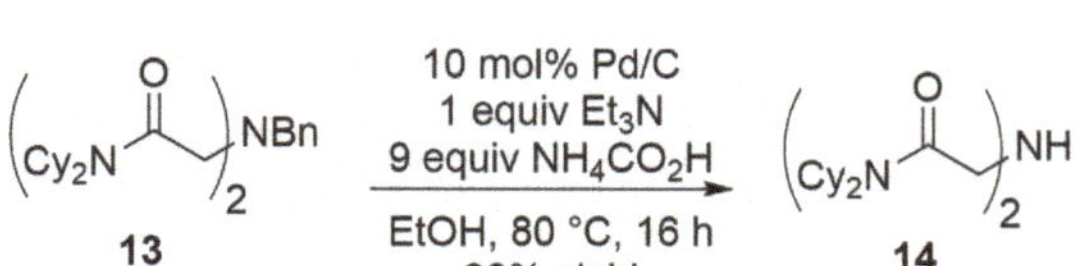

Scheme 23 Preparation of compound **14**.

6.7 Preparation of compound 16

a. 3.12 g of **15** were dissolved in 1.6 mL of 12 M HCl and 16 mL ethanol.

b. 4.36 g of $SnCl_2 \cdot (H_2O)_2$ were added and the reaction mixture was heated to 80 °C for 32 h.

c. The mixture was allowed to cool to room temperature and was then diluted with water and the pH was raised to pH 14 by addition of 3 M NaOH.
d. 15 mL of saturated $NaHCO_3$ were added and the reaction mixture was stirred for 15 min at room temperature.
e. The mixture was filtered over Celite and the filtrate was extracted with 3 × 50 mL dichloromethane.
f. The combined organic fractions were washed with 50 mL brine and dried over Na_2SO_4.
g. This mixture was filtered and concentrated under reduced pressure to afford **16** as a brown solid in a 79% yield.

^1H NMR data were consistent with the reported spectrum for this compound (Fig. 18, Scheme 24) (McSkimming et al., 2018).

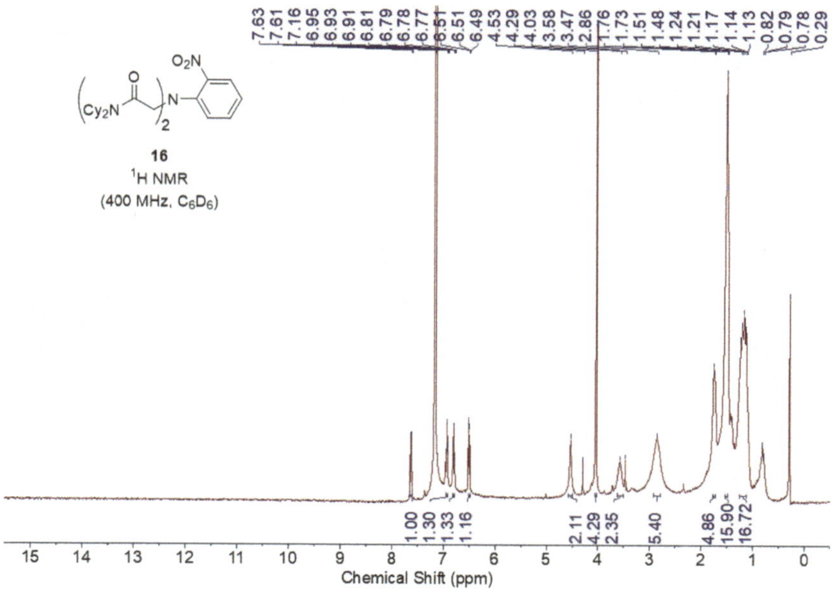

Fig. 18 ^1H NMR of **16** recorded at 400 MHz in C_6D_6. The peaks at δ 7.16 and 0.29 ppm are due to benzene and silicone grease, respectively.

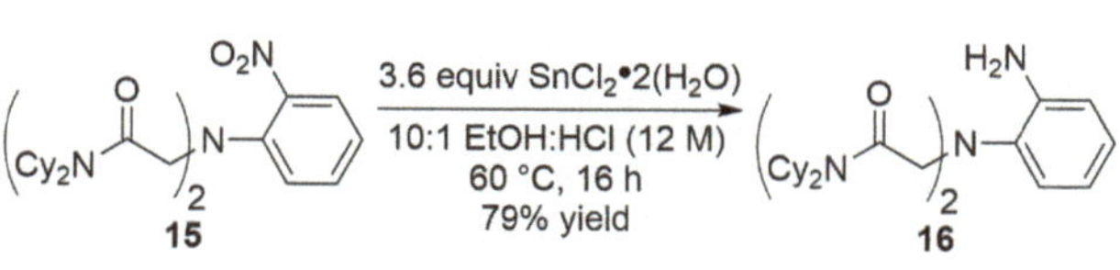

Scheme 24 Preparation of compound **16.**

Acknowledgments

We thank the University of Pennsylvania for support of this work. The Vagelos Institute for Energy Science and Technology (VIEST) is acknowledged for a fellowship to A.L.K. We also thank Dr. Robert F. Higgins for helpful discussions.

References

Bakewell, C., Platel, R. H., Cary, S. K., Hubbard, S. M., Roaf, J. M., Levine, A. C., et al. (2012). Bis(8-quinolinolato)aluminum ethyl complexes: Iso-selective initiators for *rac*-lactide polymerization. *Organometallics*, *31*(13), 4279–4736. https://doi.org/10.1021/om300307t.

Bogart, J. A., Lewis, A. J., & Schelter, E. J. (2015). DFT study of the active site of the XoxF-type natural, cerium-dependent methanol dehydrogenase enzyme. *Chemistry—A European Journal*, *21*(4), 1743–1748. https://doi.org/10.1002/chem.201405159.

Bonificio, W. D., & Clarke, D. R. (2016). Rare-earth separation using bacterial. *Environmental Science & Technology Letters*, *3*(4), 180–184. https://doi.org/10.1021/acs.estlett.6b00064.

Cady, C. W., Crabtree, R. H., & Brudvig, G. W. (2008). Functional models for the oxygen-evolving complex of photosystem II. *Coordination Chemistry Reviews*, *252*(3–4), 444–455. https://doi.org/10.1016/j.ccr.2007.06.002.

Cheisson, T., & Schelter, E. J. (2019). Rare earth elements: Mendeleev's bane, modern marvels. *Science*, *363*(6426), 489–493. https://doi.org/10.1126/science.aau7628.

Chen, Z.-F., Shi, Y.-F., Liu, Y.-C., Hong, X., Geng, B., Peng, Y., et al. (2012). TCM active ingredient oxoglaucine metal complexes: Crystal structure, cytotoxicity, and interaction with DNA. *Inorganic Chemistry*, *51*(4), 1998–2009. https://doi.org/10.1021/ic200443p.

Cotruvo, J. A., Jr. (2019). The chemistry of lanthanides in biology: Recent discoveries, emerging principles, and technological applications. *ACS Central Science*, *5*(9), 1496–1506. https://doi.org/10.1021/acscentsci.9b00642.

Cotruvo, J. A., Jr., Featherson, E. R., Mattocks, J. A., Ho, J., & Laremore, T. N. (2018). Lanmodulin: A highly selective lanthanide-binding protein from a lanthanide-utilizing bacterium. *Journal of the American Chemical Society*, *140*(44), 15056–15061. https://doi.org/10.1021/jacs.8b09842.

Culpepper, M. A., & Rosenzweig, A. C. (2012). Architecture and active site of particulate methane monooxygenase. *Critical Reviews in Biochemistry and Molecular Biology*, *47*(6), 483–492. https://doi.org/10.3109/10409238.2012.697865.

Daumann, L. J. (2019). Essential and ubiquitous: The emergence of lanthanide biochemistry. *Angewandte Chemie International Edition*, *58*(37), 12795–12802. https://doi.org/10.1002/anie.201904090.

Deng, Y. W., Ro, S. Y., & Rosenzweig, A. C. (2018). Structure and function of the lanthanide-dependent methanol dehydrogenase XoxF from the methanotroph *Methylomicrobium buryatense* 5GB1C. *Journal of Biological Inorganic Chemistry*, *23*(7), 1037–1047. https://doi.org/10.1007/s00775-018-1604-2.

Dorfner, W. L., Carroll, P. J., & Schelter, E. J. (2015). Substituted quinoline quinones as surrogates for the PQQ cofactor: An electrochemical and computational study. *Organic Letters*, *17*(8), 1850–1853. https://doi.org/10.1021/acs.orglett.5b00486.

Georg, H. (1953). *Patent No. DE869639C. German Patent and Trademark Office.*

Gnanaprakasam, B., Zhang, J., & Milstein, D. (2010). Direct synthesis of imines from alcohols and amines with liberation of H_2. *Angewandte Chemie International Edition*, *49*(8), 1468–1471. https://doi.org/10.1002/anie.200907018.

Haxel, G. B., Hedrick, J. B., & Orris, G. J. (2002). *Rare earth elements—Critical resources for high technology. US Geological Survey, Fact Sheet 087-02.*

Hibi, Y., Asai, K., Arafuka, H., Hamajima, M., Iwama, T., & Kawai, K. (2011). Molecular structure of La^{3+}-induced methanol dehydrogenase-like protein in *Methylobacterium radiotolerans*. *Journal of Bioscience and Bioengineering, 111*(5), 547–549. https://doi.org/10.1016/j.jbiosc.2010.12.017.

Huang, J., Yu, Z., & Chistoserdova, L. (2018). Lanthanide-dependent methanol dehydrogenases of XoxF4 and XoxF5 clades are differentially distributed among methylotrophic bacteria and they reveal different biochemical properties. *Frontiers in Microbiology, 9*, 1366. https://doi.org/10.3389/fmicb.2018.01366.

Huang, J., Yu, Z., Groom, J., Cheng, J.-F., Tarver, A., Yoshikuni, Y., et al. (2019). Rare earth element alcohol dehydrogenases widely occur among globally distributed, numerically abundant and environmentally important microbes. *The ISME Journal, 13*, 2005–2017. https://doi.org/10.1038/s41396-019-0414-z.

Idupulapati, N. B., & Mainardi, D. S. (2010). Quantum chemical modeling of methanol oxidation mechanisms by methanol dehydrogenase enzyme: Effect of substitution of calcium by barium in the active site. *The Journal of Physical Chemistry A, 114*(4), 1887–1896. https://doi.org/10.1021/jp9083025.

Itoh, S., Huang, X., Kawakami, H., Komatsu, M., Ohshiro, Y., & Fukuzumi, S. (1995). An active site model for calcium(II)-containing quinoproteins. *Journal of the Chemical Society, Chemical Communications, 20*, 2077–2078. https://doi.org/10.1039/C39950002077.

Itoh, S., Kawakami, H., & Fukuzumi, S. (1997). Modeling of the chemistry of quinoprotein methanol dehydrogenase. Oxidation of methanol by calcium complex of coenzyme PQQ via addition-elimination mechanism. *Journal of the American Chemical Society, 119*(2), 439–440. https://doi.org/10.1021/ja963366d.

Itoh, S., Kawakami, H., & Fukuzumi, S. (1998). Model studies on calcium-containing quinoprotein alcohol dehydrogenases. Catalytic role of Ca^{2+} for the oxidation of alcohols by coenzyme PQQ (4,5-dihydro-4,5-dioxo-1H-pyrrolo[2,3,f]quinoline-2,7,9-tricarboxylic acid). *Biochemistry, 37*(18), 6562–6571. https://doi.org/10.1021/bi9800092.

Jia, Z., Ramstad, T., & Zhong, M. (2001). Medium-throughput pK_a screening pharmaceuticals by pressure-assisted capillary electrophoresis. *Electrophoresis, 22*(6), 1112–1118. https://doi.org/10.1002/1522-2683()22:6<1112::AID-ELPS1112>3.0.CO;2-1.

Joshi, M. C., Wicht, K. J., Taylor, D., Hunter, R., Smith, P. J., & Egan, T. J. (2013). In vitro antimalarial activity, β-haematin inhibition and structure-activity relationships in a series of quinoline triazoles. *European Journal of Medicinal Chemistry, 69*, 338–347. https://doi.org/10.1016/j.ejmech.2013.08.046.

Kanady, J. S., Tsui, E. Y., Day, M. W., & Agapie, T. (2011). A synthetic model of Mn_3Ca subsite of the oxygen-evolving complex in photosystem II. *Science, 33*(6043), 733–736. https://doi.org/10.1126/science.1206036.

Kaupmees, K., Trumall, A., & Leito, I. (2014). Basicities of strong bases in water: A computational study. *Croatica Chemica Acta, 87*(4), 385–395. https://doi.org/10.5562/cca2472.

Kawahara, R., Fujita, K.-I., & Yamaguchi, R. (2012). Dehydrogenative oxidation of alcohols in aqueous media using water-soluble and reusable Cp*Ir catalysts bearing a functional bipyridine ligand. *Journal of the American Chemical Society, 134*(8), 3643–3646. https://doi.org/10.1021/ja210857z.

Kay, C. W. M., Mennenga, B., Görisch, H., & Bittl, R. (2006). Substrate binding in quinoprotein ethanol dehydrogenase from *Pseudomonas aeruginosa*; studied by electron-nuclear double resonance. *Proceedings of the National Academy of Sciences of the United States of America, 103*(14), 5267–5272. https://doi.org/10.1073/pnas.0509667103.

Keltjens, J. T., Pol, A., Reimann, J., & Op den Camp, H. J. M. (2014). PQQ-dependent methanol dehydrogenases: Rare-earth elements make a difference. *Applied Microbiology and Biotechnology, 98*(14), 6163–6183. https://doi.org/10.1007/s00253-014-5766-8.

Liu, Y.-C., Chen, Z.-F., Liu, L.-M., Peng, Y., Hong, X., Yang, B., et al. (2009). Divalent later transition metal complexes of the traditional Chinese medicine (TCM) liriodenine: Coordination chemistry, cytotoxicity, and DNA binding studies. *Dalton Transactions*, *48*, 10813–10823. https://doi.org/10.1039/b912553a.

Lumpe, H., Menke, A., Haisch, C., Mayer, P., Kabelitz, A., Buzanich, A. G., et al. (2020). The earlier the better: Structural analysis and separation of lanthanides with pyrroloquinoline quinone. *Chemistry—A European Journal*, *26*(44), 10133–10139. https://doi.org/10.1002/chem.202002653.

McSkimming, A., Cheisson, T., Carroll, P. J., & Schelter, E. J. (2018). Functional synthetic model for the lanthanide-dependent quinoid alcohol dehydrogenase active site. *Journal of the American Chemical Society*, *140*(4), 1223–1226. https://doi.org/10.1021/jacs.7b12318.

Mitome, H., Ishizuka, T., Shiota, Y., Yoshizawa, K., & Kojima, T. (2013). Heteronuclear $Ru^{II}Ag^{I}$ complexes having a pyrroloquinolinequinone derivative as a bridging ligand. *Inorganic Chemistry*, *52*(5), 2274–2276. https://doi.org/10.1021/ic302617b.

Mitome, H., Ishizuka, T., Shiota, Y., Yoshizawa, K., & Kojima, T. (2015). Controlling the redox properties of a pyrroloquinolinequinone (PQQ) derivative in a ruthenium(II) coordination sphere. *Dalton Transactions*, *44*(7), 3151–3158. https://doi.org/10.1039/C4DT03358B.

Nakagawa, T., Mitsui, R., Tani, A., Sasa, K., Tashiro, S., Iwama, T., et al. (2012). A catalytic role of XoxF1 as La^{3+}-dependent methanol dehydrogenase in *Methylobacterium extorquens* strain AM1. *PLoS One*, 7(11), e50480. https://doi.org/10.1371/journal.pone.0050480.

Nakamura, N., Kohzuma, T., Kuma, H., & Suzuki, S. (1994). Synthetic and structural studies on copper(II) complexes containing coenzyme PQQ and terpyridine. *Inorganic Chemistry*, *33*(8), 1594–1599. https://doi.org/10.1021/ic00086a007.

Park, D. M., Brewer, A., Reed, D. W., Lammers, L. N., & Jiao, Y. (2017). Recovery of rare earth elements from low-grade feedstock leachates using engineered bacteria. *Environmental Science & Technology*, *51*(22), 13471–13480. https://doi.org/10.1021/acs.est.7b02414.

Picone, N., & Op den Camp, H. J. M. (2019). Role of rare earth elements in methanol oxidation. *Current Opinion in Chemical Biology*, *49*, 39–44. https://doi.org/10.1016/j.cbpa.2018.09.019.

Pierpont, C. G., & Buchanan, R. M. (1981). Transition metal complexes of *o*-benzoquinone, *o*-semiquinone, and catecholate ligands. *Coordination Chemistry Reviews*, *38*(1), 45–87. https://doi.org/10.1016/S0010-8545(00)80499-3.

Pol, A., Barends, T. R. M., Dietl, A., Khadem, A. F., Eygensteyn, J., Jetten, M. S. M., et al. (2014). Rare earth metals are essential for methanotrophic life in volcanic mudpots. *Environmental Microbiology*, *16*(1), 255–264. https://doi.org/10.1111/1462-2920.12249.

Rudnick, R., & Gao, S. (2003). Composition of the continental crust. *Treatise on Geochemistry*, *3*, 1–64. https://doi.org/10.1016/B0-08-043751-6/03016-4.

Shannon, R. (1976). Revised effective ionic radii and systematic studies of interatomic distances in halides and chalcogenides. *Acta Crystallographica Section A*, *32*(5), 751–767. https://doi.org/10.1107/S0567739476001551.

Shiller, A. M., Chan, E. W., Joung, D. J., Redmond, M. C., & Kessler, J. D. (2017). Light rare earth element depletion during Deepwater Horizon blowout methanotrophy. *Scientific Reports*, 7(1), 10389. https://doi.org/10.1038/s41598-017-11060-z.

Suzuki, S., Sakurai, T., Itoh, S., & Ohshiro, Y. (1988). Preparation and characterization of ternary copper(II) complexes containing coenzyme PQQ and bipyridine or terpyridine. *Inorganic Chemistry*, 27(4), 591–592. https://doi.org/10.1021/ic00277a001.

Tinberg, C. E., & Lippard, S. J. (2011). Dioxygen activation in soluble methane monooxygenase. *Accounts of Chemical Research*, *44*(4), 280–288. https://doi.org/10.1021/ar1001473.

Wanner, M., Sixt, T., Klinkhammer, K.-W., & Kaim, W. (1999). First experimental structure of a 1:1 metal complex with a PQQ cofactor derivative outside dehydrogenase enzymes. *Inorganic Chemistry*, *38*(11), 2753–2755. https://doi.org/10.1021/ic981390h.

Wei, J.-H., Chen, Z.-F., Qin, J.-L., Liu, Y.-C., Li, Z.-Q., Khan, T.-M., et al. (2015). Water-soluble oxoglaucine-Y(III), Dy(III) complexes: *In vitro* and *in vivo* anticancer activities by triggering DNA damage, leading to S phase arrest and apoptosis. *Dalton Transactions*, *44*(25), 11408–11419. https://doi.org/10.1039/C5DT00926J.

Zhang, X., Reddy, S. Y., & Bruice, T. C. (2007). Mechanism of methanol oxidation by quinoprotein methanol dehydrogenase. *Proceedings of the National Academy of Sciences of the United States of America*, *104*(3), 745–749. https://doi.org/10.1073/pnas.0610126104.

Zheng, Y. J., Xia, Z., Chen, Z., Mathews, F. S., & Bruice, T. C. (2001). Catalytic mechanism of quinoprotein methanol dehydrogenase: A theoretical and x-ray crystallographic investigation. *Proceedings of the National Academy of Sciences of the United States of America*, *98*(2), 432–434. https://doi.org/10.1073/pnas.98.2.432.

CHAPTER THREE

Activity assays of methanol dehydrogenases

Sophie M. Gutenthaler, Manh Tri Phi, Helena Singer, and Lena J. Daumann*
Department of Chemistry, Ludwig-Maximilians-Universität München, München, Germany
*Corresponding author: e-mail address: lena.daumann@lmu.de

Contents

Abstract

The field of methanol dehydrogenases (MDHs) has experienced revival in the recent decade due to the observation of lanthanide-dependent MDH, in addition to widely known calcium-MDH. With the advent of lanthanide-dependent alcohol dehydrogenases, the need for reliable assays to evaluate and compare activities between different MDHs is obvious: from extremophilic to neutrophilic organisms, or with different lanthanide ions in the active site. Here we outline four assays that have been reported for Ln-MDH, discussing the advantages and disadvantages of the assays and their components. It should be noted, in 1990 Day and Anthony produced a comprehensive summary in Methods in Enzymology on the available methods for Ca-MDH assays at the time (Day & Anthony, 1990). This chapter is an updated appraisal of the most important developments in the last 30 years.

Methods in Enzymology, Volume 650
ISSN 0076-6879
https://doi.org/10.1016/bs.mie.2021.01.045

1. Assays based on artificial electron acceptors

Pyrroloquinoline quinone-dependent MDHs are responsible for the oxidation of primary alcohols such as methanol to formaldehyde in many bacteria, utilizing a specific cytochrome as an electron acceptor (EA). For kinetic investigations of MDH, the physiological electron acceptor cytochrome can be replaced with artificial components. One of the most widely used artificial dye-coupled assays is based on a protocol by Anthony from the early 1960s. The experimental conditions have been optimized and adjusted for use in physiological environments over the decades. Using only artificial components for enzymatic activity determination has two major advantages over physiological cytochrome EAs: they are commercially available, and can be used independent of the organism and its specific cytochrome (Scheme 1).

Alkylated phenazines, such as phenazine ethosulfate (PES) or phenazine methosulfate (PMS), are used as artificial primary EAs for the regeneration of the redox cofactor pyrroloquinoline quinone (PQQ) in the active site of MDH (Scheme 1) (Anthony, 2000). The oxidation of substrates can be visualized spectrophotometrically by the final EA: 2,6-dichlorophenol indophenol (DCPIP), which decolorizes on reduction at 600 nm; or with the redox dye Wurster's Blue (WB) at 610 nm (Jahn et al., 2020; Myers, 1990).

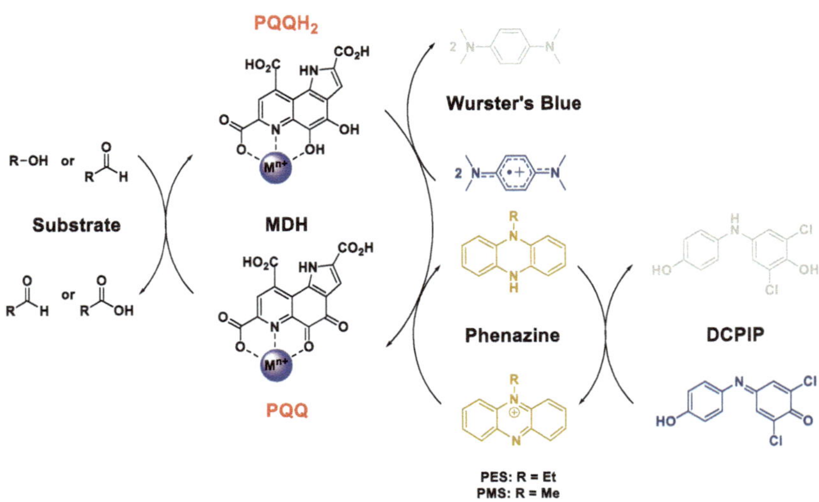

Scheme 1 Overview of the artificial dye assays for methanol dehydrogenases (Jahn et al., 2020).

If WB is used, no further EA is required since the redox indicator WB acts as both an EA and a redox dye (Anthony, 2000; Jahn et al., 2020).

Besides monitoring the enzymatic activity with exclusively artificial components, an additional method is described using the physiological electron acceptor, cytochrome c_L, coupled to the spectrophotometric output of DCPIP (Cox, Day, & Anthony, 1992). To quantify the enzymatic activity of MDH and to enable comparison between MDHs, the specific activity can be determined. Besides the amount of enzyme used, further parameters and values need to be identified to calculate the enzyme unit, *U*. The extinction coefficient ε_{600nm} for DCPIP or ε_{610nm} for WB depends on the pH as well as the temperature and buffer system. As various values for ε can be found in the literature (Jahn et al., 2020), determination of this value is recommended for each respective system based on the Beer–Lambert law. The pathlength of the cell and the volume of assay are set by the equipment and device. The initial rate of slope should be calculated after the addition of substrate; a period of 1–5 min is optimal.

$$\textit{specific activity}\left[\mu mol\ \mathrm{min}^{-1}\ mg^{-1}\right]=\frac{\textit{enzyme unit } U\left[\mu mol\ \mathrm{min}^{-1}\right]}{\textit{amount of enzyme}\left[mg\right]}$$

$$\textit{enzyme unit } U\left[\mu mol\ min^{-1}\right]=\frac{-1\cdot \textit{initial rate of slope of measurement}}{\varepsilon\left[cm^{-1}M^{-1}\right]\cdot \textit{pathlength of cell}\left[cm\right]}\cdot 10^{6}\cdot \textit{volume of assay}\left[L\right]$$

A note on the equipment: As the assay should be carried out under light exclusion for elimination of unwanted side-reactions, even if the instrument is room-light immune, measurements with an open compartment are not recommended. Performing the assay is possible with double-beam, as well as with single-beam spectrophotometers. However, diode array spectrophotometers have the distinct disadvantage that they induce phenazine dye side-reactions and decomposition. Here, no individual wavelength is selected for running enzyme kinetics, but rather the entire spectrum from, e.g., 200–800 nm interacts with the sample in the cuvette (Lente & Espenson, 2004). Furthermore, since enzymatic activity of MDH is dependent on temperature, temperature control during the measurement is indispensable for accurate and reproducible results: a thermostatted cell holder based on a Peltier element provides a convenient option. To enable the acquisition of several samples simultaneously, a microplate spectrophotometer can be used. Here, if possible, a temperature gradient of 1–2 °C should be set to prevent condensation of assay mix vapors on the microplate lid. Some plate reader instruments have a condensation control feature already built in. If this is not possible, the measurements can as well be performed without a lid, given that the assay temperatures are below 30 °C. When performing the measurements in a cuvette, the solution should be thoroughly mixed at all times. This can be obtained by using a cuvette equipped with a stirring bar. When using a plate reader for the activity determination, a shaking step should be implemented after the addition of each component.

1.1 Phenazine and DCPIP assay

A note on phenazines: Alkylated phenazines are used in the dye-coupled assay to replace the physiological EA cytochrome. Mainly phenazine ethosulfate (PES) and methosulfate (PMS) are mentioned in the literature (Ghosh & Quayle, 1979). However, PES is the preferred EA during enzymatic activity determination due to reduced radical formation, and lower tendency for dealkylation and phenazine formation (Ghosh & Quayle, 1979; Jahn et al., 2020). PMS decomposes more easily into phenazine when additives such as KCN or ammonia are added, lowering the concentration of accessible EA and increasing the amount of potential substrates for MDH similar to the PMS degradation side product of formaldehyde due to dealkylation (Jahn et al., 2020). In particular, light-induced degradation of the phenazine dyes as well as the connected radical formation poses challenges. Therefore, the preparation of EA stock solutions should be carried out under the exclusion of light, followed by a short heating of the assay mix (DCPIP and PES) to reduce radical auto-bleaching (Vu et al., 2016). The radical formation is not only light-dependent, but is also more pronounced at elevated temperatures and in an alkaline environment (Jahn et al., 2020). Furthermore, preparation of EA stock solutions in water instead of buffer and storage on ice minimizes this unwanted side-reaction (Jahn et al., 2020).

A note on DCPIP: In the absence of the enzyme or after complete consumption of the substrate, over time a slight increase of absorbance can be observed, most likely due to re-oxidation of reduced DCPIP under aerobic conditions (Naumann, Mayer, & Bannasch, 1985). Jahn et al. also reported a higher enzymatic activity of MDH as well as a reduced bleaching of the dye itself when the assay was performed under oxygen exclusion, supporting this presumption (Jahn et al., 2020). Since DCPIP is a weak acid, with a pK_a of 5.90, change of pH has an impact on the absorbance maximum of the dye and on the extinction coefficient ε (see Fig. 1A) (Armstrong, 1964). This can be visualized by comparing the absorbance

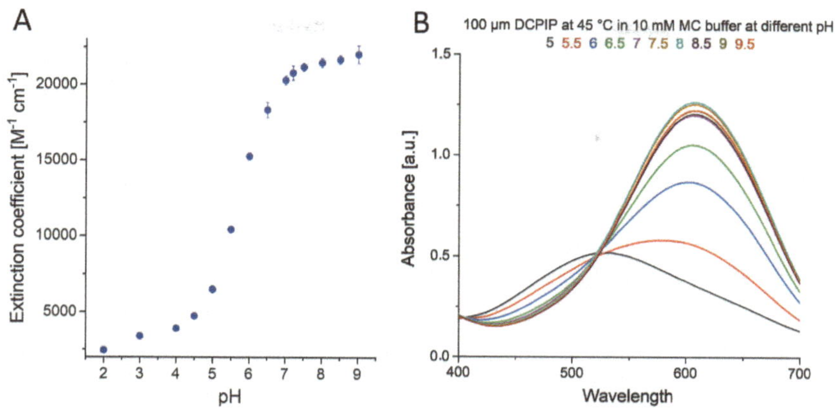

Fig. 1 (A) Extinction coefficient of DCPIP in 10 mM MC buffer (2.5 mM citrate, 2.5 mM Bis-Tris, 2.5 mM Tris, 2.5 mM CHES) at different pH. Measurements were carried out with a plate reader at 45 °C in triplicates. (B) Absorbance spectra of 100 μM DCPIP in 10 mM MC buffer at different pH at 45 °C. Spectra were collected after 2 min equilibrium time at 45 °C in a stirred quartz cuvette using a Cary 60 spectrophotometer.

spectra of DCPIP at different pHs. While spectra at pH 7–9.5 reveal a maximum at 600 nm, the maximum declines with decreasing pH value and shifts to lower wavelengths (see Fig. 1B). As a result, vastly varying extinction coefficients for DCPIP can be found in the literature, both for different buffer systems and even for similar conditions. In addition, optimal conditions for each MDH characterization need to be adjusted individually for new assay conditions (e.g., choosing a suitable buffer system, determining the pH optimum and testing if additives are required). Therefore, it is recommended to determine the extinction coefficient for the investigated system under the specific assay conditions. Furthermore, low color stability of DCPIP with ongoing storage was described which suggests that DCPIP solution should be freshly prepared (Jahn et al., 2020).

1.1.1 Protocol

Conditions: 1 mL assay volume, 1 mM PES, 100 μM DCPIP, 200 nM MDH, 50 mM MeOH (temperature, pH/buffer system of choice and optional: 15 mM NH_4Cl). Depending on the organism from which MDH is purified, the temperature for the enzymatic activity determination may vary. For example, for MDH isolated from *Methylacidiphilum fumariolicum* SolV (*M. fumariolicum* SolV) a pH of 7 and 45 °C is a reasonable compromise between enzyme stability and activity (Pol et al., 2014), while the activity of MDH from *Methylorubrum extorquens* AM1 (*M. extorquens* AM1) is monitored at pH 9 and 30 °C (Vu et al., 2016). Again, the ideal enzyme concentration depends on various factors such as the initial enzyme activity. For Eu-MDH isolated from *M. fumariolicum* SolV, 200 nM was used (Lumpe, Pol, Op den Camp, & Daumann, 2018). Light exposure should be reduced to a minimum during preparation.

1. Prepare separate fresh stock solutions of 100 mM PES (CAS: 10510-77-7, 334.39 g/mol, 33.44 mg/mL) and 2 mM DCPIP (2,6-Dichlorophenol-indophenol sodium salt dihydrate, CAS: 10510-77-7, 326.11 g/mol, 0.65 mg/mL) in ultrapure water (type 1) in amber reaction tubes and store them on ice. For a 10 mL final assay mixture at least 200 μL PES and 1 mL DCPIP are needed.
2. Prepare additive stock solutions if required; for instance, NH_4Cl should be dissolved to give a 500 mM (CAS: 12125-02-9, 53.49 g/mol, 26.74 mg/mL) solution in ultrapure water (type 1). For a 10 mL final assay mixture at least 600 μL NH_4Cl are needed.
3. For the preparation of 10 mL of assay mix solution, dilute PES to a concentration of 2 mM by adding 200 μL of the 100 mM PES solution. Add DCPIP to a concentration of 200 μM by adding 1 mL of the 2 mM DCPIP solution to the respective buffer in an amber reaction tube. If NH_4Cl or other additives are required, add them to this mixture. Then, increase the volume to 10 mL with addition of buffer. Afterward, heat the assay mixture for at least 15 min at the temperature which is specified for the assay later (e.g., 30 or 45 °C), then vortex and store on ice.
4. For the spectrophotometric readout, mix 500 μL assay mixture with 400 μL of buffer in a cuvette (e.g., 1 cm pathlength, 1400 μL volume) and let it equilibrate at the required temperature (e.g., 30 or 45 °C) for at least 2 min in the dark.
5. For 200 nM final enzyme concentration add 50 μL of MDH (4 μM in the buffer system that is used for the assay) to the cuvette and monitor the absorbance at 600 nm for 2 min. As reported by various groups, MDH reveals enzymatic activity in the absence of additional substrate, caused by the "endogenous substrate" which is not fully understood at this time. Thus, it has to be assessed for each system individually.
6. Add 50 μL of 1 M MeOH to the cuvette to yield a final concentration of 50 mM and to start the measurement. Monitor the absorbance at 600 nm. For the calculation of the enzymatic activity, the initial rate (gradient/slope) after the addition of should be determined.

1.2 Wurster's Blue assay

For the method described above, two artificial components are required for the enzymatic investigations of MDH, but kinetic measurements also can be performed by solely using the radical cation Wurster's Blue; the main advantage is that WB acts simultaneously as EA and dye. Synthesized by Wurster in 1879 (Wurster & Schobig, 1879), Duine and co-workers reported WB as potential EA for the characterization of MDH (Duine, Frank, & Westerling, 1978), 10 years after the first reported characterization of MDH isolated from *M. extorquens* AM1 by Anthony based on the DCPIP/PES assay (Anthony & Zatman, 1964). Despite the reduced number of components for the spectrophotometrical investigation of MDH when using WB, the enzymatic activity is to this day mainly monitored with the two-component system DCPIP/PES. The restrained use of the one-electron receptor could potentially be due to the fact that it previously had to be synthesized by a procedure described in the literature (Michaelis & Granick, 1943), while DCPIP and the phenazine dyes were commercially available. Since this is not the case anymore and Jahn et al. have had a closer look at the unwanted side-reactions during the assay procedure, including how to avoid them, WB might be a good alternative to the DCPIP/PES assay for investigating MDH enzymes (Jahn et al., 2020).

As described for DCPIP, it is recommended that the extinction coefficient of WB be determined for the used assay conditions, as again various coefficients can be found in the literature. Absorbance spectra collected of WB reveal two maxima at 560 and 610 nm, with the latter being commonly used for monitoring the specific activity (Fig. 2) (Frank, Dijkstra, Duine, & Balny, 1988).

The stability of WB under different storage conditions, temperatures and at various pH values was previously examined (Jahn et al., 2020). A reduced stability in alkaline buffers was reported, while at acidic pH and in water WB was confirmed to be stable. Michaelis and Granick also described the highest radical concentration at pH 4.62 (Michaelis & Granick, 1943). Since most MDH activity assays are performed at pH 9, the low stability of WB under these conditions might interfere with the kinetic investigations. To circumvent the problem of fast radical degradation, the stock solution of WB in water should be added shortly before the spectrophotometric investigations. Furthermore, the radical yield is greatly reduced with longer storage time as well as with light exposure and increasing temperature. Therefore, it is recommended that the stock solution of WB be prepared in water, immediately before the activity determination, in an amber tube and stored on ice until needed.

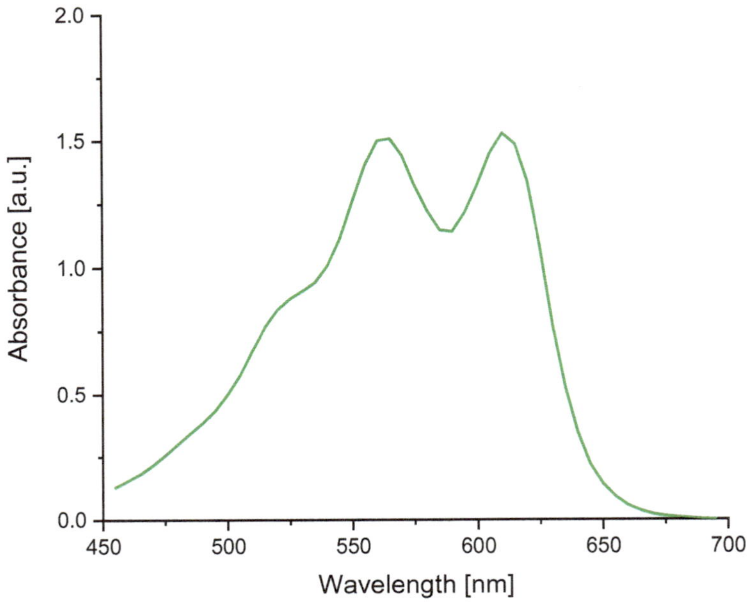

Fig. 2 Absorbance spectra of 200 μM WB in 100 mM MC buffer (2.5 mM citrate, 2.5 mM Bis-Tris, 2.5 mM Tris, 2.5 mM CHES) at pH 9 at 25 °C. Spectra were collected after 2 min equilibration time in a stirred quartz cuvette with a Cary 60 spectrophotometer.

1.2.1 Protocol

Conditions: 1 mL assay volume, 200 μM WB, 200 nM MDH, 50 mM MeOH (temperature and pH/buffer system of choice and optional: 15 mM NH_4Cl). Please see above for a discussion of temperature and pH conditions for the assay. Light exposure should be reduced to a minimum during preparation.

1. Prepare a fresh 1 mM solution of WB (Wuster's blue perchlorate, CAS: 34527-55-4, 263.69 g/moL, 0.26 mg/mL) in ultrapure water (type 1) in an amber tube. Make sure that all of the solid is dissolved by vortexing and ultrasonicating the solution and then store it on ice.
2. Prepare additive stock solutions if required, for instance NH_4Cl should be dissolved to give a 500 mM (CAS: 12125-02-9, 53.49 g/mol, 26.74 mg/mL) solution in ultrapure water (type 1). For 10 mL final assay mixture at least 600 μL of NH_4Cl are needed.
3. For the spectrophotometric readout, mix 200 μL of the 1 mM WB solution with potentially required additives and buffer to yield a final volume of 900 μL immediately before starting the activity determination and incubate for 2 min at the required temperature (e.g., 30 or 45 °C) in the dark.
4. For 200 nM final enzyme concentration, add 50 μL of MDH (4 μM in the buffer system that is used for the assay) and monitor the absorbance

at 610 nm for 2 min. As reported by various groups, MDH reveals enzymatic activity in the absence of additional substrate, caused by the "endogenous substrate" which is not fully understood at the present time. Thus, it has to be assessed for each system individually.

5. Add 50 μL of 1 M MeOH to yield a final concentration of 50 mM to start the measurement. Monitor the absorbance at 610 nm. For calculation of enzymatic activity, initial rate (gradient/slope) after substrate addition should be determined.

2. Assays based on the native cytochrome electron acceptor

In addition to the artificial assay, Anthony and co-workers introduced in the early 1980s the methanol cytochrome *c* oxidoreductase activity assay for Ca-MDH (Beardmore-Gray, O'Keeffe, & Anthony, 1983).

2.1 Spectrophotometric assay with cytochrome c_L

In contrast to the artificial assay, Ca-MDH activity is assessed by coupling its native cytochrome c_L to a secondary cytochrome *c* from bovine or equine heart. Its reduction is followed spectroscopically at 550 nm (Beardmore-Gray et al., 1983; Dijkstra, Frank, & Duine, 1989). They noted that cytochrome c_L from *Methylophilus methylotrophus* and *Paracoccus denitrificans* only interact with their respective native Ca-MDH while cytochrome c_L from AM1 is able to interact with the Ca-MDH from all three species (Beardmore-Gray et al., 1983). After further optimization, the assay mixture contained the native cytochrome c_L, secondary cytochrome *c* (from bovine or equine heart) and methanol as substrate. The assay was initiated by the addition of MDH. Recently, Chistoserdova and co-workers showed in *Methylomonas* sp. strain LW13 that a XoxG(4) protein acts as the specific native electron acceptor for its Ln-MDH (Zheng, Huang, Zhao, & Chistoserdova, 2018). Based on these findings, several groups adjusted Anthony and co-workers' previous protocol. In *M. extorquens* AM1, the native cytochrome c_L partner for Ln-MDH is referred to as XoxG which was biochemically and structurally characterized by Featherston et al. (2019). The native cytochrome c_L in *M. fumariolicum* SolV is a fusion protein and is denominated as c_{GJ}. The fusion protein consists of a XoxG cytochrome and periplasmic binding protein XoxJ. Protein-based activity assays for MDH from *M. extorquens* AM1 and *M. fumariolicum* SolV have been reported and consist of MDH, secondary cytochrome *c* (from bovine or equine heart), methanol as substrate and are initiated by addition of their respective native cytochrome (Fig. 3)

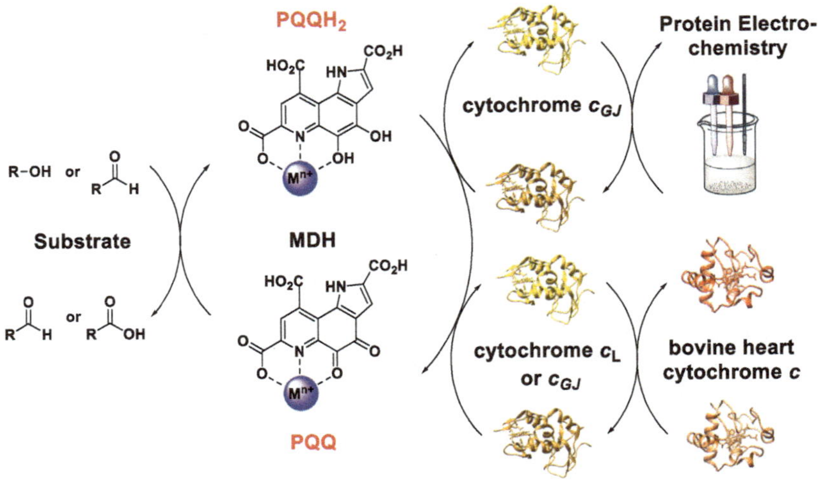

Fig. 3 Overview of assay methods involving the natural electron acceptor of MDH. Lower assay shows the spectrophotometric method where cytochrome reduction is coupled to a second cytochrome from equine or bovine heart. For the MDH/cytochrome c_{GJ} pair of *M. fumariolicum* SolV an assay using protein electrochemistry has also been reported. The protein structures depicted here were generated with Chimera (Pettersen et al., 2004), PDB 6ONQ and 6FF5. Please note, for cytochrome c_{GJ} from *M. fumariolicum* SolV no crystal structure has been reported so far and the structure of the cytochrome from *M. extorquens* AM1 has been used in its place.

(Featherston et al., 2019; Kalimuthu, Daumann, Pol, Op den Camp, & Bernhardt, 2019; Versantvoort et al., 2019). Independently of strain, the native cytochrome c_L is re-oxidized by the secondary electron acceptor equine/bovine heart cytochrome *c*, whose α-band intensity increases upon reduction by cytochrome c_L. The reduction can thus be followed spectrophotometrically at 550 nm. The amount of reduced cytochrome *c* per minute is assessed by subtracting the absorbance of the oxidized form from the fully reduced cytochrome to obtain an extinction coefficient of 19.5 mM^{-1} cm^{-1} at pH 7.2 (Day & Anthony, 1990; Versantvoort et al., 2019). Cytochrome *c* can be fully reduced by the addition of 100 equivalents of sodium dithionite.

Note on salt content and additives: The electron transfer from MDH to cytochrome c_L is based on electrostatic interaction between lysine residues on MDH and carboxyl residues on cytochrome c_L (Cox et al., 1992). It is known that this single "docking" site is affected by the ionic strength of the buffer. Anthony and co-workers have shown that increasing the concentration of NaCl from 0 to 50 mM drastically decreases the affinity of cytochrome c_L to Ca-MDH which is reflected by a sixfold increase of the

K_M value. The effect of NaCl can be reversed by increasing the concentration of cytochrome c_L (Cox et al., 1992). Furthermore, they observed that ethylenediaminetetraacetic acid (EDTA) and other tetracarboxylates (e.g., EGTA and CDTA) could interfere with the protein–protein interaction by interaction of their carboxy groups with lysine residues on the "docking" site of MDH. Recent work by Op den Camp and co-workers support these findings and showed that the activity readout of MDH (assessed with the cytochrome assay) decreases by 50% when a NaCl concentration of higher than 100 mM is used (Versantvoort et al., 2019). Since protein purifications often involve high NaCl gradient for elution of the target protein, buffer exchange or dialysis step of MDH and cytochrome c_L before usage in protein-based activity assays is recommended.

2.1.1 Protocol

Conditions: 100 µL assay volume (in 96 well-plate), 100 nM of XoxF-MDH, 50 µM cytochrome *c* from equine or bovine heart, 20 mM MeOH and 0–10 µM cytochrome c_L (pH 7.0–7.2, buffer system of choice). The appropriate amount of cytochrome c_L can vary across different systems. The K_M value of XoxF-XoxG needs to be considered when designing these assays. The reaction was incubated for 20 min at 30 °C for XoxG (*M. extorquens* AM1) and for 2 min at 45 °C for c_{GJ} (*M. fumariolicum* SolV) before absorbance was measured at 550 nm.

1. Prepare a fresh stock solution of 0.5 mM cytochrome *c* from equine heart (CAS: 9007-43-6, 12,384 g/mol, 6.2 mg/mL) in buffer in reaction tubes. Store on ice until needed. For 1 mL final assay mix at least 100 µL cytochrome *c* are needed.
2. Prepare a stock solution of 1 M MeOH (32.04 g/mol, 40 µL/mL) in ultrapure water (type 1) in reaction tubes. For 1 mL final assay mixture at least 10 µL MeOH are needed.
3. For the preparation of 90 µL assay mixture, combine 68 µL of buffer, 10 µL of 0.5 mM cytochrome c from equine heart, 10 µL of a 1 µM MDH solution (in the used buffer system) and 2 µL of a 1 M MeOH solution. Store on ice until needed. It is recommended to pre-heat the plate in the plate reader and to store the buffer at room temperature.
4. If a final concentration of 1 µM c_L is required, the 10 µL cytochrome c_L mixture is prepared by diluting cytochrome c_L to a concentration of 10 µM in buffer. Store on ice until needed.
5. Incubate both mixtures in separate wells of a 96-well plate. The assay is initiated by addition of the cytochrome c_L solution to the assay mixture.

The average gradient (slope) after cytochrome c_L addition is used to calculate the rate of reduction. The pathlength of the well plate has to be measured as well.

$$\textit{specific activity}\left[\mu mol \min^{-1} mg^{-1}\right] = \frac{\textit{enzyme unit } U\left[\mu mol \min^{-1}\right]}{\textit{amount of enzyme}\left[mg\right]}$$

$$\textit{enzyme unit } U\left[\mu mol \min^{-1}\right] = \frac{\textit{initial rate of slope of measurement}}{\varepsilon\left[cm^{-1}M^{-1}\right] \bullet \textit{pathlength of cell}\left[cm\right]} \bullet 10^6 \bullet \textit{volume of assay}\left[L\right]$$

2.2 Electrochemical assay

Kalimuthu et al. developed a method that allows the co-adsorption of Eu-MDH and its cytochrome c_{GJ} from *M. fumariolicum* SolV with the biopolymer chitosan onto an Au working electrode (Kalimuthu et al., 2019). The Au electrode replaces the bovine/equine heart cytochrome *c* as the secondary electron acceptor and the activity is determined by measuring the current. With that electrode in hand the effect of ammonium ions, methanol concentration, formaldehyde concentration, pH and temperature can be assayed electrochemically. Here, activating agents such as ammonium ions or glycine ethyl ester (GEE) are not necessary, nor is the addition of cyanide as protecting agent. The transformation of a quasi-reversible wave of c_{GJ} into a sigmoidal wave upon addition of either methanol or formaldehyde and MDH displays a homogeneous (Eu-MDH to cytochrome c_{GJ}) and heterogeneous (cytochrome c_{GJ} to Au electrode) electron transfer process. Screening across the pH range of 5.4–9.5 identified the pH optimum at pH 7.3, which is in accordance with results from dye-linked assays. With similarities to the dye-linked assays, Eu-MDH and c_{GJ} display increased activity with higher temperature. To the best of our knowledge, there are no additional publications on bioelectrochemistry of Ln-MDHs. For a precise procedure we refer the reader to Kalimuthu et al. (2019).

3. Role of activators, additives, and buffers

3.1 Ammonium salts

Ammonium chloride is usually needed as activator in MDH assays with artificial electron acceptors (Afolabi et al., 2001; Anthony, 2000; Day & Anthony, 1990; Frank et al., 1988; Goodwin & Anthony, 1996; Harris & Davidson, 1993). It acts as an activator at low concentrations by increasing

V_{max} (Goodwin & Anthony, 1996), while an inhibitory effect can be observed at higher concentrations (Afolabi et al., 2001; Harris & Davidson, 1993). Anthony and Zatman suggested in 1964 that the free base ammonia and not the ammonium ion is the active form as they observed that at pH 9 the activation was more effective than at pH 7, even if higher amounts of ammonium chloride were added at neutral pH (Anthony & Zatman, 1964). The conclusion is based on the fact that the pK_a of NH_3/NH_4^+ in water is 9.21 which means that the predominant species at pH7 is ammonium and at pH9 more free base is available (Hall, 1957). Thus, to completely shift the equilibrium to ammonia, pH values higher than pH9.21 would be needed.

The inhibitory effect of added ammonium chloride was suggested by molecular modeling to be due to the ammonium ion, preventing methanol from binding properly to the active site of MDH (Reddy & Bruice, 2004). However, the full mechanism of activation by ammonia is still not fully understood. It was proposed that ammonia acts by activating the reduction of PQQ by the substrate (Afolabi et al., 2001; Frank et al., 1988; Goodwin & Anthony, 1996). Furthermore, Reddy and Bruice suggested in 2004, based on molecular modeling, that the formation of an MDH-PQQ-(NH)-methanol intermediate might be involved (Reddy & Bruice, 2004). Moreover, it must be noted that ammonia can induce the decomposition of artificial electron acceptors such as PMS and PES leading to phenazine which is problematic as the concentration of the electron acceptor is decreased and formaldehyde is formed as a side product which can act as a substrate for the studied MDH system. To circumvent this problem, it is recommended that PES be used rather than PMS (as this behavior is less pronounced for PES), to keep the light-exposure to a minimum, to prepare the stock solution in water instead of buffer and to heat them briefly at the temperature used for the assay prior to use in order to minimize the mentioned side-reactions (Jahn et al., 2020; Vu et al., 2016).

If the assay is performed in the presence of natural electron acceptors (the respective cytochrome c_L or c_{GJ}), no ammonia needs to be added as an activator, and the assay can be performed at lower pH (Anthony, 2000; Day & Anthony, 1990; Featherston et al., 2019; Kalimuthu et al., 2019; Versantvoort et al., 2019). Furthermore, not all MDHs require the addition of an activator. For example, the Ln-MDH isolated from *M. fumariolicum* SolV doesn't need ammonia as activator (albeit the activity with it is slightly higher) and can as well be assayed at neutral pH with artificial EAs (Pol et al., 2014).

3.2 Glycine ethyl ester (GEE)

As alternative activator to ammonium salts, amines such as methylamine or esters of glycine and alanine have been proposed in the literature (Anthony, 1986). Martinez-Gomez and coworkers reported that methylamine was an even better activator than ammonia for the ethanol dehydrogenase (ExaF) isolated from *M. extorquens* (Good et al., 2016). Hothi et al. stated in 2005 that GEE might be a more beneficial activator for PQQ-dependent MDHs as they observed higher activity with GEE at lower concentrations than with ammonium chloride and a less pronounced inhibition at higher concentrations (Hothi, Sutcliffe, & Scrutton, 2005). However, it should be considered that GEE can degrade under assay conditions depending on the used pH, temperature and activator concentration (e.g., at pH 8.6 ethanol is rapidly produced in a 10 mM solution, Fig. 4); since esters are known to hydrolyze under acidic or basic conditions producing a carboxylic acid and an alcohol. Therefore, it is recommended that the stability of the used GEE be checked (e.g., by ^{1}H NMR spectroscopy) under assay conditions in the absence of MDH as the substrate ethanol might be generated.

3.3 Potassium cyanide (KCN)

KCN is routinely added to MDH assays as it was found to suppress the endogenous MDH activity and the reduction of the used artificial electron acceptor as well as to prevent enzyme inactivation (Duine & Frank, 1980). Endogenous activity refers to the activity observed for MDHs in the absence of added substrate, which is believed to be caused by low concentrations of contaminating aldehydes and alcohols presumably present at low concentrations in used reagents (Anthony & Zatman, 1964; Day & Anthony, 1990; Duine et al., 1978; Duine & Frank, 1980). However, depending on the used assay conditions, it should be kept in mind that the main species in the solution under investigation might not be cyanide but HCN. The pK_a of HCN is 9.21 (Beck, 1987), which means that at pH 7.2 the predominant species is HCN which can outgas of the solution, especially if the assay is performed at elevated temperatures (e.g., at 45 °C used for investigating the Ln-MDH of *M. fumariolicum* SolV (Jahn et al., 2020)) (Arun et al., 2005). At pH 9 the predominant species is still HCN, but to a much lesser extent. Above pH 9.21 the predominate species is cyanide. Depending on the chosen assay conditions (high temperature, neutral pH, high KCN concentration), HCN (highly toxic!) might outgas of the investigated solution. The assay should therefore be performed in a well-ventilated area. It should further be noted that the KCN can also induce the decomposition of PMS and PES (compare

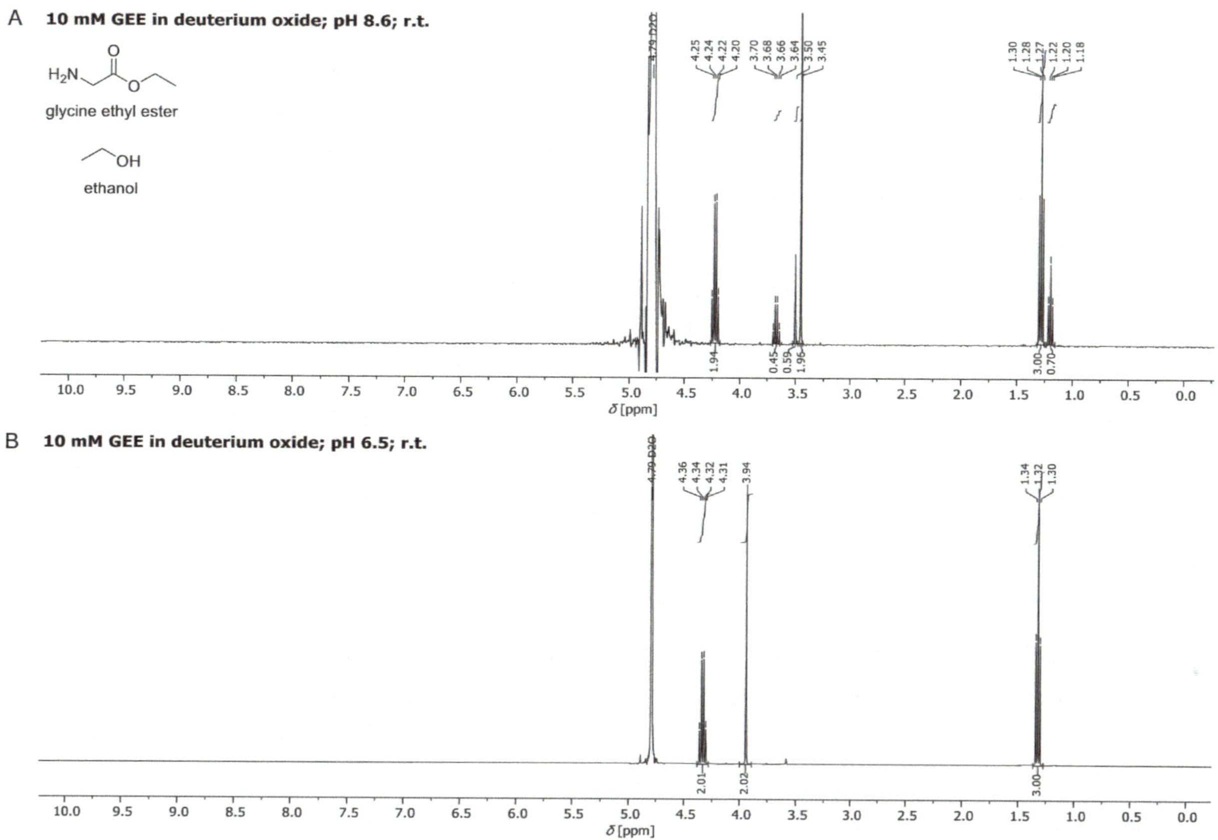

Fig. 4 (A) ^{1}H NMR spectrum of a GEE sample that was immediately measured after the pH adjustment to 8.6. In addition to the signals of GEE (δ [ppm] = 4.23 (q, 2H), 3.45 (s, 2H), 1.28 (t, 3H)), signals that can be assigned to ethanol (δ [ppm] = 3.67 (q, 2H), 1.20 (t, 3H)) are observed as well. (B) ^{1}H NMR spectrum of a GEE sample that was adjusted to pH 6.5 and measured after 24 h storage at room temperature. No ethanol could be detected in this sample; the different shifts for the NMR signals of the GEE are due to the different pH value. All samples were measured on a Bruker 400 MHz instrument.

with ammonia), leading to the same troublesome side products phenazine and formaldehyde (Jahn et al., 2020). The same recommendations, as stated above, apply.

3.4 Lanthanide salts

If assaying Ln-dependent alcohol dehydrogenases (ADHs), sometimes $LnCl_3$ are added for the saturation of the apo-enzyme (concentrations 0–20 μM were used for MDH isolated from *M. fumariolicum* SolV, and up to 100 μM for ethanol dehydrogenase isolated from *M. extorquens* AM1) (Good et al., 2016; Lumpe et al., 2018). However, it should be noted that this can impact the electrostatic interaction of MDH with its cytochrome. Additionally, when preparing $LnCl_3$ stock solutions it should be kept in mind that $LnCl_3$ readily hydrolyzes at alkaline pH forming $Ln(OH)_3$ precipitates and can also react with carbon dioxide to form $Ln_2(CO_3)_3$ precipitates. Due to the hygroscopic nature of $LnCl_3$ salts, the water content of the used $LnCl_3 \times n\ H_2O$ salt for preparing the stock solutions should be determined regularly by, e.g., elemental analysis. All things considered, it is recommended to prepare the stock solutions freshly and use them for a maximum of 1 week, stored in plastic containers (contamination with other ions can occur if stored in glass) at a low pH (addition of, e.g., HCl to prevent hydrolysis, not necessary if the stock is used freshly) and at fairly high concentrations (e.g., 1–10 mM) (Evans, 1990). The stock solutions should always be checked for precipitates before use.

3.5 Pyrroloquinoline quinone (PQQ)

In some reports, PQQ is added to ethanol dehydrogenase assay mixtures to restore activity (Good et al., 2016). Whether this works for MDHs is currently unknown and the literature on reconstitution is scarce. However, it should be noted in any case, that PQQ can readily precipitate with lanthanides in aqueous solutions (Lumpe et al., 2020). This might be disadvantageous, if $LnCl_3$ salts are added as well, for the saturation of the apo-Ln-MDH.

3.6 Buffers

As there is no such thing as a perfect buffer, the pros and cons of every buffer system should be precisely considered in respect to the envisioned experimental setting. For example, if the redox chemistry of a system is under investigation (MDHs assay) the usage of PIPES and HEPES might be inadvisable as both buffers are known for radical formation. In Table 1 an

Table 1 Selected considerations of different buffers frequently used in MDH assays.

Buffer	pH change upon heating	Metal binding ability	Known side reaction/other issues
Acetate 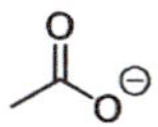	Yes* (Fukada & Takahashi, 1998)	Yes, e.g., Y(III), Cu(II), Zn(II), Cd(II), Pb(II) and Ln(III) (Kolat & Powell, 1962)	
Citrate 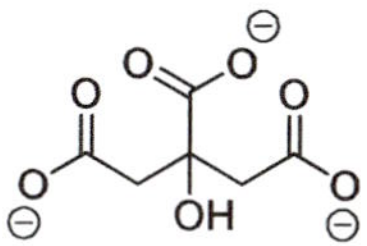	None (Ellis & Morrison, 1982)	Yes, e.g., alkaline earth metal ions; transitions metal ions (Wyrzykowski & Chmurzyński, 2010; Wyrzykowski, Czupryniak, Ossowski, & Chmurzyński, 2010) and Ln(III) (Nikonorov, 2010)	
Phosphate 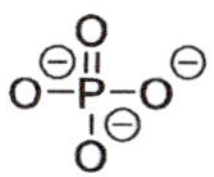	Yes* (Fukada & Takahashi, 1998)	Yes, e.g., alkali and alkaline earth metal ions (Van Wazer & Callis, 1958)	Possible precipitation of Ln(III) (Jordan et al., 2018; Liu & Byrne, 1997)
Bis-Tris 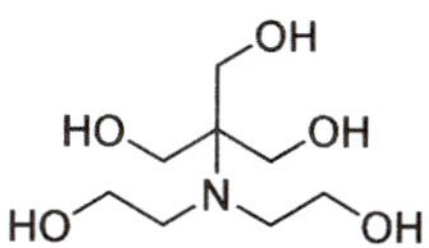	Yes** (Ellis & Morrison, 1982)	Yes, e.g., alkali, alkaline earth and transition metal ions (Sigel, Scheller, & Prijs, 1982) and Ln(III) (Jin Oh et al., 1998)	
HEPES 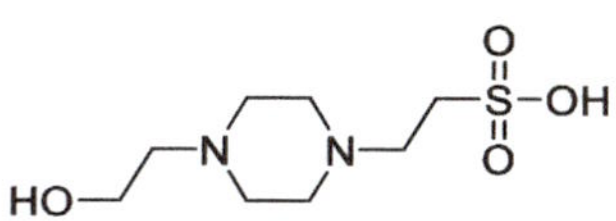	Yes** (Fukada & Takahashi, 1998)	Yes, mostly weak binding of e.g. Cu(II) (Sokołowska & Bal, 2005), Zn(II), Pb(II), Cd(II) (Zeinab Mohamed Anwar, 2005) and Ln(III) (Anwar & Azab, 2001; Soares & Conde, 2000) some reports do not observe any complexation of Zn(II) and Cd(II) ➔ opinion in the literature not unified	Radical formation, not suitable for redox-active systems (Grady, Chasteen, & Harris, 1988)

Continued

Table 1 Selected considerations of different buffers frequently used in MDH assays.—cont'd

Buffer	pH change upon heating	Metal binding ability	Known side reaction/other issues
PIPES 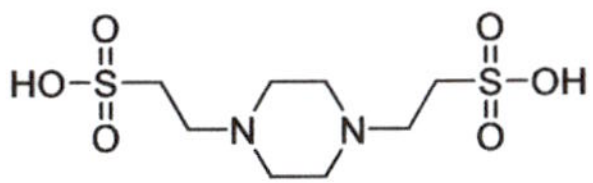	Yes* (Fukada & Takahashi, 1998)	Yes, e.g., Ln(III), (Azab, Al-Deyab, Anwar, Abd El-Gawad, & Kamel, 2011) Co(II) and Ni(II) (Wyrzykowski, Pilarski, Jacewicz, & Chmurzyński, 2013)	Radical formation, not suitable for redox-active systems (Grady et al., 1988)
Tris 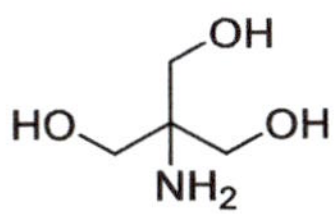	Yes*** (Ellis & Morrison, 1982)	Yes, alkali, alkaline earth metal ions (Sigel et al., 1982), some transition metals (Xiao et al., 2020), Eu(III) and Tb(III) (Pfefferlé & Bünzli, 1989)	Imine condensations with aldehydes (Hoffman, Frey, Smith, & Auble, 2015)
CAPS 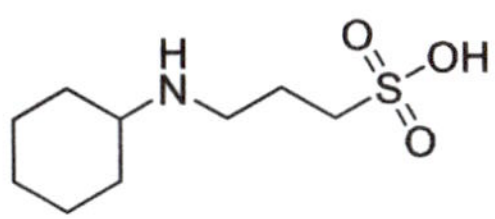	Yes*** (Fukada & Takahashi, 1998)	Yes, weak binding of, e.g., Cu(II), Pb(II), Cd(II) and Zn(II) (Ferreira, Pinto, Soares, & Soares, 2015)	
MOPS 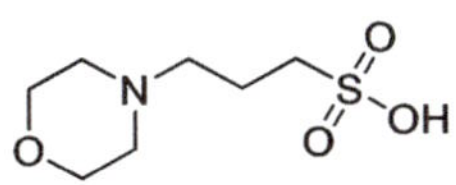	Yes** (Fukada & Takahashi, 1998)	Yes, mild or weak binding of e.g. Cu(II), Pb(II), Zn(II), Cd(II) (Xiao et al., 2020)	

All Values from Fukada and Takahashi (1998) were obtained in 0.1 M KCl and from Ellis and Morrison (1982) with ionic strength 0.1. * = pH change per 5 °C change <0.05, ** = 0.10 > pH change per 5 °C change >0.05, *** = pH change per 5 °C change >0.10; pH change refers to the temperature range from 20 to 45 °C.

overview of different buffers, frequently used in MDH assays, and some selected considerations, e.g., about susceptibility to pH change upon temperature changes or potential metal complexation, are given without claim of completeness. For a comprehensive overview of different buffer systems we refer the reader to Ferreira et al. (2015).

4. Safety considerations

Potassium cyanide and likewise its acid, HCN, are potent inhibitors of cellular respiration and thus highly toxic. Appropriate care should be taken when handling and disposing of cyanide containing stock solutions. The Wurster's blue radical cation usually contains a perchlorate counterion. Perchlorates can potentially form explosive mixtures with metal salts or organic compounds. Appropriate care should be taken when handling the Wurster's blue salt.

References

Afolabi, P. R., Mohammed, F., Amaratunga, K., Majekodunmi, O., Dales, L., Gill, R., et al. (2001). Site-directed mutagenesis and X-ray crystallography of the PQQ-containing quinoprotein methanol dehydrogenase and its electron acceptor, cytochrome cL. *Biochemistry*, *40*(33), 9799–9809. https://doi.org/10.1021/bi002932l.

Anthony, C. (1986). Bacterial oxidation of methane and methanol. In A. H. Rose, & D. W. Tempest (Eds.), *Vol. 27. Advances in microbial physiology* (pp. 113–210). Academic Press.

Anthony, C. (2000). Methanol dehydrogenase, a PQQ-containing quinoprotein dehydrogenase. In A. Holzenburg, & N. S. Scrutton (Eds.), *Enzyme-catalyzed electron and radical transfer: Subcellular biochemistry* (pp. 73–117). Boston, MA: Springer US.

Anthony, C., & Zatman, L. J. (1964). Microbial oxidation of methanol. 2. The methanol-oxidizing enzyme of Pseudomonas sp. M27. *Biochemical Journal*, *92*, 614–621.

Anwar, Z. M. (2005). Complexation equilibria of Zn(II), Pb(II) and Cd(II) with reduced glutathione (GSH) and biologically important zwitterionic buffers. *Journal of the Chinese Chemical Society (Taipei, Taiwan)*, *52*(5), 863–871. https://doi.org/10.1002/jccs.200500121.

Anwar, Z. M., & Azab, H. A. (2001). Ternary complexes formed by trivalent lanthanide ions, nucleotides, and biological buffers. *Journal of Chemical and Engineering Data*, *46*(3), 613–618. https://doi.org/10.1021/je000187+.

Armstrong, J. M. (1964). The molar extinction coefficient of 2,6-dichlorophenol indophenol. *Biochimica et Biophysica Acta*, *86*, 194–197.

Arun, P., Moffett, J. R., Ives, J. A., Todorov, T. I., Centeno, J. A., Namboodiri, M. A. A., et al. (2005). Rapid sodium cyanide depletion in cell culture media: Outgassing of hydrogen cyanide at physiological pH. *Analytical Biochemistry*, *339*(2), 282–289. https://doi.org/10.1016/j.ab.2005.01.015.

Azab, H. A., Al-Deyab, S. S., Anwar, Z. M., Abd El-Gawad, I. I., & Kamel, R. M. (2011). Comparison of the coordination tendency of amino acids, nucleobases, or mononucleotides toward the monomeric and dimeric lanthanide complexes with biologically important compounds. *Journal of Chemical and Engineering Data*, *56*(5), 2613–2625. https://doi.org/10.1021/je200099n.

Beardmore-Gray, M., O'Keeffe, D. T., & Anthony, C. (1983). The methanol: Cytochrome c oxidoreductase activity of methylotrophs. *Microbiology*, *129*(4), 923–933.

Beck, M. T. (1987). Critical survey of stability constants of cyano complexes. *Pure and Applied Chemistry*, *59*(12), 1703–1720. https://doi.org/10.1351/pac198759121703.

Cox, J. M., Day, D. J., & Anthony, C. (1992). The interaction of methanol dehydrogenase and its electron acceptor, cytochrome cL in methylotrophic bacteria. *Biochimica et Biophysica Acta, Protein Structure and Molecular Enzymology*, *1119*(1), 97–106. https://doi.org/10.1016/0167-4838(92)90240-E.

Day, D. J., & Anthony, C. (1990). Methanol dehydrogenase from methylobacterium extorquens AM1. In M. E. Lidstrom (Ed.), *Vol. 188. Methods in enzymology* (pp. 210–216). Academic Press.

Dijkstra, M., Frank, J., Jr., & Duine, J. A. (1989). Studies on electron transfer from methanol dehydrogenase to cytochrome cL, both purified from Hyphomicrobium X. *Biochemical Journal*, *257*(1), 87–94. https://doi.org/10.1042/bj2570087.

Duine, J. A., & Frank, J., Jr. (1980). Studies on methanol dehydrogenase from Hyphomicrobium X. Isolation of an oxidized form of the enzyme. *Biochemical Journal*, *187*(1), 213–219. https://doi.org/10.1042/bj1870213.

Duine, J. A., Frank, J., & Westerling, J. (1978). Purification and properties of methanol dehydrogenase from Hyphomicrobium X. *Biochimica et Biophysica Acta (BBA)—Enzymology*, *524*(2), 277–287. https://doi.org/10.1016/0005-2744(78)90164-X.

Ellis, K. J., & Morrison, J. F. (1982). Buffers of constant ionic strength for studying pH-dependent processes. *Methods in Enzymology*, *87*, 405–426. https://doi.org/10.1016/s0076-6879(82)87025-0.

Evans, C. H. (1990). *Biochemistry of the lanthanides*. New York: Springer Science & Business Media.

Featherston, E. R., Rose, H. R., McBride, M. J., Taylor, E., Boal, A. K., & Cotruvo, J. J. (2019). Biochemical and structural characterization of XoxG and XoxJ and their roles in activity of the lanthanide-dependent methanol dehydrogenase, XoxF. *ChemBioChem*, *20*, 2360–2372. https://doi.org/10.1002/cbic.201900184.

Ferreira, C. M. H., Pinto, I. S. S., Soares, E. V., & Soares, H. M. V. M. (2015). (Un)suitability of the use of pH buffers in biological, biochemical and environmental studies and their interaction with metal ions—A review. *RSC Advances*, *5*(39), 30989–31003. https://doi.org/10.1039/C4RA15453C.

Frank, J., Dijkstra, M., Duine, J. A., & Balny, C. (1988). Kinetic and spectral studies on the redox forms of methanol dehydrogenase from Hyphomicrobium X. *European Journal of Biochemistry*, *174*(2), 331–338. https://doi.org/10.1111/j.1432-1033.1988.tb14102.x.

Fukada, H., & Takahashi, K. (1998). Enthalpy and heat capacity changes for the proton dissociation of various buffer components in 0.1 M potassium chloride. *Proteins: Structure, Function, and Bioinformatics*, *33*(2), 159–166. https://doi.org/10.1002/(sici)1097-0134(19981101)33:2<159::Aid-prot2>3.0.Co;2-e.

Ghosh, R., & Quayle, J. R. (1979). Phenazine ethosulfate as a preferred electron acceptor to phenazine methosulfate in dye-linked enzyme assays. *Analytical Biochemistry*, *99*, 112–117.

Good, N. M., Vu, H. N., Suriano, C. J., Subuyuj, G. A., Skovran, E., & Martinez-Gomez, N. C. (2016). Pyrroloquinoline quinone ethanol dehydrogenase in *Methylobacterium extorquens* AM1 extends lanthanide-dependent metabolism to multicarbon substrates. *Journal of Bacteriology*, *198*(22), 3109–3118. https://doi.org/10.1128/jb.00478-16.

Goodwin, M. G., & Anthony, C. (1996). Characterization of a novel methanol dehydrogenase containing a Ba2+ ion at the active site. *Biochemical Journal*, *318*(2), 673–679. https://doi.org/10.1042/bj3180673.

Grady, J. K., Chasteen, N. D., & Harris, D. C. (1988). Radicals from "good's" buffers. *Analytical Biochemistry*, *173*(1), 111–115. https://doi.org/10.1016/0003-2697(88)90167-4.

Hall, H. K. (1957). Correlation of the base strengths of amines. *Journal of the American Chemical Society*, *79*(20), 5441–5444. https://doi.org/10.1021/ja01577a030.

Harris, T. K., & Davidson, V. L. (1993). A new kinetic model for the steady-state reactions of the quinoprotein methanol dehydrogenase from Paracoccus denitrificans. *Biochemistry, 32*(16), 4362–4368. https://doi.org/10.1021/bi00067a028.

Hoffman, E. A., Frey, B. L., Smith, L. M., & Auble, D. T. (2015). Formaldehyde crosslinking: A tool for the study of chromatin complexes. *Journal of Biological Chemistry, 290*(44), 26404–26411. https://doi.org/10.1074/jbc.R115.651679.

Hothi, P., Sutcliffe, M. J., & Scrutton, N. S. (2005). Kinetic isotope effects and ligand binding in PQQ-dependent methanol dehydrogenase. *Biochemical Journal, 388*(Pt. 1), 123–133. https://doi.org/10.1042/BJ20041731.

Jahn, B., Jonasson, N. S. W., Hu, H., Singer, H., Pol, A., Good, N. M., et al. (2020). Understanding the chemistry of the artificial electron acceptors PES, PMS, DCPIP and Wurster's Blue in methanol dehydrogenase assays. *JBIC, Journal of Biological Inorganic Chemistry, 25*, 199–212. https://doi.org/10.1007/s00775-020-01752-9.

Jin Oh, S., Choi, Y.-S., Hwangbo, S., Chul Bae, S., Kang Ku, J., & Won Park, J. (1998). Structure and phosphodiesterase activity of Bis-Tris coordinated lanthanide (III) complexes. *Chemical Communications (Cambridge, United Kingdom)*, (20), 2189–2190. https://doi.org/10.1039/A806021E.

Jordan, N., Demnitz, M., Lösch, H., Starke, S., Brendler, V., & Huittinen, N. (2018). Complexation of trivalent lanthanides (Eu) and actinides (cm) with aqueous phosphates at elevated temperatures. *Inorganic Chemistry, 57*(12), 7015–7024. https://doi.org/10.1021/acs.inorgchem.8b00647.

Kalimuthu, P., Daumann, L. J., Pol, A., Op den Camp, H. J. M., & Bernhardt, P. V. (2019). Electrocatalysis of a europium-dependent bacterial methanol dehydrogenase with its physiological electron-acceptor cytochrome cGJ. *Chemistry—A European Journal, 25*(37), 8760–8768. https://doi.org/10.1002/chem.201900525.

Kolat, R. S., & Powell, J. E. (1962). Acetate complexes of the rare earth and several transition metal ions. *Inorganic Chemistry, 1*(2), 293–296. https://doi.org/10.1021/ic50002a019.

Lente, G., & Espenson, J. H. (2004). Photoreduction of 2,6-dichloroquinone in aqueous solution: Use of a diode array spectrophotometer concurrently to drive and detect a photochemical reaction. *Journal of Photochemistry and Photobiology A: Chemistry, 163*(1), 249–258. https://doi.org/10.1016/j.jphotochem.2003.12.005.

Liu, X., & Byrne, R. H. (1997). Rare earth and yttrium phosphate solubilities in aqueous solution. *Geochimica et Cosmochimica Acta, 61*(8), 1625–1633. https://doi.org/10.1016/S0016-7037(97)00037-9.

Lumpe, H., Menke, A., Haisch, C., Mayer, P., Kabelitz, A., Yusenko, K. V., et al. (2020). The earlier the better: Structural analysis and separation of lanthanides with pyrroloquinoline quinone. *Chemistry—A European Journal, 26*(44), 10133–10139. https://doi.org/10.1002/chem.202002653.

Lumpe, H., Pol, A., Op den Camp, H. J. M., & Daumann, L. J. (2018). Impact of the lanthanide contraction on the activity of a lanthanide-dependent methanol dehydrogenase—A kinetic and DFT study. *Dalton Transactions, 47*, 10463–10472. https://doi.org/10.1039/C8DT01238E.

Michaelis, L., & Granick, S. (1943). The polymerization of the free radicals of the Wurster dye type: The dimeric resonance bond. *Journal of the American Chemical Society, 65*(9), 1747–1755. https://doi.org/10.1021/ja01249a026.

Myers, A. (1990). Use of DCPIP in a colorimetric method to investigate electron transport in crude heart mitochondrial extracts. *Journal of Biological Education, 24*(2), 123–127. https://doi.org/10.1080/00219266.1990.9655123.

Naumann, R., Mayer, D., & Bannasch, P. (1985). Voltammetric measurements of the kinetics of enzymatic reduction of 2,6-dichlorophenolindophenol in normal and neoplastic hepatocytes using glucose as substrate. *Biochimica et Biophysica Acta (BBA)—Molecular Cell Research, 847*(1), 96–100. https://doi.org/10.1016/0167-4889(85)90158-2.

Nikonorov, V. V. (2010). Determination of the stability constants of lanthanide complexes with oxyacids using capillary electrophoresis. *Journal of Analytical Chemistry (Translation of Zhurnal Analiticheskoi Khimii)*, *65*(4), 359–365. https://doi.org/10.1134/S1061934810040040.

Pettersen, E. F., Goddard, T. D., Huang, C. C., Couch, G. S., Greenblatt, D. M., Meng, E. C., et al. (2004). UCSF chimera—A visualization system for exploratory research and analysis. *Journal of Computational Chemistry*, *25*, 1605–1612. https://doi.org/10.1002/jcc.20084.

Pfefferlé, J.-M., & Bünzli, J.-C. G. (1989). Interaction between the buffer Tris and the eu(III) ion: Luminescence and potentiometric investigation. *Helvetica Chimica Acta*, *72*(7), 1487–1494. https://doi.org/10.1002/hlca.19890720709.

Pol, A., Barends, T. R. M., Dietl, A., Khadem, A. F., Eygensteyn, J., Jetten, M. S. M., et al. (2014). Rare earth metals are essential for methanotrophic life in volcanic mudpots. *Environmental Microbiology*, *16*(1), 255–264. https://doi.org/10.1111/1462-2920.12249.

Reddy, S. Y., & Bruice, T. C. (2004). Mechanisms of ammonia activation and ammonium ion inhibition of quinoprotein methanol dehydrogenase: A computational approach. *Proceedings of the National Academy of Sciences of the United States of America*, *101*(45), 15887–15892. https://doi.org/10.1073/pnas.0407209101.

Sigel, H., Scheller, K. H., & Prijs, B. (1982). Metal ion/buffer interactions. Stability of alkali and alkaline earth ion complexes with triethanolamine (tea), 2-amino-2(hydroxymethyl)-1,3-propanediol (tris)and 2-[bis(2-hydroxyethyl)-amino] 2(hydroxymethyl)-1,3-propanediol (Bistris) in aqueous and mixed solvents. *Inorganica Chimica Acta*, *66*, 147–155. https://doi.org/10.1016/S0020-1693(00)85805-3.

Soares, H. M. V. M., & Conde, P. C. F. L. (2000). Electrochemical investigations of the effect of N-substituted aminosulfonic acids with a piperazinic ring pH buffers on heavy metal processes which may have implications on speciation studies. *Analytica Chimica Acta*, *421*(1), 103–111. https://doi.org/10.1016/S0003-2670(00)01028-X.

Sokołowska, M., & Bal, W. (2005). Cu(II) complexation by "non-coordinating" N-2-hydroxyethylpiperazine-N′-2-ethanesulfonic acid (HEPES buffer). *Journal of Inorganic Biochemistry*, *99*(8), 1653–1660. https://doi.org/10.1016/j.jinorgbio.2005.05.007.

Van Wazer, J. R., & Callis, C. F. (1958). Metal complexing by phosphates. *Chemical Reviews (Washington, DC, United States)*, *58*(6), 1011–1046. https://doi.org/10.1021/cr50024a001.

Versantvoort, W., Pol, A., Daumann, L. J., Larrabee, J. A., Strayer, A. H., Jetten, M. S. M., et al. (2019). Characterization of a novel cytochrome cGJ as the electron acceptor of XoxF-MDH in the thermoacidophilic methanotroph Methylacidiphilum fumariolicum SolV. *Biochimica et Biophysica Acta, Proteins and Proteomics*, *1867*(6), 595–603. https://doi.org/10.1016/j.bbapap.2019.04.001.

Vu, H. N., Subuyuj, G. A., Vijayakumar, S., Good, N. M., Martinez-Gomez, N. C., & Skovran, E. (2016). Lanthanide-dependent regulation of methanol oxidation Systems in *Methylobacterium extorquens* AM1 and their contribution to methanol growth. *Journal of Bacteriology*, *198*(8), 1250–1259. https://doi.org/10.1128/JB.00937-15.

Wurster, C., & Schobig, E. (1879). Ueber die Einwirkung oxydirender Agentien auf Tetramethylparaphenylendiamin. *Berichte der Deutschen Chemischen Gesellschaft*, *12*(2), 1807–1813. https://doi.org/10.1002/cber.187901202156.

Wyrzykowski, D., & Chmurzyński, L. (2010). Thermodynamics of citrate complexation with Mn2+, Co2+, Ni2+ and Zn2+ ions. *Journal of Thermal Analysis and Calorimetry*, *102*(1), 61–64. https://doi.org/10.1007/s10973-009-0523-4.

Wyrzykowski, D., Czupryniak, J., Ossowski, T., & Chmurzyński, L. (2010). Thermodynamic interactions of the alkaline earth metal ions with citric acid. *Journal of Thermal Analysis and Calorimetry*, *102*(1), 149–154. https://doi.org/10.1007/s10973-010-0970-y.

Wyrzykowski, D., Pilarski, B., Jacewicz, D., & Chmurzyński, L. (2013). Investigation of metal–buffer interactions using isothermal titration calorimetry. *Journal of Thermal Analysis and Calorimetry*, *111*(3), 1829–1836. https://doi.org/10.1007/s10973-012-2593-y.

Xiao, C.-Q., Huang, Q., Zhang, Y., Zhang, H.-Q., & Lai, L. (2020). Binding thermodynamics of divalent metal ions to several biological buffers. *Thermochimica Acta*, *691*, 178721. https://doi.org/10.1016/j.tca.2020.178721.

Zheng, Y., Huang, J., Zhao, F., & Chistoserdova, L. (2018). Physiological effect of XoxG(4) on lanthanide-dependent methanotrophy. *MBio*, *9*(2), e02430-02417. https://doi.org/10.1128/mBio.02430-17.

CHAPTER FOUR

Expression, purification and properties of the enzymes involved in lanthanide-dependent alcohol oxidation: XoxF4, XoxF5, ExaF/PedH, and XoxG4

Jing Huang[a], Yue Zheng[b], Joseph D. Groom[c], Zheng Yu[d], and Ludmila Chistoserdova[c,*]

[a]Department of Parasitology, Xiangya School of Medicine, Central South University, Changsha, Hunan, China
[b]CAS Key Laboratory of Urban Pollutant Conversion, Institute of Urban Environment, Chinese Academy of Sciences, Xiamen, China
[c]Department of Chemical Engineering, University of Washington, Seattle, WA, United States
[d]Department of Microbiology, Xiangya School of Medicine, Central South University, Changsha, Hunan, China
[*]Corresponding author: e-mail address: milamilachis@gmail.com

Contents

Methods in Enzymology, Volume 650
ISSN 0076-6879
https://doi.org/10.1016/bs.mie.2021.01.041

Abstract

In this chapter we describe logistics, protocols and conditions for expression, purification and characterization of Ln^{3+}-dependent alcohol dehydrogenases representing three distinct phylogenetic clades of these enzymes, classified as XoxF4, XoxF5 and ExaF/PedH. We present data on the biochemical properties of a dozen enzymes, all generated by our group, in a comparative fashion. These enzymes display a range of properties in terms of substrate and metal specificities, pH and ammonium requirement, as well as catalytic constants. In addition, we describe a single novel cytochrome, XoxG4, that likely serves as a natural electron acceptor from XoxF5 in methanotrophs of the Gammaproteobacteria class.

1. Introduction

Lanthanide (Ln^{3+})-dependent alcohol dehydrogenases have emerged recently as important and relatively widely distributed enzymes (Chistoserdova, 2016; Chistoserdova & Kalyuzhnaya, 2018; Cotruvo, 2019; Daumann, 2019). While it has been proposed that these enzymes are more catalytically efficient than their calcium (Ca^{2+})-dependent counterparts (Keltjens, Pol, Reimann, & Op den Camp, 2014), the question remains whether the two enzymes act independently or in unison. In many organisms, the two alternative forms are present and operating simultaneously, their relative expression subject to multiple regulatory mechanisms (Yu, Zheng, Huang, & Chistoserdova, 2019). As Ln^{3+}-dependent microbial metabolism is still an emerging area, more data are needed on the properties and the activities of the Ln^{3+}-dependent enzymes. Notably, Ln^{3+}-dependent enzymes are extremely diverse, often showing less amino acid conservation with members of a different clade than with their Ca^{2+}-dependent counterparts (Chistoserdova, 2011; Chistoserdova & Kalyuzhnaya, 2018; Keltjens et al., 2014). Here we describe protocols for expression, purification and analysis of several types of such enzymes. Specifically, we focus on methanol dehydrogenases of two phylogenetic affiliations, XoxF4 and XoxF5, and ethanol dehydrogenase classified as ExaF/PedH. Finally, we describe the properties and the activity of one associated cytochrome, XoxG4.

2. Expression hosts and expression conditions

2.1 Plasmid-based expression

2.1.1 Gene synthesis and expression vector construction

We originally attempted to express several Ln^{3+}-dependent alcohol dehydrogenases in *Escherichia coli*, using commercial pET vectors, but these constructs did not result in production of active enzymes, even when cells were grown in the presence of pyrroloquinoline quinone (PQQ) and Ln^{3+}, the natural cofactors, or when the expressed proteins were incubated with these cofactors in attempts to reconstitute active holoenzymes (not shown). An alternative approach was to utilize a broad host range vector. To express Ln^{3+}-dependent alcohol dehydrogenases, Huang et al. (2019) opted to modify the pCM80-PxoxF vector that contained a native promoter for the *mxaF* gene from *Methylorubrum extorquens* AM1 (Marx & Lidstrom, 2001), and to express proteins of interest in *M. extorquens* AM1. The modifications introduced into the original vector included a 411 bp synthetic DNA fragment, presumed to contain a native promoter for a *xoxF* gene in the same organism (between PmxaF and the *lacZ* reporter gene), a synthetic ribosome-binding site, a unique *Nco*I restriction site for cloning genes of interest, a TEV protease cleavage site, and an 8 × His-tag sequence (Huang et al., 2019). The select genes encoding Ln^{3+}-dependent alcohol dehydrogenases were codon-optimized for expression in *M. extorquens* AM1, using an empirically derived codon usage table. Synthetic DNA molecules were generated by Twist Biosciences and cloned into the *Nco*I site of the pCM80-PxoxF vector using the Gibson Assembly method (NEBuilder HiFi, NEB). The *E. coli* Top10 transformants were plated on lysogeny broth (LB) agar plates supplemented with tetracycline ($10\,\mu g\,mL^{-1}$). All gene constructs were sequence-verified using the PacBio RSII sequencing platform (Pacific Biosciences). It is important to note that not all constructs could be successfully cloned into *E. coli*. In some cases, deletions in the gene of interest occurred, suggesting that certain variants of Ln^{3+}-dependent alcohol dehydrogenases were toxic to this host, even when transcribed from foreign promoters. For the list of successful expression constructs described in this paper see Table 1.

2.1.2 Plasmid transfer into host strain

M. extorquens AM1 (a rifampicin-resistant variant) was chosen as an alternative host. This organism naturally produces PQQ, which serves as a

Table 1 Enzymes, expression vectors and expression hosts.

Enzyme	Enzyme source	Plasmid/chromosome insert; promoter	Expression host	Antibiotic resistance	Reference
XoxF4-1	*Methylotenera mobilis* JLW8	Chromosome; native	*M. mobilis* JLW8	Gentamicin	Huang, Yu, and Chistoserdova (2018)
XoxF4-2	*Methylotenera mobilis* JLW8	Chromosome; native	*M. mobilis* JLW8	Kanamycin	Huang et al. (2018)
XoxF5-1	*Methylorubrum extorquens* AM1	pCM80-PxoxF; PxoxF5-1	*M. extorquens* AM1	Tetracycline	Huang et al. (2019)
XoxF5-2	*Methylorubrum extorquens* AM1	pCM80-PxoxF; PxoxF5-1	*M. extorquens* AM1	Tetracycline	Huang et al. (2019)
XoxF5	*Sinorhizobium meliloti* 5A14	pCM80-PxoxF; PxoxF5-1	*M. extorquens* AM1	Tetracycline	Huang et al. (2019)
XoxF5	*Bradyrhizobium diazoefficiens* USDA 110	pCM80-PxoxF; PxoxF5-1	*M. extorquens* AM1	Tetracycline	Huang et al. (2019)
XoxF5	*Rhodovulum kholense* DSM 19783	pCM80-PxoxF; PxoxF5-1	*M. extorquens* AM1	Tetracycline	Huang et al. (2019)
XoxF5	*Tistlia consotensis* DSM 21585	pCM80-PxoxF; PxoxF5-1	*M. extorquens* AM1	Tetracycline	Huang et al. (2019)
XoxF5	*Grimontia marina* CECT 8713	pCM80-PxoxF; PxoxF5-1	*M. extorquens* AM1	Tetracycline	Huang et al. (2019)
XoxF5	*Methyloversatilis discipulorum* FAM1	pCM80-PxoxF; PxoxF5-1	*M. extorquens* AM1	Tetracycline	Huang et al. (2019)
XoxF5	*Methylomonas* sp. LW13	Chromosome; native	*Methylomonas* sp. LW13	Kanamycin	Zheng, Huang, Zhao, and Chistoserdova (2018)
ExaF/PedH	*Methylopila* sp. M107	pCM80-PxoxF; PxoxF5-1	*M. extorquens* AM1	Tetracycline	Huang et al. (2019)
XoxG4	*Methylomonas* sp. LW13	Chromosome; native	*Methylomonas* sp. LW13	Kanamycin	Zheng et al. (2018)

cofactor for a calcium-dependent methanol dehydrogenase (MDH), MxaFI, two Ln^{3+}-dependent MDH enzymes, classified as XoxF5, and a Ln^{3+}-dependent ethanol dehydrogenase (EDH), classified as ExaF/PedH (Good et al., 2016). A quadruple mutant defective in all these enzymes was utilized, unable to grow on either methanol or ethanol (Vu et al., 2016). Plasmid constructs were transferred from *E. coli* Top10 to the host via three-way conjugation as follows. The donor *E. coli* Top10 cells carrying an expression construct were mixed with the recipient, *M. extorquens* AM1 and a helper strain, *E. coli* 2073, in approximately equal proportions. As controls, matings were performed in the absence of the helper strain, to ensure that the progenies only resulted from the complete mixtures, as the helper strain was essential for plasmid transfer. The matings were carried out on nutrient agar (Thermo Fisher Scientific, Waltham, MA, USA) plates overnight, at 30 °C, after which appropriate dilutions of the biomass were plated onto a selective minimal medium (Hypho; Delaney et al., 2013) containing succinate (0.2% w/v) as a substrate and tetracycline ($10\,\mu g\,mL^{-1}$) and rifampicin ($50\,\mu g\,mL^{-1}$) for, respectively, selecting for the plasmid-borne function and counter-selecting for *E. coli*.

2.1.3 Growth complementation

Constructs expressing an active Ln^{3+}-dependent alcohol dehydrogenase can be selected based on their ability to grow on methanol (0.5% v/v) or ethanol (0.15% v/v) when supplemented with (1–30 μM) of lighter Ln^{3+}, such as La^{3+}, Ce^{3+}, Nd^{3+}, or a mixture of these metals. Growth rates differed for different constructs, so selective plates were incubated (at room temperature; ~25 °C) for up to 6 weeks. In our hands, some constructs (e.g., construct carrying *xoxF5* from *Rhodovulum kholense* DSM 19783) required lower temperature (20 °C) to demonstrate optimal growth and higher MDH activity (Huang et al., 2019). Some of the constructs did not result in mutant phenotype complementation, suggesting that either inactive proteins were produced, or deleterious mutations arose after the plasmid entered the *M. extorquens* AM1 host (not tested). One of the constructs that did not result in complementation, carrying *xoxF5* from *Methyloversatilis discipulorum* FAM1, was shown to produce an active enzyme. We hypothesized that, in this case, no functional electron transfer chain was available in the host strain to receive electrons from this enzyme (Huang et al., 2019). Thus, selection by functional complementation is not a universal approach for characterizing Ln^{3+}-dependent enzymes.

2.2 Expression from chromosomal inserts

We were also successful expressing Ln^{3+}-dependent enzymes in their native hosts, by either replacing respective native genes with His-tagged constructs (Huang et al., 2018), or by adding a His-tag-coding sequence, along with the kanamycin (Km) resistance gene to the chromosome, for selection (Zheng et al., 2018). In each case, the respective genes were amplified with upstream fragments of approximately 600 bp at the 5′ termini, with an addition of a His-tag encoding six histidine residues at the 3′ termini, and these fragments were assembled with downstream fragments of approximately 600 bp by fusion PCR, three to four fragments dependent on the specific construct, as has been described elsewhere (Shevchuk et al., 2004). Briefly, the fragments to be amplified were designed to overlap by 30–40 nucleotides, and these were incorporated as parts of the primer design. The reaction mixture for fusion PCR was as follows: 9.5 μL water, 5 μL Phusion buffer (5 ×) (NEB, USA), 2 μL deoxynucleoside triphosphate (dNTP) mix (2.5 mM each), 8 μL gel-purified fragments (about 50 ng per fragment), and 0.5 μL Phusion DNA polymerase (NEB, USA). The assembled PCR products were further amplified using nested primers at a high annealing temperature. The resulting PCR products were purified using the QIAquick PCR purification kit (Qiagen, USA). The correctness of all the constructs was verified by PCR amplification, followed by sequencing. After DNA purification, the products were directly electroporated into the respective native host (Yan, Chu, Puri, Fu, & Lidstrom, 2016). Prior to electroporation, host cells were harvested from plates or from liquid media. Cells were resuspended in 50 mL sterilized Milli-Q H_2O at room temperature, and collected by centrifugation at 4700 × *g* at 4 °C for 10 min. Pellets were washed in 10 mL sterilized Milli-Q H_2O, at room temperature, and centrifuged at 4700 × *g* at 4 °C for 10 min. The resulting pellets were resuspended in 100–150 μL sterilized Milli-Q H_2O and placed on ice. The products of the assembled fragments were directly electroporated into pre-treated cells.

After electroporation, cells were pre-cultivated overnight in liquid medium with either methane (*Methylomonas* sp. LW13; Zheng et al., 2018) or methanol (*Methylotenera mobilis* JLW8; Huang et al., 2018), without antibiotics, after which cells were harvested and transferred onto plates containing the respective antibiotic (50 μg mL^{-1} Km or 10 μg mL^{-1} gentamicin (Gm)). Single colonies were selected, and the desired mutation was verified by sequencing a respective PCR-amplified fragment of the chromosomal DNA. Specifically, wild type *Methylomonas* sp. LW13 was used to add the His-tag and the Km gene sequence to express XoxF5 and XoxG4, and recombinant clones were selected on plates supplemented with Km, under methane, in the presence of Ce^{3+} (30 μM; Zheng et al., 2018). The double *xoxF4* mutant of *M. mobilis* JLW8 was used to insert the His-tagged versions of *xoxF4-1* and *xoxF4-2*, and recombinant clones were selected by complementation for growth on methanol, on solid media supplied with methanol and Ce^{3+} (30 μM), in the presence of, respectively, Gm or Km (Huang et al., 2018).

We also employed *Methylomonas* sp. LW13 as a host for heterologous expression of the divergent presumed *xoxG* genes (Zheng et al., 2018). The assembled constructs contained the presumed *xoxG* gene, the Km resistance gene cassette, and the two regions flanking the presumed *xoxG*. The heterologously expressed genes belonged to *Methylotenera versatilis* 7 (*xoxG1*), *Methylophilus methylotrophus* Q8 (*xoxG2*), *Methylosinus* sp. PW1 (*xoxG3*), and *Methylosarcina lacus* LW14 (*xoxG4*), with His tags encoding six histidine residues added to the 3′ terminus of each target gene (Zheng et al., 2018). Details of chromosomal insert constructs are summarized in Table 1.

3. Protein purification and analysis protocols

3.1 Protein expression in *M. extorquens* AM1

For proteins expressed in *M. extorquens* AM1, generally, 300 mL cultures were grown in shake flasks with succinate as a substrate (0.2% w/v) to late exponential phase and collected by centrifugation at 4700 × *g* for 10 min. Pellets were transferred to 100 mL of fresh minimal medium supplemented with methanol (0.5% v/v) and Ce^{3+} (30 μM), and these were incubated overnight at 30 °C with shaking at 200 rpm to an OD_{600} of 0.6–0.8. Cells were harvested by centrifugation at 4700 × *g* for 15 min at 4 °C, and pellets were used immediately or kept at −80 °C. For some proteins (e.g., XoxF5 from *R. kholense* DSM 19783), best results were achieved when expression was carried out as above, but at 20 °C, for 2 days.

3.2 Protein expression in *M. mobilis* JLW8

Each of the two native XoxF4 enzymes of *M. mobilis* JLW8 were expressed separately in a double mutant that lacked both enzymes and thus unable to grow on methanol. The recombinant strains producing active enzymes were selected based on the reversal of the methanol-negative mutant phenotype, on plates supplemented with methanol and Ce^{3+} (30 μM, Huang et al., 2018). Constructs expressing these enzymes were grown (1000 mL) at 30 °C with shaking at 200 rpm to an OD_{600} of 0.6–0.8.

3.3 Protein expression in *Methylomonas* sp. LW13

The *Methylomonas* sp. LW13 recombinants expressing its native enzymes, XoxF5 and XoxG4 were selected under methane, by Km resistance, as the Km gene was included into the constructs, in the presence of Ce^{3+} (30 μM; Zheng et al., 2018). Cultures (1000 mL) for protein recovery were incubated under methane, with Ce^{3+}, at 30 °C with shaking at 200 rpm to an OD_{600} of 0.6–0.8.

3.4 Purification protocol

Cell pellets were resuspended in 4 mL of 100 mM Tris-HCl (pH 8.0, 8.5 or 9.0) buffer containing 3.5 mM β-mercaptoethanol (β-ME) and passed twice through a French Pressure Cell (Sim-Aminco) at 1.0×10^8 Pa. Cell debris was removed by centrifugation at 20,800 × *g* for 15 min at 4 °C. The supernatants were mixed with 5 volumes of the start buffer (100 mM Tris-HCl pH 9.0, 150 mM NaCl, 5 mM imidazole, 1 mM β-ME, a Pierce™ protease inhibitor tablet (Thermo Fisher Scientific)) and 1 volume of pre-balanced Ni-NTA agarose (Qiagen), and these mixtures were shaken for 10 min at 4 °C, to enhance the specificity of binding. Then the mixtures were loaded onto empty PD-10 columns (GE Healthcare). After two successive wash steps with 5 volumes of the start buffer and 3 volumes of the wash buffer (100 mM Tris-HCl pH 9.0, 150 mM NaCl, 30 mM imidazole, 1 mM β-ME), the elution step was carried out using the elution buffer (100 mM Tris-HCl pH 9.0, 150 mM NaCl, 250 mM imidazole, 1 mM β-ME). As one exception, a higher concentration of NaCl, up to 350 mM, was required for optimal results with purification of XoxF5 from *R. kholense* DSM 19783, in all the purification buffers.

Protein samples were desalted and concentrated by a series of dilution/concentration steps (100 mM Tris-HCl pH 9.0, 1 mM dithiothreitol; DTT), using 50 kDa Amicon Ultra centrifugal filter units (Millipore), until the concentration of imidazole reached below 1 μM.

The purified proteins were analyzed by 12.5% SDS-PAGE running with 1 × Tris/Glycine/SDS buffer (BIO-RAD). Protein concentrations were determined by the bicinchoninic acid Protein Assay kit (Pierce, Thermo Fisher Scientific), with serum albumin as a standard.

3.5 Enzyme activity assay

3.5.1 Quantitative assay

Alcohol dehydrogenase activity was measured by monitoring the phenazine methosulfate (PMS)-mediated reduction of 2,6-dichlorophenolindophenol (DCPIP; $\varepsilon_{600} = 21.0\,mM^{-1}\,cm^{-1}$). Initially, all assays were carried out at pH 9.0 in accordance with the classic assay (Anthony & Zatman, 1967), the standard reaction mixture containing: 100 mM Tris-HCl buffer pH 9.0, 45 mM NH_4Cl, 1 mM PMS, 150 μM DCPIP, 10 mM methanol and 10–20 μL of crude cell extract (20–35 $mg\,mL^{-1}$ protein) or 5–15 μL of pure protein preparation (0.5–5 $mg\,mL^{-1}$ protein). Assays were performed at room temperature (approximately 26 °C), in a total volume of 0.8 mL in plastic cuvettes (1 cm path length). One unit (U) of specific enzyme activity was defined as 1 μM DCPIP reduced per minute (determined at 600 nm) and was expressed as U per milligram of protein. While this assay is straightforward in cases of the purified enzymes, in crude extracts it measures cumulative activities of multiple enzymes, both Ln^{3+}-dependent and non-Ln^{3+}-dependent, if multiple enzymes are active.

3.5.2 Activity visualization

Some Ln^{3+}-dependent alcohol dehydrogenases can be visualized, either in liquid or in gels. In the former case, a quick evaluation of the presence of activity can be carried out. In the latter case, the Ln^{3+}-dependent enzyme can be separated in space from a Ca^{2+}-dependent enzyme and their respective activities can be evaluated, if only qualitatively. For example, Zheng et al. (2018) utilized gel staining to demonstrate that enzymes with alternative metal specificities, Ln^{3+} versus Ca^{2+}, were both active between Ln^{3+} concentrations of 0.3–60 μM La^{3+}. The assay can be carried out with either isoelectrofocusing gel (not published) or a gradient PAGE electrophoresis, for example 4–25% (Zheng et al., 2018). The staining mixture is essentially the same as in the quantitative assay above, but the DCPIP is substituted with a dye such as nitroblue tetrazolium (1 mM, Sigma-Aldrich, USA) as an electron acceptor, which results in precipitation of the reduced dye. While attractive, this method is limited at this time. While successfully used with Ca^{2+}-dependent MDH enzymes (Chistoserdova & Lidstrom, 1997) as well as with XoxF5-type enzymes (Zheng et al., 2018), in-gel staining failed to produce positive results for XoxF4-type enzymes, whether purified or as parts of crude extracts (unpublished data). We do not have a reasonable explanation for this failure at the time, but it highlights the differences in properties among different groups of XoxF enzymes.

3.6 Metal analysis

Metals were analyzed by inductively coupled plasma mass spectrometry (ICP-MS) on an ELAN DRC-e (PerkinElmer, Shelton, CT, USA). Standard curves were generated with a dilution series of elements (La^{3+}, Ce^{3+}, Nd^{3+}, Ca^{2+}), with concentrations of 0, 500, 1000, 1500, and 3000 nM in 1% nitric acid (Thermo Fisher Scientific, Waltham, MA, USA). Using this equipment, amounts greater than 3000 nM resulted in a non-linear standard curve. Ammonia was used in the Dynamic Reaction Cell (DRC) mode on the ELAN DRC-e to measure Ca^{2+}, because O_2 generates non-calcium compounds that interfere with Ca^{2+} measurements.

To measure metal content in proteins, 50–300 μL of purified enzyme (0.5–4 mg mL^{-1}) was incubated with nitric acid (5–10%) at 90 °C for 1 h and diluted to 5 mL with water, to achieve an approximate expected metal concentration that fell within the range of the standard curve. Metal content was calculated as the mole to mole percentage of the respective metal per subunit of enzyme (mol%).

4. Enzyme properties

4.1 pH optimum

To determine optimum pH values, standard buffers were used as follows: 100 mM sodium/potassium phosphate (pH range 6.0–8.0), 100 mM Tris-HCl (pH range 7.5–9.0), 100 mM sodium carbonate/sodium bicarbonate (pH range 9.0–10.0). All the XoxF5 enzymes as well as the single ExaF/PedH enzyme we characterized revealed a pH optimum between 9.0 and 9.5. One of the XoxF4 enzymes revealed a similar pH optimum with a peak at 9.0, while the second revealed a somewhat lower pH optimum, with a peak at 8.5 (Huang et al., 2018, 2019; Table 2).

4.2 Ammonium requirement

The effect of ammonium was examined in the standard assay including or omitting 45 mM NH_4Cl, at a range of pH values. Ammonium was strictly required for all the XoxF5 enzymes in the in vitro activity assay. Ammonium was not required for the activity of the XoxF4-2, and was not strictly required for the XoxF4-1 or ExaF/PedH enzymes, while having a stimulatory effect on the latter two (Huang et al., 2018, 2019; Table 2).

Table 2 Properties of Ln^{3+}-dependent alcohol dehydrogenases.

Protein source	Enzyme	V_{max} ($U mg^{-1}$)	K_m (μM)	K_{eff} ($s^{-1} mM^{-1}$)	Preferred substrate	Ln^{3+} range	pH optimum	NH_4 required	Expression host
Bradyrhizobium diazoefficiens USDA 110	XoxF5	2.7 ± 0.1	15 ± 2	371 ± 4	Methanol	La^{3+}-Nd^{3+}	9.0–9.5	R	*M. extorquens* AM1
Grimontia marina CECT 8713	XoxF5	14.9 ± 0.2	24 ± 2	1296 ± 175	Methanol	La^{3+}-Nd^{3+}	9.0–9.5	R	*M. extorquens* AM1
Methyloversatilis discipulorum FAM1	XoxF5	3.1 ± 0.1	15 ± 2	434 ± 51	Methanol	La^{3+}-Nd^{3+}	9.0–9.5	R	*M. extorquens* AM1
Methylorubrum extorquens AM1	XoxF5-1	12.5 ± 0.2	21 ± 1	1252 ± 135	Methanol	La^{3+}-Nd^{3+}	9.0–9.5	R	*M. extorquens* AM1
Methylorubrum extorquens AM1	XoxF5-2	3.2 ± 0.1	12 ± 2	564 ± 110	Methanol	La^{3+}-Nd^{3+}	9.0–9.5	R	*M. extorquens* AM1
Rhodovulum kholense DSM 19783	XoxF5	1.6 ± 0.03	17 ± 1	194 ± 24	Methanol	La^{3+}-Nd^{3+}	9.0–9.5	R	*M. extorquens* AM1
Sinorhizobium meliloti 5A14	XoxF5	4.4 ± 0.1	23 ± 2	408 ± 44	Methanol	La^{3+}-Nd^{3+}	9.0–9.5	R	*M. extorquens* AM1
Tistlia consotensis DSM 21585	XoxF5	5.3 ± 0.3	30 ± 6	366 ± 46	Methanol	La^{3+}-Nd^{3+}	9.0–9.5	R	*M. extorquens* AM1
Methylomonas sp. LW13	XoxF5	5.97 ± 0.94	39 ± 11	342 ± 99	Methanol	La^{3+}-Nd^{3+}	9.0–9.5	R	*Methylomonas* sp. LW13
Methylotenera mobilis JLW8	XoxF4-1	0.65 ± 0.56	55 ± 32	27 ± 20	Methanol	Ln^{3+}-Dy^{3+}	9.0	S	*M. mobilis* JLW8
Methylotenera mobilis JLW8	XoxF4-2	0.78 ± 0.3	42 ± 18	42 ± 17	Methanol	Ln^{3+}-Sm^{3+}	8.5	N	*M. mobilis* JLW8
Methylopila sp. M107	ExaF/PedH	0.45 ± 0.01	16 ± 3	57 ± 2	Ethanol	Ln^{3+}-Yb^{3+}	9.0–9.5	S	*M. extorquens* AM1

Kinetic constants are provided for the preferred substrate. *N*, not required; *R*, required; *S*, stimulatory.

4.3 Substrate range

Substrate specificities were determined for several alcohols (methanol, ethanol, 1-propanol, 2-propanol, butanol, hexanol) and aldehydes (formaldehyde, acetaldehyde, propionaldehyde), using the optimal assay conditions for each enzyme, where a substrate was supplied at 10 mM (saturating concentration). All substrates were purchased from Sigma-Aldrich. Formaldehyde was prepared by incubating paraformaldehyde in water at 95 °C for approximately 1 h, until all the powder has been completely dissolved.

All the enzymes revealed high activities with the saturating concentrations of methanol, ethanol, 1-propanol, 1-butanol, 1-hexanol. Of the aldehydes tested, the highest activity was apparent for formaldehyde (Huang et al., 2018, 2019).

4.4 Enzyme kinetics

Kinetic parameters were determined using the respective optimal assay conditions, using varying concentrations of select substrates (0.001–5 mM). The values for K_m and V_{max} were obtained using the Michaelis-Menten equation, with a GraphPad Software (GraphPad Inc., USA). Averages of kinetic constants were determined for each enzyme based on three to five independent experiments, and the respective enzyme efficiencies (K_{eff}) were calculated from these values (Huang et al., 2018, 2019; Table 2).

The two XoxF4 enzymes showed similar V_{max} values for a range of alcohols and for formaldehyde. However, the K_m values were significantly lower for methanol, ethanol, propanol, and hexanol than for butanol and formaldehyde. The enzyme efficiencies for XoxF4 were also similar for methanol, ethanol, propanol, and hexanol, while being somewhat lower for butanol and significantly lower for formaldehyde (Huang et al., 2018).

The XoxF5 enzymes also demonstrated affinities for a broad range of alcohols, and, similarly to the XoxF4 enzymes, their affinities for formaldehyde were significantly lower than for alcohols. One exception was XoxF5 from *R. kholense* DSM 19783, revealing poor affinities for propanol, butanol and hexanol (Huang et al., 2018, 2019). In general, V_{max} values for the XoxF5 enzymes were significantly higher than those for the XoxF4 enzymes, and the respective K_m values were somewhat lower, resulting in significant differences in enzyme efficiencies (Table 2).

It is interesting to point out that the two XoxF5 enzymes from the expression host organism, *M. extorquens* AM1, while 88% identical and 94% similar, demonstrated different kinetic properties. XoxF5-1, expressed under its own promoter, revealed a much higher V_{max} rates with all substrates employed in the study, as well as much higher K_{eff} values compared to XoxF5-2, and it showed the highest affinities toward methanol and hexanol as substrates, while XoxF5-2 revealed the highest affinity for methanol (Huang et al., 2019).

The ExaF/PedH enzyme revealed higher affinities for ethanol and propanol, compared to other alcohols, while the affinity for formaldehyde, again, was much lower than for alcohols.

4.5 Metal specificity

Metal specificity was determined in growth experiments, both in liquid and on plates, supplying different Ln^{3+} at 30 μM concentration. All the XoxF5 enzymes revealed similar metal dependence, being active only in the presence of the lightest Ln^{3+} (La^{3+} to Nd^{3+}). XoxF4-2 was able to utilize a broader range of the light Ln^{3+}, La^{3+} to Sm^{3+}, and XoxF4-1 could utilize an even broader, range, including heavier lanthanides, La^{3+} through Dy^{3+}, with the growth rate diminishing toward heavier Ln^{3+} (Huang et al., 2018, 2019; Table 2). ExaF/PedH from *Methylopila* sp. M107 revealed activity with an even broader range of Ln^{3+}, allowing for growth of the expressing strain with all the Ln^{3+} tested (through Yb^{3+}), with diminishing growth toward the heaviest Ln^{3+} (Huang et al., 2019; Table 2). Remarkably, this enzyme appeared to be promiscuous toward metal dependence, showing (reduced) growth on methanol/ethanol and (reduced) enzymatic activity even in the absence of Ln^{3+} (Huang et al., 2019). Note, however, that low activity in the absence of Ln^{3+} has previously been reported for the purified XoxF5-1 from *M. extorquens* AM1 (Schmidt, Christen, Kiefer, & Vorholt, 2010), and low activity of both XoxF4 enzymes in *M. mobilis* JLW8 was implicated from growth experiments (Mustakhimov, Kalyuzhnaya, Lidstrom, & Chistoserdova, 2013).

We measured metal contents in protein preparations from cultures grown in the presence of both Ln^{3+} and Ca^{2+} and found that all proteins bound Ln^{3+} (up to 0.9 mol per mol of the enzyme protomer). Several enzymes were purified from cultures grown with cocktails of lighter Ln^{3+}, and in each of these cases, the lightest of the Ln^{3+} were preferentially bound (Huang et al., 2019).

4.6 Enzyme preparation stability and storage

Enzyme stability was determined by three activity measurements, at times 0, 4 and 24 h after purification, while kept on ice, in 100 mM Tris-HCl (pH 9.0) buffer containing 1 mM dithiothreitol and 10% glycerol, in at least two independent experiments. While different deactivation trajectories were noted for different enzymes, all appeared relatively unstable even in the presence of glycerol and dithiothreitol, losing 12–80% of initial activity within 24 h (Huang et al., 2019). The enzymes showed UV–Vis absorption spectra typical of PQQ-linked enzymes, with a maximum of 355 nm and a slight shoulder at 400 nm. The PQQ peak decreased somewhat over time, potentially indicative of protein denaturation (Huang et al., 2019). Thus, it is recommended that enzyme analyses are carried out within hours after obtaining pure protein preparations, and the preparations for future use should be immediately frozen at −80 °C.

5. Functional characterization of XOXG4

5.1 Heterologous expression of alternative cytochromes

As Ln^{3+}-dependent alcohol dehydrogenases are represented by proteins widely diverging in their sequences (Chistoserdova, 2011; Chistoserdova & Kalyuzhnaya, 2018), it appears that the electron transfer chains also vary widely. At least four divergent cytochromes have been identified among alpha-, beta- and gammaproteobacterial methylotrophs that were highly expressed during growth on methane (Yu, Beck, & Chistoserdova, 2017). In addition, a plastocyanin-encoding gene has been identified as part of a Xox5-encoding gene cluster in the *Grimontia marina* CECT 8713 genome (Huang et al., 2019). To start to investigate the functions and the properties of these novel proposed electron acceptors, we tested whether alternative XoxG proteins (XoxG1, XoxG2, or XoxG3) could functionally substitute for the XoxG4 proteins that are typical of gammaproteobacterial methanotrophs, by expressing these alternative variants in the ΔXoxG4 mutant of *Methylomonas* sp. LW13, under the native promoter of *xoxG4*. Of the four tested constructs, only the one expressing a XoxG4 homolog was able to complement growth of the XoxG4 mutant (in the presence of Ln^{3+}; Zheng et al., 2018).

5.2 Biochemical demonstration of reduction of XoxG4 by XoxF5

We tested whether XoxG4 is indeed a *c*-type cytochrome and whether it can be directly reduced by XoxF5. We expressed and purified the His-tagged versions of both proteins, belonging to *Methylomonas* sp. LW13. The purified XoxG4, after reduction by dithionite, showed two strong absorption peaks, around 525 and 552 nm, as is typical of cytochromes *c*. We further recorded absorbance by XoxG4, at 550 nm, before and after the addition of the purified XoxF5 and observed an increase in absorbance, indicative of the XoxF5-dependent reduction of XoxG4. The increase in the concentration of XoxF5 correlated with the increase in the absorbance at 550 nm (Zheng et al., 2018). This demonstrated direct redox interaction between XoxF5 and XoxG4.

6. Summary

We here presented logistics, protocols and conditions for expression, purification and characterization of Ln^{3+}-dependent alcohol dehydrogenases from a variety of bacterial sources. We also presented data on the biochemical properties of a dozen enzymes generated by our group, in a comparative fashion, pinpointing some contrasting characteristics. We also briefly described properties of a novel cytochrome that likely serves as a natural electron acceptor from XoxF5 MDH enzymes, acting in methanotrophs of the Gammaproteobacteria class.

Acknowledgment

This material is based upon work supported by the US Department of Energy, Office of Science, Office of Biological and Environmental Research under award DE-SC-0016224.

References

Anthony, C., & Zatman, L. J. (1967). The microbial oxidation of methanol. The prosthetic group of the alcohol dehydrogenase of *Pseudomonas* sp. M27: A new oxidoreductase prosthetic group. *Biochemical Journal*, *104*, 960–969.

Chistoserdova, L. (2011). Modularity of methylotrophy, revisited. *Environmental Microbiology*, *13*, 2603–2622.

Chistoserdova, L. (2016). Lanthanides: New life metals? *World Journal of Microbiology and Biotechnology*, *32*, 138.

Chistoserdova, L., & Kalyuzhnaya, M. G. (2018). Current trends in methylotrophy. *Trends in Microbiology*, *26*, 703–714.

Chistoserdova, L., & Lidstrom, M. E. (1997). Molecular and mutational analysis of a DNA region separating two methylotrophy gene clusters in *Methylobacterium extorquens* AM1. *Microbiology*, *143*, 1729–1736.

Cotruvo, J. A. (2019). The chemistry of lanthanides in biology: Recent discoveries, emerging principles, and technological applications. *ACS Central Science*, *5*, 1496–1506.

Daumann, L. J. (2019). Essential and ubiquitous: The emergence of lanthanide metallobiochemistry. *Angewandte Chemie International Edition*, *58*, 12795–12802.

Delaney, N. F., Kaczmarek, M. E., Ward, L. M., Swanson, P. K., Lee, M.-C., & Marx, C. J. (2013). Development of an optimized medium, strain and high-throughput culturing methods for *Methylobacterium extorquens*. *PLoS One*, *8*, e62957.

Good, N. M., Vu, H. N., Suriano, C. J., Subuyuj, G. A., Skovran, E., & Martinez-Gomez, N. C. (2016). Pyrroloquinoline quinone ethanol dehydrogenase in *Methylobacterium extorquens* AM1 extends lanthanide-dependent metabolism to multicarbon substrates. *Journal of Bacteriology*, *198*, 3109–3118.

Huang, J., Yu, Z., & Chistoserdova, L. (2018). Lanthanide-dependent methanol dehydrogenases of XoxF4 and XoxF5 clades display differential distribution among methylotrophic bacteria and different biochemical properties. *Frontiers in Microbiology*, *9*, 1366.

Huang, J., Yu, Z., Groom, J., Cheng, J.-F., Traver, A., Yoshikuni, Y., et al. (2019). Rare earth element alcohol dehydrogenases widely occur among globally distributed, numerically abundant and environmentally important microbes. *The ISME Journal*, *13*, 2005–2017.

Keltjens, J. T., Pol, A., Reimann, J., & Op den Camp, H. J. M. (2014). PQQ-dependent methanol dehydrogenases: Rare-earth elements make a difference. *Applied Microbiology and Biotechnology*, *98*, 6163–6183.

Marx, C. J., & Lidstrom, M. E. (2001). Development of improved versatile broad-host-range vectors for use in methylotrophs and other Gram-negative bacteria. *Microbiology*, *147*, 2065–2075.

Mustakhimov, I., Kalyuzhnaya, M. G., Lidstrom, M. E., & Chistoserdova, L. (2013). Insights into denitrification in Methylotenera mobilis from denitrification pathway and methanol metabolism mutants. *Journal of Bacteriology*, *195*, 2207–2211.

Schmidt, S., Christen, P., Kiefer, P., & Vorholt, J. A. (2010). Functional investigation of methanol dehydrogenase-like protein XoxF in *Methylobacterium extorquens* AM1. *Microbiology*, *156*, 2575–2586.

Shevchuk, N. A., Bryksin, A. V., Nusinovich, Y. A., Cabello, F. C., Sutherland, M., & Ladisch, S. (2004). Construction of long DNA molecules using long PCR-based fusion of several fragments simultaneously. *Nucleic Acids Research*, *32*, e19.

Vu, H. N., Subuyuj, G. A., Vijayakumar, S., Good, N. M., Martinez-Gomez, N. C., & Skovran, E. (2016). Lanthanide-dependent regulation of methanol oxidation systems in *Methylobacterium extorquens* AM1 and their contribution to methanol growth. *Journal of Bacteriology*, *98*, 1250–1259.

Yan, X., Chu, F., Puri, A. W., Fu, Y., & Lidstrom, M. E. (2016). Electroporation-based genetic manipulation in type I methanotrophs. *Applied and Environmental Microbiology*, *82*, 2062–2069.

Yu, Z., Beck, D. A. C., & Chistoserdova, L. (2017). Natural selection in synthetic communities highlights the roles of *Methylococcaceae* and *Methylophilaceae* and suggests differential roles for alternative methanol dehydrogenases in methane consumption. *Frontiers in Microbiology*, *8*, 2392.

Yu, Z., Zheng, Y., Huang, J., & Chistoserdova, L. (2019). Systems biology meets enzymology: Recent insights into communal metabolism of methane and the role of lanthanides. *Current Issues in Molecular Biology*, *33*, 183–196.

Zheng, Y., Huang, J., Zhao, F., & Chistoserdova, L. (2018). Physiological effect of XoxG (4) on lanthanide-dependent methanotrophy. *mBio*, *9*, e02430-17.

CHAPTER FIVE

Expression, purification and testing of lanthanide-dependent enzymes in *Methylorubrum extorquens* AM1

Nathan M. Good and N. Cecilia Martinez-Gomez*
Department of Plant and Microbial Biology, University of California, Berkeley, CA, United States
*Corresponding author: e-mail address: cecimartinez@berkeley.edu

Contents

Abstract

With mounting evidence of the importance of lanthanide metals in biology and among diverse bacterial phyla, a platform for high-throughput microbial growth for expression and purification of lanthanide-dependent enzymes is increasingly important. Presented in this chapter is a stream-lined approach for growth of the model methylotrophic bacterium *Methylorubrum extorquens* AM1 for the expression of lanthanide-dependent enzymes. Growth is optimized for both high-throughput phenotypic characterization facilitating *in vivo* studies, as well as for scaled-up batch cultivation for enzyme purification allowing for *in vitro* enzymatic studies. Both approaches have been shown to be important to understanding the function and structure of these enzymes. Expression systems have been designed for production of enzymes with and without lanthanide metals, allowing for detection of lanthanide dependence. The protocol described herein is expected to accelerate the discovery of novel lanthanide-dependent enzymes and our understanding of the role of these metals in the greater biological world.

Methods in Enzymology, Volume 650
ISSN 0076-6879
https://doi.org/10.1016/bs.mie.2021.02.001

1. Introduction: A platform for expression of Ln^{3+}-dependent enzymes

The discovery that some lanthanide elements (Ln^{3+}; atomic numbers 57 [La^{3+}] through 63 [Eu^{3+}], excluding 61 [Pm^{3+}]), function with pyrroloquinoline quinone (PQQ) as cofactors of alcohol dehydrogenases (ADHs), enzymes that catalyze the oxidation of primary and secondary alcohols, has led to the emergence of the novel field of lanthanide biochemistry and added the Ln^{3+} to the biological table of elements (Chistoserdova, 2016; Daumann, 2019; Good et al., 2020, 2016; Hibi et al., 2011; Keltjens, Pol, Reimann, & Op den Camp, 2014; Pol et al., 2014). Ln^{3+}-dependent ADHs are currently represented by XoxF-type methanol dehydrogenases (MDHs), of which there are at least five distinct phylogenetic clades (Deng, Ro, & Rosenzweig, 2018; Good et al., 2020; Huang, Yu, & Chistoserdova, 2018; Nakagawa et al., 2012; Pol et al., 2014; Vu et al., 2016; Wang et al., 2019); and ExaF-type/PedH-type ethanol dehydrogenases (EDHs) (Good et al., 2016; Wehrmann, Billard, Martin-Meriadec, Zegeye, & Klebensberger, 2017). XoxF-MDHs have been identified and characterized from methylotrophic *Alpha-*, *Beta-*, and *Gammaproteobacteria*, *Verrucomicrobia*, and the NC10 phylum bacterium "*Candidatus* Methylomirabilis oxyfera" (Deng et al., 2018; Good et al., 2020; Huang et al., 2018, 2019; Keltjens et al., 2014; Yanpirat et al., 2020). ExaF EtDH, the first reported Ln^{3+}-dependent enzyme with a preference for a multi-carbon substrate, was also discovered in a methylotroph, *Methylorubrum* (previously *Methylobacterium*) *extorquens* AM1 (Good et al., 2016). However, the hunt for Ln^{3+}-dependent enzymes outside of known methylotrophic bacteria has resulted in the discoveries of previously unknown methylotrophs (Wang et al., 2019) and the usage of Ln^{3+} by non-methylotrophic bacteria (Wegner, Gorniak, Riedel, Westermann, & Küsel, 2019; Wehrmann et al., 2017). In addition, the prosthetic group PQQ is important for mammals (humans included) (Killgore et al., 1989), and plants (Choi et al., 2008). PQQ-dependent enzymes, or quinoproteins, have been discovered in archaea and eukaryotes (Matsumura et al., 2014; Takeda et al., 2015) as well. Further, in addition to alcohol dehydrogenases, bacterial quinoproteins include sugar-oxidizing enzymes broadening the substrates of these enzymes (Anthony, 2001; Goodwin & Anthony, 1998; Matsushita, Toyama, Yamada, & Adachi, 2002). Together, these findings indicate that the use of Ln^{3+} extends outside of methylotrophy and may likely have a physiological role in other domains of life.

Detailed protocols for working with Ln^{3+}-dependent enzymes, while currently limited to the bacterial representatives available, can provide guidance for studying candidates in novel bacterial species and potentially higher organisms. The organism that has been used most extensively as a platform for the studying of Ln^{3+}-dependent enzymes to date is the model methylotrophic bacterium *M. extorquens* AM1. This can be attributed to the relative ease of growing an aerobic mesophile (Peel & Quayle, 1961), the availability of abundant genetic tools for generating gene knock-out mutants and expressing recombinant enzymes (Marx & Lidstrom, 2001, 2002; von Borzyskowski, Lennart, Weishaupt, Vorholt, & Erb, 2015), and the genomic capacity for producing at minimum three Ln^{3+}-dependent enzymes (Good et al., 2016, 2020; Nakagawa et al., 2012; Vu et al., 2016; Vuilleumier et al., 2009). The genome of *M. extorquens* AM1 contains two *xoxF* genes encoding proteins, XoxF1 and XoxF2, that are 90% identical (Vuilleumier et al., 2009). XoxF1 is the primary MDH produced in the presence of Ln^{3+}, though XoxF2 is capable of supporting growth on its own (Vu et al., 2016). *M. extorquens* AM1 can also produce ExaF EtDH, for which ethanol is the preferred substrate, though the enzyme is capable of oxidizing formaldehyde and supporting growth alone with methanol (Good, Moore, Suriano, & Cecilia Martinez-Gomez, 2019; Good et al., 2016). Additionally, *M. extorquens* AM1 produces the Ca^{2+}-dependent MDH MxaFI in the absence of Ln^{3+}, allowing for growth of the strain with methanol as the sole carbon and energy source. Addition of Ln^{3+} to the growth medium promotes a change from MxaFI to XoxF as the primary MDH. The understanding of this regulatory switch in methanol oxidation systems is fundamental to expressing Ln^{3+}-dependent enzymes in *M. extorquens* AM1.

This chapter will focus on the application *M. extorquens* AM1 as a platform for the expression, purification, and testing of Ln^{3+}-dependent enzymes using methanol as the growth substrate. We will cover the design of genetic constructs for expression, generation of expression strains, scale up of culture conditions for protein production and purification, and high-throughput growth analysis for *in vivo* testing. As a whole, this chapter will serve as a template for both *in vivo* and *in vitro* studies of Ln^{3+}-dependent enzymes, both native and non-native, produced in *M. extorquens* AM1.

2. Expression of Ln^{3+}-dependent enzymes

Unlike common bacterial platforms for protein expression, such as *Escherichia coli*, *M. extorquens* AM1 has the robust capacity to grow with and utilize Ln^{3+}, making it a valuable platform for expression of Ln^{3+}-dependent enzymes, both native and non-native. Ln^{3+}-dependent methanol growth of *M. extorquens* AM1 has been demonstrated for the lanthanide series from La^{3+} through Sm^{3+}, excluding Pm^{3+} (Vu et al., 2016). Sensing, transport, and even intracellular storage has been experimentally validated as well (Mattocks, Ho, & Cotruvo, 2019; Roszczenko-Jasińska et al., 2020), though a complete understanding of the proteins involved in trafficking, Ln^{3+} delivery and insertion into MDH is currently not well understood. The use of *M. extorquens* AM1 enables a wide range of Ln^{3+} that can be tested with a candidate enzyme. Additionally, the availability of Ln^{3+}-sensitive and -insensitive promoters, as well as constitutive and inducible promoters (Chubiz, Purswani, Carroll, & Marx, 2013), offers a high degree of flexibility in expression (Fig. 1).

2.1 Design of expression constructs

Construct design for expression of Ln^{3+}-dependent enzymes requires careful consideration of several factors including promoter sensitivity, promoter strength, and additional regulatory components. The *xox1* promoter (P*xox1*) is the best-understood Ln^{3+}-dependent promoter to date, and for that reason it will be used as an exemplary promoter for construct design. Exogenous Ln^{3+} at concentrations as low as 100 nM have been shown to be sufficient to effect the switch from P*mxa* to P*xox1* expression (Vu et al., 2016). P*xox1* is not equally sensitive to all Ln^{3+} though, and expression decreases precipitously with Ln^{3+} beyond samarium. Even a 10-fold increase in exogenous Sm^{3+} is unable to increase P*xox1* expression significantly. Promoter fusion studies of P*xox1* with a Ln^{3+} profile correlate very well with methanol growth phenotypes. At full expression, P*xox1* is considered to be a relatively "strong" promoter, with expression around 60% that of *Pmxa*. It is important to remark that P*xox1* and P*mxa* are oppositely regulated by Ln^{3+} even in the presence of Ca^{2+}. The promoter strength of P*xox1* is further supported by biochemical assays showing similar methanol oxidation activities in cell-free extracts from cultures grown on methanol with and without exogenous Ln^{3+}, and again these activities correlate with very similar growth rates. P*xox1* in an expression construct should include at least 300 bp upstream of the XoxF1 translation start site. This additional sequence includes a predicted promoter and other potential regulatory region(s) that are required for Ln^{3+}-inducible expression (Good et al., 2016; Nakagawa et al., 2012; Skovran, Palmer, Rountree, Good, & Lidstrom, 2011; Vu et al., 2016). Genetic constructs have been designed for expression and study of native Ln^{3+}-dependent enzymes in *M. extorquens* AM1. Both XoxF1 and ExaF have been expressed *via* P*xox1* in permissive Ln^{3+} conditions (Good et al., 2019, 2016).

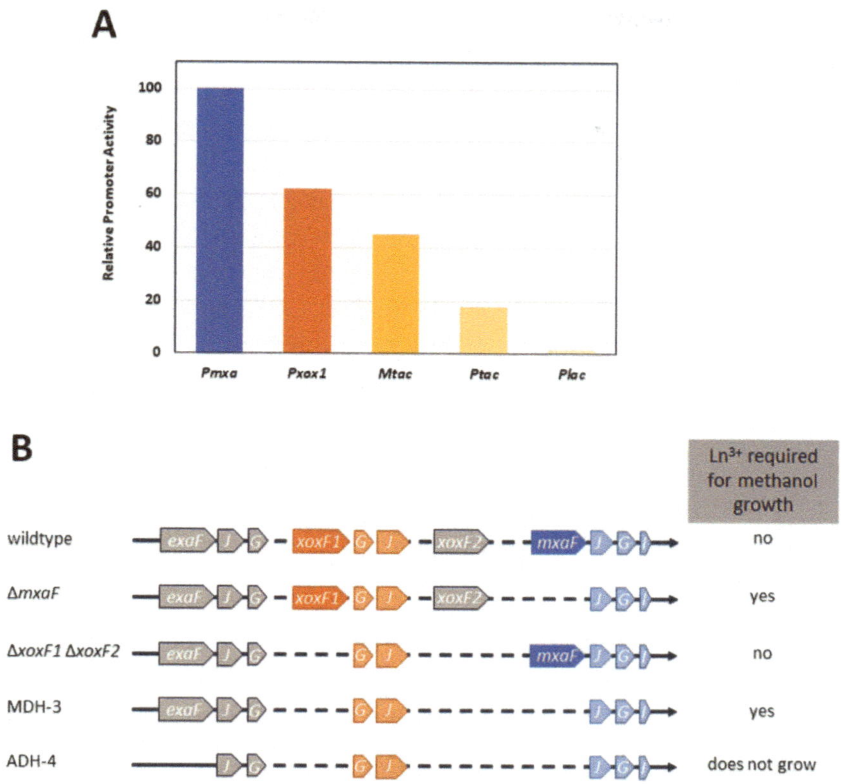

Fig. 1 Promoters and genetic backgrounds for expression in *M. extorquens* AM1. (A) Comparison of promoter strengths with methanol in *M. extorquens* AM1. Expression levels are adapted from Zhang, FitzGerald, and Lidstrom (2005), Marx and Lidstrom (2001), Vu et al. (2016), Roszczenko-Jasińska et al. (2020) and normalized to *Pmxa*. Promoters shown in yellow are constitutive in *M. extorquens* AM1 because the *lac* repressor is not produced natively. (B) Genetic backgrounds for expression and testing of Ln^{3+}-dependent enzymes. Wildtype and the Δ*xoxF1* Δ*xoxF2* double knockout mutant will grow on methanol in the absence or presence of Ln^{3+}. The ADH-4 strain is incapable of methanol growth under laboratory conditions because it cannot produce a functional oxidation system.

The requirement of Ln^{3+} for expression of Ln^{3+}-dependent enzymes, as we know them, presents a challenge for diagnostic studies of metal dependency. In order to truly test whether or not an enzyme requires a Ln^{3+} ion for catalytic function the enzyme in question must be assayed for activity when it is produced in both the presence and absence of exogenous Ln^{3+}. Regulatory constraints can be circumvented by generating Ln^{3+}-sensitive constructs with P*xox1* and Ln^{3+}-insensitive constructs with P*mxa*. ExaF was successfully produced from two such constructs (Good et al., 2016).

In another study, clever design using both the P*xox1* and P*mxa* promoters in tandem to accommodate expression in both the presence or absence of Ln^{3+} allowed for the study of non-native XoxF enzymes using *M. extorquens* AM1 for protein production (Huang et al., 2019). In this way, the number of constructs needed was effectively reduced by half and *M. extorquens* AM1 was shown to be a suitable platform for recombinant Ln^{3+}-dependent enzyme production. In a separate study *Mtac*, a modified *tac* promoter for higher expression in *M. extorquens* AM1, which is insensitive to Ln^{3+}, was used to study native XoxF1 and ExaF both with and without addition of Ln^{3+} to the growth medium (Good et al., 2020; Zhang et al., 2005). Both of the approaches described above successfully decoupled the *mxa-xox1* inverse regulation induced by exogenous Ln^{3+}.

2.1.1 Protocol 1: Generating expression strains

2.1.1.1 Media/Reagents/Equipment (Table 1)

Table 1 Media/reagents/equipment.

Media	Sterile nutrient broth
	Growth medium (MP minimal medium [Delaney et al., 2013]; see notes)
Selection medium	MP medium with 15 mM succinic acid and 50 μg/mL kanamycin sulfate
Reagents	Expression construct(s)
	M. extorquens background strain(s)
	Ice-cold sterile ddH_2O
	Ice-cold sterile 10% glycerol
	Electroporation cuvettes
	14-mL sterile polypropylene culture tubes
	1.5-mL polypropylene eppendorf tubes
Equipment	Tabletop centrifuge
	Electroporator
	Shaker
	Incubator at 30 °C

2.1.1.2 Steps

1. Inoculate 2 mL of basal succinate growth medium with the *M. extorquens* AM1 background strain(s).
2. Incubate the culture tube at 30 °C, shaking at 200 rpm for 16–24 h until the cultures have reached late exponential growth phase (OD_{600} 1.8–2.0).
3. Incubate culture on ice for 20 min
4. Pre-incubate a 1.5-mL eppendorf tube on ice
5. Transfer 1 mL of culture to a sterile, ice-cold 1.5-mL eppendorf tube
6. Pellet cells by centrifugation at 1000 × *g* for 10 min. Decant the supernatant
7. Wash cells with 1 mL ice-cold sterile ddH_2O
8. Pellet cells by centrifugation at 1000 × *g* for 10 min. Decant the supernatant
9. Wash cells with 1 mL ice-cold sterile ddH_2O
10. Pellet cells by centrifugation at 1000 × *g* for 10 min. Decant the supernatant
11. Remove any remaining supernatant by pipetting
12. Resuspend the cell pellet by gently pipetting 40 μL of 10% glycerol. Keep cells on ice
13. Pre-incubate a sterile electroporation cuvette on ice for 10 min
14. Add 100 ng of purified plasmid DNA to cells
15. Incubate on ice for 5 min
16. Transfer cells and plasmid DNA to the pre-incubated electroporation cuvette
17. Electroporate at 2500 V
18. Incubate on ice for 5 min
19. Add 1 mL of sterile nutrient broth. Incubate at 30 °C shaking at 200 rpm for 16–24 h
20. Pellet cells by centrifugation at 1000 × *g* for 10 min. Decant the supernatant
21. Remove any residual nutrient broth by pipetting
22. Gently resuspend the cell pellet with 50 μL of sterile basal medium (no carbon source).
23. Transfer 50 μL of cell suspension to a petri dish with solidified succinate growth medium with 50 μg/mL kanamycin sulfate (see notes).
24. Spread the cell suspension evenly over the surface of the solidified growth medium
25. Incubate the petri dish at 30 °C for 4–5 days or until single pink colonies appear

26. Re-streak 4–8 single colonies on succinate growth medium
27. Incubate the petri dish at 30 °C for 3–5 days
28. Verify individual colonies for plasmid sequence by PCR

2.1.1.3 A note on generating expression strains

MP minimal medium has been tested and shown to provide robust high-throughput growth results (Delaney et al., 2013). Similarly, robust growth results have been obtained in large-scale cultures. Before the lanthanide effect was discovered, pCM80 was originally designed as an expression construct. A modified version for expression and His-tag purification of protein was reported in (Schmidt, Christen, Kiefer, & Vorholt, 2010). If using pCM80 (carrying P*mxa*), 20 µg/mL working concentration of tetracycline should be used in place of kanamycin for plasmid retention. However, the use of tetracycline as an antibiotic for selection has proven to induce cobalt limitation and, as a result, decrease growth rate of *M. extorquens* (Kiefer et al., 2009).

3. Purification of recombinant Ln^{3+}-dependent enzymes

Purification is essential to conducting detailed biochemical characterizations of Ln^{3+}-dependent enzymes. A number of crucial studies have investigated catalytic properties, metal coordination, and the crystal structure of enzymes produced from *M. extorquens* AM1, and several important observations have been made through this work. First, metallation of Ln^{3+} -dependent enzymes is metal species dependent. Molar ratios of 0.8–1.0 La^{3+}:1.0 protomer (Good et al., 2020; Huang et al., 2019; Nakagawa et al., 2012) and 0.5–0.9 Nd^{3+}:1.0 protomer (Good et al., 2019; Huang et al., 2019) have been reported for purified XoxF1 MDH produced in culture medium with a single Ln^{3+} species. The reduction in Nd^{3+} coordination suggested discrepancies in either metal transport or insertion/coordination compared to La^{3+}, though this requires more investigation. Interestingly, reconstitution of XoxF1 with La^{3+} or Nd^{3+} was unsuccessful in our hands, indicating a dedicated metal insertion system is required *in vivo*. Unsuccessful reconstitution was also reported for distantly related XoxF enzymes from *Methylotenera mobilis* JLW8 and *Methylomonas* sp. LW13, but XoxF from the extremophile *Methylacidaphilum fumariolicum* SolV (Huang et al., 2018; Jahn et al., 2018) was successfully reconstituted by adding a second Ln^{3+} to partially apo enzyme purified with Eu^{3+}. Reconstitution of ExaF from *M. extorquens* AM1 was also unsuccessful in our hands (Good et al., 2016). These are important considerations when producing heterologously expressed enzymes since production in *E. coli* would require reconstitution with Ln^{3+} after purification, while enzymes produced in *M. extorquens* AM1 would not as the bacterium is already capable of transporting and inserting Ln^{3+}. This is further evidenced by the first study producing heterologously expressed Ln^{3+}-dependent enzymes from multiple organisms using *M. extorquens* AM1 as a platform. In addition to XoxF1 and XoxF2 from *M. extorquens* AM1, seven heterologous enzymes were produced, kinetically characterized, and shown to be Ln^{3+}-dependent (Huang et al., 2019). An eighth enzyme of the ExaF/PedH EtDH class from *Methylophila* sp. 107 was identified as promiscuous for Ln^{3+} or Ca^{2+}. Interestingly, in our work XoxF1 purified in the absence of Ln^{3+} showed that the enzyme was capable of coordinating Ca^{2+} as well, though in this case it was catalytically inactive (Good et al., 2020). Work with both La^{3+} and Ca^{2+} also supported the presence of a Ln^{3+} insertion system independent from the Ca^{2+} insertion system (Amaratunga, Goodwin, David, & Anthony, 1997). Second, *in vitro* studies of catalysis by Ln^{3+}-dependent enzymes led to the discovery of the most efficient PQQ-dependent EtDH to date, ExaF (Good et al., 2016). A comparison of catalytic efficiencies shows a ~4.5-fold higher efficiency than the next highest EtDH, the Ca^{2+}-dependent enzyme from *Pseudomonas* BB1 (Dijkstra, Van Den Tweel, De Bont, Frank, & Duine, 1985), and a ~220-fold higher efficiency than the only other Ln^{3+}-dependent EtDH, reported from *P. putida* KT2440 (Wehrmann et al., 2017). Third, crystallization of enzymes can provide key insights into their structure and cofactor coordination. The crystal structure of XoxF1, along with site-directed mutagenesis of the active site, was key to validating the necessity of an essential aspartate residue for Ln^{3+} coordination (Good et al., 2020). A second structure of XoxF1 MDH was resolved and found to not contain the coenzyme PQQ. This was a surprising finding since it was speculated that insertion of the PQQ-La cofactor complex leads to dimeric interfacing (Featherston et al., 2019), yet the enzyme remained dimerized after the loss of PQQ. These results attest to the importance of working with purified preps to address critical questions about the structure and function of Ln^{3+}-dependent enzymes.

3.1.1 Protocol 2: Culture for protein purification

3.1.1.1 A note on Ln^{3+} stocks

Stocks should be prepared in plastic tubes as metal ions can leach from glassware. It is recommended that stocks be used within 1 week or be remade. Before use, stocks should be checked for precipitation and remade if necessary.

3.1.1.2 Media/Reagents/Equipment (Table 2)

Table 2 Media/reagents/equipment.

Growth medium	MP medium with 15 mM succinic acid and 50 μg/mL kanamycin sulfate
Reagents	20 mM $LnCl_3$ stock(s)
	100% methanol
	Kanamycin sulfate
Equipment	250-mL shake flask with cap; sterile
	2.8-L Fernbach flask; sterile
	Shaker
	Incubator at 30 °C

3.1.1.3 Steps

1. Make 1.5 L of basal growth medium and transfer to a 2.8-L Fernbach flask
2. Autoclave to sterilize. *Important: If using MP growth medium, $CaCl_2$ should be added from a sterile stock after autoclaving to avoid formation of calcium carbonate*
3. Add 150 μL of $LnCl_3$ stock to one tube of methanol growth medium to generate methanol + Ln growth medium. The final concentration of Ln^{3+} is 20 μM
4. After the media has cooled to room temperature, add 7.5 mL of methanol (125 mM final concentration).
5. Add 75 mg of kanamycin sulfate (see notes). Mix until dissolved
6. Transfer 2 mL of cells of the desired expression strain, pre-grown in succinate growth medium, to 50 mL succinate growth medium with kanamycin in a sterile 250-mL shake flask
7. Incubate at 30 °C, shaking at 200 rpm for ~24 h
8. Measure OD_{600} of culture to ensure culture is in late-exponential growth phase (OD_{600} 1.6–2.0) before proceeding
9. Transfer 50 mL of culture to 1.5 L of expression growth medium with 125 mM methanol (prepared in steps 1–5).

10. Incubate at 30 °C, shaking at 200 rpm for ~24 h
11. Measure OD_{600} of culture to ensure culture is in late-exponential growth phase with methanol (OD_{600} 1.6–2.0) before proceeding
12. *Optional: Without additional methanol substrate, cultures should reach an* OD_{600} *of* ~ *2.4. Cultures can be fed additional methanol to increase culture yield if desired. Addition of 7.5 mL methanol can boost the final culture yield to* OD_{600} ~ *4.5*
13. Once the culture is approaching maximal yield, harvest cells by centrifugation at 21,000 × *g* at 4 °C for 10 min
14. Discard the supernatant
15. Transfer the cell pellet(s) to a clean 50 mL conical tube and incubate on ice if continuing or flash freeze in liquid nitrogen and store at −80 °C

3.1.2 Protocol 3: Immobilized ion affinity chromatography

3.1.2.1 Buffers/Reagents/Equipment (Table 3)

Table 3 Buffers/reagents/equipment.

Buffers	25 mM Tris-HCl (pH 8.0) with 150 mM NaCl and 5 mM imidazole
	25 mM Tris-HCl (pH 8.0) with 150 mM NaCl and 50 mM imidazole
	25 mM Tris-HCl (pH 8.0) with 150 mM NaCl and 400 mM imidazole
Reagents	Ni-nitrilotriacetic acid (NTA) Superflow resin
	100% methanol
	Kanamycin sulfate
Equipment	Gravity flow column
	PD-10 column
	Cell disruptor
	Ultracentrifuge
	Spin cells
	Magnetic stir plate
	Small glass beaker
	Magnetic stir bar

3.1.2.2 Steps

1. Resuspend cell pellets in two volumes of 25 mM Tris-HCl (pH 8.0) with 150 mM NaCl and 5 mM imidazole. Keep on ice
2. Disrupt cells at 25,000 p.s.i
3. Transfer cell lysate to spin cell(s).
4. Remove cell debris by ultracentrifugation for 1 h at 4 °C at 100,000 × *g*
5. Decant supernatant into a clean, small glass beaker. Be careful so as to not pour over cell debris
6. In the small glass beaker, mix cell extract with 2 mL of Ni-nitrilotriacetic acid (NTA) Superflow resin pre-equilibrated with 25 mM Tris-HCl (pH 8.0) with 150 mM NaCl and 5 mM imidazole
7. Incubate mixing gently for 1 h at 4 °C
8. Transfer cell extract and resin to an empty gravity flow column
9. Wash the resin with 10 mL 25 mM Tris-HCl (pH 8.0) with 150 mM NaCl and 50 mM imidazole
10. Elute protein by addition of 25 mM Tris-HCl (pH 8.0) with 150 mM NaCl and 400 mM imidazole, collecting 1 mL fractions
11. Desalt protein fractions by loading onto a PD-10 column and eluting with 25 mM Tris-HCl (pH 8.0), 150 mM NaCl
12. Confirm protein enrichment *via* SDS-PAGE analysis

3.1.2.3 A note on enzyme storage

In the event that a purified enzyme must be stored before being used, it is important that proper storage conditions be identified and maintained to avoid degradation and loss of catalytic activity. For example, XoxF1 MDH can be stored in pH 8.0 Tris or HEPES buffer with 10% glycerol or methanol at −80 °C for 12 months or longer with little loss of enzyme activity. It may be preferable to store the enzyme with methanol, as this is the primary substrate for XoxF MDH. On the other hand, ExaF EtDH rapidly undergoes degradation when frozen in the same conditions. If, however, the enzyme is stored at 4 °C, it remains active for up to 3 months. We suggest, therefore, that when working with new enzymes, aliquots are stored at 4 °C and −80 °C and then tested for activity to help determine optimal storage conditions.

4. Growth phenotyping with Ln^{3+}-dependent enzymes

Understanding how Ln^{3+}-dependent enzymes function *in vivo* requires a biological system that is amenable to laboratory cultivation and genetic manipulation, such that the function of individual enzymes can be dissected. A suite of gene deletion mutant strains of *M. extorquens* AM1 have been generated for exactly these purposes and provide necessary genetic backgrounds for testing individual Ln^{3+}-dependent enzymes *in vivo*:Δ*mxaF* (Skovran et al., 2011), Δ*xoxF1* (Skovran et al., 2011), Δ*xoxF2* (Skovran et al., 2011), Δ*xoxF1* Δ*xoxF2* (Skovran et al., 2011), Δ*mxaF* Δ*xoxF1* Δ*xoxF2* [MDH-3] (Good et al., 2016; Vu et al., 2016), Δ*mxaF* Δ*xoxF1* Δ*xoxF2* Δ*exaF* [ADH-4] (Good et al., 2016). The flexibility of *M. extorquens* AM1 also allows for the comparisons of *in vitro* enzyme activity with *in vivo* growth data, which have been shown to have high correlativity regarding primary substrates (Good et al., 2016; Vu et al., 2016).

M. extorquens has also been effectively utilized as an expression platform to dissect secondary enzyme activities *in vivo* (Fig. 2). Formaldehyde oxidation by the EtDH ExaF was detected with pure enzyme and in cell-free extract. In wildtype *M. extorquens* AM1, formaldehyde, a toxic intermediate, is generated from methanol oxidation and then converted to formate through the oxidative tetrahydromethanopterin pathway. When the oxidative tetrahydromethanopterin pathway was genetically disrupted, a condition where formaldehyde would accumulate to toxic levels unless a secondary oxidation system was available; ExaF was able to alleviate toxicity by oxidizing formaldehyde, allowing the strain to grow normally (Good et al., 2019). In the same study, formaldehyde oxidation by XoxF MDH was not seen *in vivo*, as formaldehyde levels remained at toxic levels when the oxidative tetrahydromethanopterin pathway was disrupted. Therefore, even though XoxF MDH has the capacity for formaldehyde oxidation *in vitro*, this activity is not observed *in vivo*, indicating the presence of an unknown effector or regulatory mechanism.

Finally, *M. extorquens* AM1 has been effectively utilized for the expression of site-directed mutant variant enzymes of XoxF and ExaF. Both enzymes

A

enzyme or cell-free extract	V_{max} methanol (U-mg^{-1})	V_{max} formaldehyde (U-mg^{-1})
XoxF1-La	5.72 ± 0.13	5.00 ± 0.07
wild-type (XoxF1)	0.081 ± 0.003	not detected
ExaF-La	6.62 ± 0.12	8.13 ± 0.06
MDH-3 (ExaF)	0.045 ± 0.02	0.058 ± 0.04

B

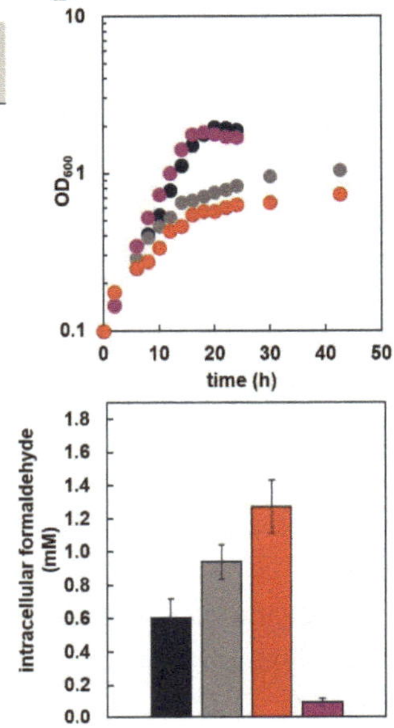

C

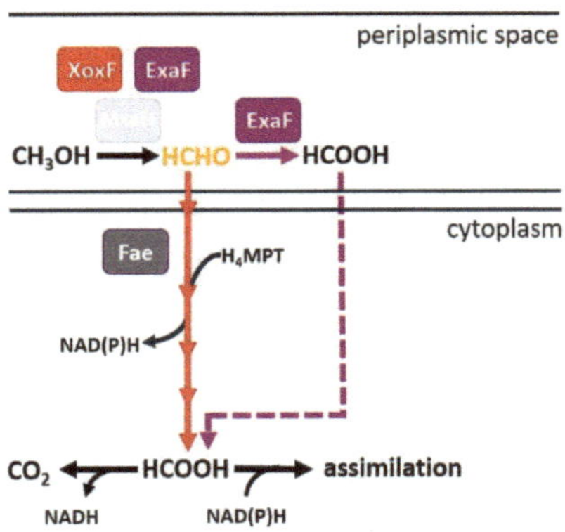

Fig. 2 Dissection of secondary catalytic activities for Ln^{3+}-dependent enzymes. (A) Catalytic activities of enzymes purified with La^{3+} or cell-free extracts prepared from culture of strains with XoxF1 (wildtype, red) or ExaF (MDH-3, purple) as the primary methanol oxidation system. Both enzymes exhibit robust activities with methanol and formaldehyde when in pure form, but formaldehyde oxidation is only detected in the MDH-3, ExaF-expressing, cell-free extract. (B) Alleviation of growth inhibition by toxic formaldehyde produced from methanol oxidation. Strains are co-fed succinate and methanol and grown with La^{3+} (top). Wildtype (black) grows normally, but a mutant strain unable to produce formaldehyde-activating enzyme, Fae (gray), exhibits growth inhibition: The strain grows on succinate, but eventually formaldehyde from methanol oxidation accumulates to toxic levels. A more severe growth inhibition is seen in a *fae* mutant background producing XoxF1 but not ExaF (red). Complete alleviation of growth inhibition is seen in a *fae* mutant background producing ExaF but not XoxF1 (purple). Intracellular formaldehyde concentrations correlate with growth patterns (bottom). When only XoxF1 is available, the strain accumulates more formaldehyde than *fae* alone, indicating that ExaF contributes to formaldehyde oxidation in the *fae* mutant. When only ExaF is available, no growth inhibition is seen indicating efficient formaldehyde oxidation. (C) Model showing *in vivo* formaldehyde oxidation activity of ExaF (purple) and the requirement of the tetrahydromethanopterin pathway when XoxF1 is the primary methanol oxidation system (red).

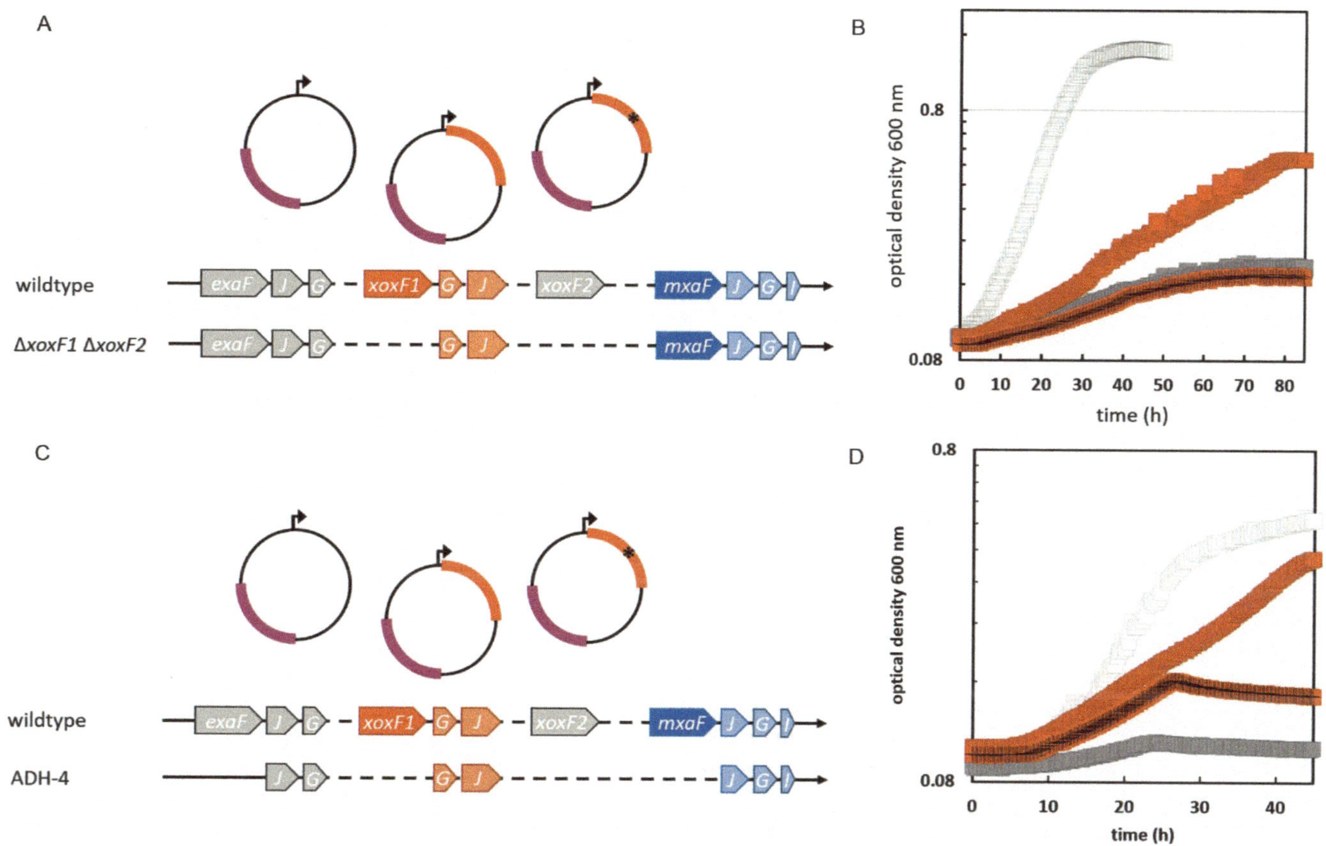

Fig. 3 See figure legend on next page.

were shown to contain a specific aspartate residue that is necessary for Ln^{3+} coordination and enzymatic function (Good et al., 2020). In addition to structural studies and enzyme activity assays, we showed the necessity of the Asp residue *via* growth phenotyping by expressing each enzyme variant in genetic backgrounds that would allow us to assess *in vivo* function (Fig. 3). By coupling *in vivo* growth analysis with *in vitro* assays, we have been able to test and discern structure and function requirements of Ln^{3+}-dependent enzymes. This approach could be used for high-throughput variant analysis, such as deep-mutational scanning, enzyme engineering or directed evolution.

Below is a protocol for high-throughput growth phenotype analysis that can be used for rapid testing of multiple strains and constructs.

4.1.1 Protocol 4: High-throughput growth analysis

4.1.1.1 A note on microplate setup and growth

For best reproducibility, comparative phenotyping media should be distributed to the 48-well microplate in a mirrored pattern to reduce artifacts due to unequal oxygenation. We also recommend using a repeating pipettor to reduce the risk of contamination of the microplate and growth media. If a non-repeater micropipette is used for this step, a new pipette tip should be used for each transfer. We strongly discourage the use of 96-well plates for high-throughput growth of *M. extorquens* AM1, and instead recommend 48-well plates for best results. For more detailed information see (Delaney et al., 2013).

Fig. 3 Expression of Ln^{3+}-dependent enzyme variants. (A) Expression setup for testing XoxF1 (orange) and XoxF1 D320A (orange star) enzymes. Genes were expressed by the constitutive *Mtac* promoter to produce enzymes in the Δ*xoxF1* Δ*xoxF2* double MDH mutant strain. This strain retains the *exaF* and accessory genes (gray) and *mxaFI* and accessory genes (blue). Constructs carry the kanamycin resistance cassette for selective plasmid maintenance (purple). (B) Growth, as a proxy of enzyme function, was tested with La^{3+} and methanol. With La^{3+} in the growth medium, the XoxF1 variant is the primary methanol oxidation enzyme. Production of wild-type XoxF1 MDH (orange) partially restores growth compared to the wildtype (white) and Δ*xoxF1* Δ*xoxF2* double mutant (gray) strains carrying empty plasmids. The D320A variant of XoxF1 MDH has lost catalytic function and no growth restoration is observed (orange star). (C) A similar expression scheme was used to test the function of ExaF EtDH variants. The *Mtac* promoter was used once again to express *exaF* alleles to produce wild-type ExaF (orange) and ExaF D319S (orange star). In this scheme, the Δ*mxaF* Δ*xoxF1* Δ*xoxF2* Δ*exaF* [ADH-4] mutant strain was used as a genetic background to eliminate residual expression of *exaF* from a genomic copy of the gene. (D) Growth with ethanol and La^{3+} was tested and again partial restoration was observed for the strain producing ExaF relative to the wildtype (white) and ADH-4 mutant strain (gray) carrying empty plasmids. The D319S variant of ExaF (orange star) is not functional and no growth is observed.

4.1.1.2 Media/Reagents/Equipment (Table 4)

Table 4 Media/reagents/equipment.

Media	Basal growth medium (MP)
	MP medium with 15 mM succinic acid and 50 μg/mL kanamycin sulfate
Reagents	20 mM $LnCl_3$ stock(s) (freshly prepared; see notes)
	100% methanol
	M. extorquens AM1 strain(s) of interest as isolated colonies
	Sterile 1.5-mL polypropylene eppendorf tubes
	Kanamycin sulfate
Equipment	14-mL sterile polypropylene culture tubes
	Benchtop microplate spectrophotometer
	48-well cell culture microplate
	Vortexer
	Shaker
	Incubator at 30 °C

4.1.1.3 Steps

1. Distribute 35 mL of basal growth medium to sterile 50-mL conical tubes
2. Add 100 μL of methanol to one tube of basal growth medium to generate methanol growth medium (50 mM final concentration). Vortex briefly
3. Distribute half of the methanol medium (17.5 mL) into a new 50 mL conical tube
4. Add 1.75 μL of $LnCl_3$ stock to one tube of methanol growth medium to generate methanol + Ln growth medium. The working final concentration of Ln is 2 μM. Vortex briefly
5. Add 1.75 μL of sterile ddH_2O to the other tube of methanol growth medium. Vortex briefly
6. Transfer 2 mL of succinate growth medium with kanamycin to sterile 14 mL polyethylene culture tube

7. Inoculate by transferring 1 colony of each *M. extorquens* AM1 expression or control strain to the growth medium in the tube
8. Incubate the culture tube at 30 °C, shaking at 200 rpm for ~24 h
9. Measure OD_{600} of culture to ensure culture is in late-exponential growth phase before proceeding
10. Transfer 1 mL of mid-exponential growth phase culture to a sterile 1.5-mL eppendorf tube
11. Pellet cells by centrifugation at $1000 \times g$ for 10 min
12. Decant supernatant and remove any residual growth medium by pipetting
13. Add 1 mL of basal growth medium (no carbon source or Ln).
14. Gently wash the growth medium over the cell pellet until the cells are fully resuspended
15. Pellet cells by centrifugation at $1000 \times g$ for 10 min
16. Decant supernatant and remove any residual growth medium by pipetting
17. Resuspend the cell pellet in a volume of basal growth medium such that the cell suspension is concentrated 6.5-fold
18. Pre-incubate the microplate reader. The microplate needs to be capable of maintaining an internal temperature of 30 °C, shaking at a high rate (see below) for prolonged time periods (24–96 h), and reading light scatter at 600 nm in 48-well microplate format in 15-min time intervals
19. Transfer 640 µL of phenotyping growth medium to each well of a 48-well microplate (see notes).
20. Add 10 µL of concentrated *M. extorquens* cell suspension to each microplate well. We recommend using a repeating pipettor to reduce the risk of contamination of the microplate and growth media. If a non-repeater micropipette is used for this step, a new pipette tip should be used for each transfer
21. Cover the microplate with the lid
22. Pre-incubate the plate reader to 30 °C
23. Place the microplate in the reader
24. Shake the microplate at 548 rpm
25. Read absorbance at 600 nm every 15 min until all growth curves have reached stationary phase

5. Final remarks

The expanding field of lanthanide biochemistry seeks to address fundamental questions about how these metals, critical for our technocentric modern life, are important for microscopic and perhaps even macroscopic life. We are only just beginning to grasp the breadth of impacts lanthanides have on biological systems, from transport and storage to gene regulation and enzyme cofactors. In this chapter, we have laid out a series of protocols for expressing enzymes using *M. extorquens* AM1 as a platform. We have shown how this platform can be used to ask biological questions ranging from enzyme structure and function to enzyme regulation. Our hope is that by sharing these protocols we will facilitate the discovery or new Ln^{3+}-dependent enzymes, functions and organisms.

References

Amaratunga, K., Goodwin, P. M., David, C., & Anthony, C. (1997). The methanol oxidation genes mxaFJGIR(S)ACKLD in Methylobacterium Extorquens. *FEMS Microbiology Letters*, *146*(1), 31–38. https://doi.org/10.1016/S0378-1097(96)00399-0.

Anthony, C. (2001). Pyrroloquinoline quinone (PQQ) and quinoprotein enzymes. *Antioxidants & Redox Signaling*, *3*(5), 757–774. https://doi.org/10.1089/15230860152664966.

Chistoserdova, L. (2016). Lanthanides: New life metals? *World Journal of Microbiology and Biotechnology*, *32*(8), 138. https://doi.org/10.1007/s11274-016-2088-2.

Choi, O., Kim, J., Kim, J.-G., Jeong, Y., Moon, J. S., Park, C. S., et al. (2008). Pyrroloquinoline Quinone is a plant growth promotion factor produced by Pseudomonas Fluorescens B16. *Plant Physiology*, *146*(2), 657–668. https://doi.org/10.1104/pp.107.112748.

Chubiz, L. M., Purswani, J., Carroll, S. M., & Marx, C. J. (2013). A novel pair of inducible expression vectors for use in Methylobacterium Extorquens. *BMC Research Notes*, *6*(1), 183. https://doi.org/10.1186/1756-0500-6-183.

Daumann, L. J. (2019). Essential and ubiquitous: The emergence of lanthanide Metallobiochemistry. *Angewandte Chemie*, *58*(37), 12795–12802. https://doi.org/10.1002/anie.201904090.

Delaney, N. F., Kaczmarek, M. E., Ward, L. M., Swanson, P. K., Lee, M.-C., & Marx, C. J. (2013). Development of an optimized medium, strain and high-throughput culturing methods for Methylobacterium Extorquens. *PLoS One*, *8*(4), e62957. https://doi.org/10.1371/journal.pone.0062957.

Deng, Y. W., Ro, S. Y., & Rosenzweig, A. C. (2018). Structure and function of the lanthanide-dependent methanol dehydrogenase XoxF from the Methanotroph Methylomicrobium Buryatense 5GB1C. *Journal of Biological Inorganic Chemistry: JBIC: A Publication of the Society of Biological Inorganic Chemistry*, *23*(7), 1037–1047. https://doi.org/10.1007/s00775-018-1604-2.

Dijkstra, M., Van Den Tweel, W. J. J., De Bont, J. A. M., Frank, J., & Duine, J. A. (1985). Monomeric and dimeric quinoprotein alcohol dehydrogenase from alcohol- grown pseudomonas BB1. *Journal of General Microbiology*, *131*, 3163–3169.

Featherston, E. R., Rose, H. R., McBride, M. J., Taylor, E. M., Boal, A. K., & Cotruvo, J. A., Jr. (2019). Biochemical and structural characterization of XoxG and XoxJ and their roles in lanthanide-dependent methanol dehydrogenase activity. *Chembiochem: A European Journal of Chemical Biology*, *20*(18), 2360–2372. https://doi.org/10.1002/cbic.201900184.

Good, N. M., Fellner, M., Demirer, K., Hu, J., Hausinger, R. P., & Cecilia Martinez-Gomez, N. (2020). Lanthanide-dependent alcohol dehydrogenases require an essential aspartate residue for metal coordination and enzymatic function. *The Journal of Biological Chemistry*, *295*(24), 8272–8284. https://doi.org/10.1074/jbc.RA120.013227.

Good, N. M., Moore, R. S., Suriano, C. J., & Cecilia Martinez-Gomez, N. (2019). Contrasting in vitro and in vivo methanol oxidation activities of lanthanide-dependent alcohol dehydrogenases XoxF1 and ExaF from Methylobacterium Extorquens AM1. *Scientific Reports*, *9*(1), 4248. https://doi.org/10.1038/s41598-019-41043-1.

Good, N. M., Vu, H. N., Suriano, C. J., Subuyuj, G. A., Skovran, E., & Cecilia Martinez-Gomez, N. (2016). Pyrroloquinoline quinone ethanol dehydrogenase in Methylobacterium Extorquens AM1 extends lanthanide-dependent metabolism to multicarbon substrates. *Journal of Bacteriology*, *198*(22), 3109–3118. https://doi.org/10.1128/JB.00478-16.

Goodwin, P. M., & Anthony, C. (1998). The biochemistry, physiology and genetics of PQQ and PQQ-containing enzymes. *Advances in Microbial Physiology*, *40*, 1–80. https://doi.org/10.1146/annurev.mi.23.100169.002543.

Hibi, Y., Asai, K., Arafuka, H., Hamajima, M., Iwama, T., & Kawai, K. (2011). Molecular structure of La3+-induced methanol dehydrogenase-like protein in Methylobacterium Radiotolerans. *Journal of Bioscience and Bioengineering*, *111*(5), 547–549. https://doi.org/10.1016/j.jbiosc.2010.12.017.

Huang, J., Yu, Z., & Chistoserdova, L. (2018). Lanthanide-dependent methanol dehydrogenases of XoxF4 and XoxF5 clades are differentially distributed among methylotrophic bacteria and they reveal different biochemical properties. *Frontiers in Microbiology*, *9*(June), 1366. https://doi.org/10.3389/fmicb.2018.01366.

Huang, J., Yu, Z., Groom, J., Cheng, J.-F., Tarver, A., Yoshikuni, Y., et al. (2019). Rare earth element alcohol dehydrogenases widely occur among globally distributed, numerically abundant and environmentally important microbes. *The ISME Journal*, *13*(8), 2005–2017. https://doi.org/10.1038/s41396-019-0414-z.

Jahn, B., Pol, A., Lumpe, H., Barends, T. R. M., Dietl, A., Hogendoorn, C., et al. (2018). Similar but not the same: First kinetic and structural analyses of a methanol dehydrogenase containing a europium ion in the active site. *Chembiochem*, *19*, 1147–1153. March https://doi.org/10.1002/cbic.201800130.

Keltjens, J. T., Pol, A., Reimann, J., & Op den Camp, H. J. M. (2014). PQQ-dependent methanol dehydrogenases: Rare-earth elements make a difference. *Applied Microbiology and Biotechnology*, *98*(14), 6163–6183. https://doi.org/10.1007/s00253-014-5766-8.

Kiefer, P., Buchhaupt, M., Christen, P., Kaup, B., Schrader, J., & Vorholt, J. A. (2009). Metabolite profiling uncovers plasmid-induced cobalt limitation under methylotrophic growth conditions. *PLoS One*, *4*(11), e7831. https://doi.org/10.1371/journal.pone.0007831.

Killgore, J., Smidt, C., Duich, L., Romero-Chapman, N., Tinker, D., Reiser, K., et al. (1989). Nutritional importance of pyrroloquinoline quinone. *Science*, *245*(4920), 850–852. https://doi.org/10.1126/science.2549636.

Marx, C. J., & Lidstrom, M. E. (2001). Development of improved versatile broad-host-range vectors for use in methylotrophs and other gram-negative bacteria. *Microbiology*, *147*(Pt 8), 2065–2075. https://doi.org/10.1099/00221287-147-8-2065.

Marx, C. J., & Lidstrom, M. E. (2002). Broad-host-range cre-lox system for antibiotic marker recycling in gram-negative Bacteria. *BioTechniques*, *33*(5), 1062–1067. https://doi.org/10.2144/02335rr01.

Matsumura, H., Umezawa, K., Takeda, K., Sugimoto, N., Ishida, T., Samejima, M., et al. (2014). Discovery of a eukaryotic pyrroloquinoline quinone-dependent oxidoreductase belonging to a new auxiliary activity family in the database of carbohydrate-active enzymes. *PLoS One*, *9*(8), e104851. https://doi.org/10.1371/journal.pone.0104851.

Matsushita, K., Toyama, H., Yamada, M., & Adachi, O. (2002). Quinoproteins: Structure, function, and biotechnological applications. *Applied Microbiology and Biotechnology*, *58*, 13–22. https://doi.org/10.1007/s00253-001-0851-1.

Mattocks, J. A., Ho, J. V., & Cotruvo, J. A., Jr. (2019). A selective, protein-based fluorescent sensor with Picomolar affinity for rare earth elements. *Journal of the American Chemical Society*, *141*(7), 2857–2861. https://doi.org/10.1021/jacs.8b12155.

Nakagawa, T., Mitsui, R., Tani, A., Sasa, K., Tashiro, S., Iwama, T., et al. (2012). A catalytic role of XoxF1 as La3+-dependent methanol dehydrogenase in Methylobacterium Extorquens strain AM1. *PLoS One*, 7(11), e50480. https://doi.org/10.1371/journal.pone.0050480.

Peel, D., & Quayle, J. R. (1961). Microbial growth on C1 compounds. I. Isolation and characterization of pseudomonas AM 1. *Biochemical Journal*, *81*(December), 465–469. https://doi.org/10.1042/bj0810465.

Pol, A., Barends, T. R. M., Dietl, A., Khadem, A. F., Eygensteyn, J., Jetten, M. S. M., et al. (2014). Rare earth metals are essential for Methanotrophic life in volcanic mudpots. *Environmental Microbiology*, *16*(1), 255–264. https://doi.org/10.1111/1462-2920.12249.

Roszczenko-Jasińska, P., Vu, H. N., Subuyuj, G. A., Crisostomo, R. V., Cai, J., Lien, N. F., et al. (2020). Gene products and processes contributing to lanthanide homeostasis and methanol metabolism in Methylorubrum Extorquens AM1. *Scientific Reports*, *10*(1), 12663. https://doi.org/10.1038/s41598-020-69401-4.

Schmidt, S., Christen, P., Kiefer, P., & Vorholt, J. A. (2010). Functional investigation of methanol dehydrogenase-like protein XoxF in Methylobacterium Extorquens AM1. *Microbiology*, *156*(Pt 8), 2575–2586. https://doi.org/10.1099/mic.0.038570-0.

Skovran, E., Palmer, A. D., Rountree, A. M., Good, N. M., & Lidstrom, M. E. (2011). XoxF is required for expression of methanol dehydrogenase in Methylobacterium Extorquens AM1. *Journal of Bacteriology*, *193*(21), 6032–6038. https://doi.org/10.1128/JB.05367-11.

Takeda, K., Matsumura, H., Ishida, T., Samejima, M., Ohno, H., Yoshida, M., et al. (2015). Characterization of a novel PQQ-dependent quinohemoprotein pyranose dehydrogenase from Coprinopsis Cinerea classified into auxiliary activities family 12 in carbohydrate-active enzymes. *PLoS One*, *10*(2), e0115722. https://doi.org/10.1371/journal.pone.0115722.

von Borzyskowski, S., Lennart, M. R.-E., Weishaupt, R., Vorholt, J. A., & Erb, T. J. (2015). A set of versatile brick vectors and promoters for the assembly, expression, and integration of synthetic operons in Methylobacterium Extorquens AM1 and other alphaproteobacteria. *ACS Synthetic Biology*, *4*(4), 430–443. https://doi.org/10.1021/sb500221v.

Vu, H. N., Subuyuj, G. A., Vijayakumar, S., Good, N. M., Cecilia Martinez-Gomez, N., & Skovran, E. (2016). Lanthanide-dependent regulation of methanol oxidation Systems in Methylobacterium Extorquens AM1 and their contribution to methanol growth. *Journal of Bacteriology*, *198*(8), 1250–1259. https://doi.org/10.1128/JB.00937-15.

Vuilleumier, S., Chistoserdova, L., Lee, M.-C., Bringel, F., Lajus, A., Yang, Z., et al. (2009). Methylobacterium genome sequences: A reference blueprint to investigate microbial metabolism of C1 compounds from natural and industrial sources. *PLoS One*, *4*(5), e5584. https://doi.org/10.1371/journal.pone.0005584.

Wang, L., Suganuma, S., Hibino, A., Mitsui, R., Tani, A., Matsumoto, T., et al. (2019). Lanthanide-dependent methanol dehydrogenase from the legume symbiotic nitrogen-fixing bacterium Bradyrhizobium diazoefficiens strain USDA110. *Enzyme and Microbial Technology*, *130*(November), 109371. https://doi.org/10.1016/j.enzmictec.2019.109371.

Wegner, C.-E., Gorniak, L., Riedel, S., Westermann, M., & Küsel, K. (2019). Lanthanide-dependent Methylotrophs of the family Beijerinckiaceae: Physiological and genomic insights. *Applied and Environmental Microbiology*, *86*(1), 1–18. https://doi.org/10.1128/AEM.01830-19.

Wehrmann, M., Billard, P., Martin-Meriadec, A., Zegeye, A., & Klebensberger, J. (2017). Functional role of lanthanides in enzymatic activity and transcriptional regulation of pyrroloquinoline quinone-dependent alcohol dehydrogenases in pseudomonas putida KT2440. *MBio*, *8*(3), e00570–17. https://doi.org/10.1128/mBio.00570-17.

Yanpirat, P., Nakatsuji, Y., Hiraga, S., Fujitani, Y., Izumi, T., Masuda, S., et al. (2020). Lanthanide-dependent methanol and formaldehyde oxidation in Methylobacterium Aquaticum strain 22A. *Microorganisms*, *8*(6), 822. https://doi.org/10.3390/microorganisms8060822.

Zhang, M., FitzGerald, K. A., & Lidstrom, M. E. (2005). Identification of an upstream regulatory sequence that mediates the transcription of Mox genes in Methylobacterium Extorquens AM1. *Microbiology*, *151*(Pt 11), 3723–3728. https://doi.org/10.1099/mic.0.28243-0.

CHAPTER SIX

Heterologous expression, purification, and characterization of proteins in the lanthanome

Emily R. Featherston, Joseph A. Mattocks, Jonathan L. Tirsch, and Joseph A. Cotruvo Jr.*
Department of Chemistry, The Pennsylvania State University, University Park, PA, United States
*Corresponding author: e-mail address: juc96@psu.edu

Contents

Abstract

Recent work has revealed that certain lanthanides—in particular, the more earth-abundant, lighter lanthanides—play essential roles in pyrroloquinoline quinone (PQQ) dependent alcohol dehydrogenases from methylotrophic and non-methylotrophic

Methods in Enzymology, Volume 650
ISSN 0076-6879
https://doi.org/10.1016/bs.mie.2021.02.004

bacteria. More recently, efforts of several laboratories have begun to identify the molecular players (the lanthanome) involved in selective uptake, recognition, and utilization of lanthanides within the cell. In this chapter, we present protocols for the heterologous expression in *Escherichia coli*, purification, and characterization of many of the currently known proteins that comprise the lanthanome of the model facultative methylotroph, *Methylorubrum extorquens* AM1. In addition to the methanol dehydrogenase XoxF, these proteins include the associated *c*-type cytochrome, XoxG, and solute binding protein, XoxJ. We also present new, streamlined protocols for purification of the highly selective lanthanide-binding protein, lanmodulin, and a solute binding protein for PQQ, PqqT. Finally, we discuss simple, spectroscopic methods for determining lanthanide- and PQQ-binding stoichiometry of proteins. We envision that these protocols will be useful to investigators identifying and characterizing novel members of the lanthanome in many organisms.

1. Introduction

Since the initial discovery in 2011 (Fitriyanto et al., 2011; Hibi et al., 2011), the lighter and most abundant members (La, Ce, Pr, Nd, Sm, and Eu) of the lanthanide (Ln) series—the 15 elements from La to Lu—have been shown to play essential catalytic and functional roles in biology, most notably in methylotrophic bacteria. Methylotrophic bacteria use reduced one-carbon compounds (e.g., methanol) as their sole carbon and energy source (Anthony, 1982; Chistoserdova, Kalyuzhnaya, & Lidstrom, 2009). The discovery that lanthanides are essential for catalysis in the most broadly distributed class of pyrroloquinoline quinone (PQQ)-dependent methanol dehydrogenases (MDHs), XoxF, has changed the field of methylotrophy as well as bioinorganic chemistry more generally. Although the XoxF-type lanthanide-dependent MDHs were discovered decades after their calcium-dependent orthologs (Anthony & Williams, 2003), lanthanide-dependent methanol oxidation may be the ancestral process (Chistoserdova, 2011; Chistoserdova & Kalyuzhnaya, 2018; Keltjens, Pol, Reimann, & Op den Camp, 2014; Picone & Op den Camp, 2019). It is now clear that methylotrophic (and some non-methylotrophic) bacteria that utilize lanthanides occupy diverse habitats (Huang et al., 2019), from volcanic mudpots (Pol et al., 2014) to the phyllosphere (Ochsner et al., 2019) to freshwater and marine environments (Krause et al., 2017; Taubert et al., 2015).

The utilization of lanthanides by MDHs, other alcohol dehydrogenases (ADHs), and likely other enzymes, necessitates the involvement of many more proteins and other biomolecules in selective acquisition, trafficking, regulation, and utilization of these metal ions, which we have termed the "lanthanome" (Cotruvo, 2019), the study of which is "lanthanomics." XoxF is encoded in an operon alongside genes encoding a *c*-type cytochrome, XoxG, the physiological electron acceptor for XoxF, as well as a periplasmic solute binding protein (SBP) of currently unknown function, XoxJ. Many lanthanide-dependent ADHs have been characterized, not limited to MDHs but also ethanol dehydrogenases, ExaF and PedH (Good et al., 2016; Wehrmann, Billard, Martin-Meriadec, Zegeye, & Klebensberger, 2017). Apart from these core ADH proteins, the first native lanthanide-binding protein to be discovered was lanmodulin (LanM), which uses metal-binding motifs similar to those of Ca^{II}-binding proteins to instead bind lanthanides with high affinity and high selectivity (Cook, Featherston, Showalter, & Cotruvo, 2019; Cotruvo, Featherston, Mattocks, Ho, & Laremore, 2018). This discovery led to identification of a gene cluster, of which *lanM* is a part, encoding several proteins involved in uptake and utilization of lanthanides in the model methylotroph, *Methylorubrum extorquens* (Mattocks, Ho, & Cotruvo, 2019; Ochsner et al., 2019), also later identified by transposon mutagenesis (Roszczenko-Jasińska et al., 2020). Together, these reports confirmed the involvement of metallophore-mediated lanthanide uptake suggested by early studies (Gu, Haque, Dispirito, & Semrau, 2016; Keltjens et al., 2014). Another protein in this cluster, LanD (META1p1781 or LutD) was also shown to directly bind lanthanides (Mattocks et al., 2019). In addition to genes involved in lanthanide metallophore (lanthanophore) uptake, this cluster may include other lanthanoenzymes and other lanthanide-binding proteins. In addition, one of the relatively small number of genes upregulated by lanthanides in *M. extorquens* has been characterized as a PQQ-binding SBP, putatively involved in PQQ uptake (Ho & Cotruvo, 2019). Identities of proteins potentially involved in PQQ transport from the cytosol (Roszczenko-Jasińska et al., 2020) and metal-PQQ chaperoning in the periplasm (Featherston & Cotruvo, 2021) have also been speculated. Important regulatory mechanisms have yet to be fully elucidated (Akberdin et al., 2018; Chu, Beck, & Lidstrom, 2016; Farhan Ul Haque et al., 2015; Gu & Semrau, 2017; Skovran, Raghuraman, & Martinez-Gomez, 2019; Wehrmann, Berthelot, Billard, & Klebensberger, 2018). Therefore, the lanthanome has expanded greatly in recent years and will surely continue to do so as more organisms and proteins involved in lanthanide biochemistry are characterized (Chistoserdova, 2019; Featherston & Cotruvo, 2021).

In this chapter, we present detailed methods for the heterologous expression, purification, and characterization of most of the presently confirmed members of the lanthanome of *M. extorquens* AM1—XoxG, XoxJ, LanM, and PqqT—other than the alcohol dehydrogenases. Although strategies exist to purify these proteins from overexpression in *M. extorquens* AM1 (see chapters "Expression, purification and properties of the enzymes involved in lanthanide-dependent alcohol oxidation: XoxF4, XoxF5, ExaF/PedH, and XoxG4" by Huang et al. and "Expression, purification and testing of lanthanide-dependent enzymes in *Methylorubrum extorquens* AM1" by Good and Martinez-Gomez, in this volume), using *Escherichia coli* as the expression host may be preferable for some proteins: *E. coli* grow more quickly than *M. extorquens*, yields of proteins (once codon-optimized for expression in *E. coli*) are high, periplasmic extraction is easier in *E. coli* than in *M. extorquens*, and the required cells, vectors, and techniques are already in the possession of almost every biochemist. The high yields also facilitate the use of constructs without purification tags, which can affect protein function and are inefficient to remove using proteases. Finally, we also provide additional notes for each protein to provide context for its involvement in lanthanide biochemistry and to guide future studies in this new and rapidly changing field.

2. Purification and assay of native XoxF from *M. extorquens* AM1

Unlike the other proteins described in this chapter, in order to be active, XoxF must be purified from a suitable methylotrophic bacterium that can express the appropriate cofactor insertion machinery—at least until the nature of this machinery is better understood and can be replicated in another expression host. Here, we focus on important issues regarding expression, metalation, and assay of these enzymes.

2.1 Notes on expression and purification

Detailed protocols for growth of *M. extorquens* AM1 (ATCC 14718, NCIB 9133) in MP medium can be found in Delaney et al. (2013) and in chapters "Discovery of lanthanide-dependent methylotrophy and screening methods for lanthanide-dependent methylotrophs" by Tani et al., "Expression, purification and properties of the enzymes involved in lanthanide-dependent alcohol oxidation: XoxF4, XoxF5, ExaF/PedH, and XoxG4" by Huang et al., and "Expression, purification and testing of lanthanide-dependent enzymes in *Methylorubrum extorquens* AM1" by Good and Martinez-Gomez in the present volume. Protocols for purification of the native, untagged protein from *M. extorquens* have been provided by Nakagawa and co-workers [(Nakagawa et al., 2012; Wang et al., 2020) and chapter "Discovery of lanthanide-dependent methylotrophy and screening methods for lanthanide-dependent methylotrophs" by Tani et al. in the present volume] and by our laboratory (Cotruvo et al., 2018; Featherston et al., 2019). [For XoxF from *M. fumariolicum*, we refer the reader to Pol et al. (2014).] These purification strategies are generally similar, including ammonium sulfate fractionation, cation exchange and hydrophobic interaction chromatography, and desalting.

Lanthanide-dependent ADHs are unusual among enzymes in that they can utilize multiple metal ions efficiently. The number of distinct lanthanides supporting catalysis in vitro and/or in vivo differs from enzyme to enzyme: e.g., La-Nd, La-Sm/Eu, or La-Gd, according to the XoxF clade in which an enzyme is found (Fitriyanto et al., 2011; Huang, Yu, & Chistoserdova, 2018; Lumpe, Pol, Op den Camp, & Daumann, 2018; Pol et al., 2014; Vu et al., 2016; Wehrmann et al., 2017). In the case of *M. extorquens* AM1, XoxF1 can be activated in vivo with La, Ce, Pr, and Nd, and poorly or not at all with Sm (Featherston et al., 2019; Good, Moore, Suriano, & Martinez-Gomez, 2019; Wang et al., 2020). The reasons for the inability to utilize later lanthanides are likely manifold—selectivity of the lanthanophore-mediated mechanism of uptake for early lanthanides (Mattocks et al., 2019), selectivity of trafficking and cofactor insertion machinery (Deblonde et al., 2020), stability of XoxF (Wang et al., 2020), and reduced efficiency of electron transfer to XoxG in Nd-XoxF vs. La-XoxF (Featherston et al., 2019). Related to XoxF purification, it is not necessarily true that production of XoxF for biochemical studies is equally effective with the four lightest lanthanides, La-Nd, when they are added individually to the growth media. Whereas researchers reporting metal contents of XoxF expressed in the presence of La, either from endogenous levels (Featherston et al., 2019; Nakagawa et al., 2012) or recombinantly in a methylotroph (Good et al., 2019; Huang et al., 2018), show roughly stoichiometric La incorporation, Nd incorporation is more variable. Purifying XoxF from endogenous levels, we observed similar protein yields and A_{280nm}/A_{359nm} ratio (Fig. 1, a measure of cofactor loading) for La, Ce, and Nd forms of XoxF, as well as similar activities (V_{max}) when measured using XoxG (Featherston et al., 2019). By contrast, plasmid-based expression of XoxF in the presence of Nd led to substoichiometric Nd insertion into XoxF (Good et al., 2019). Although the situation is clearly complex, one possible interpretation of these results is that plasmid-based XoxF overexpression in *M. extorquens* might cause Nd import to become limiting for

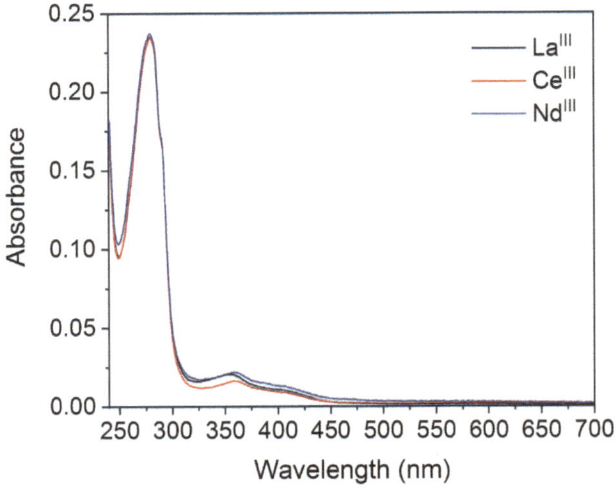

Fig. 1 UV–visible absorption spectra of La-, Ce-, and Nd-XoxFs purified from *M. extorquens*. PQQ-containing MDHs have characteristic absorption features around 350–360nm with a shoulder at ~410nm. The A_{280nm}/A_{359nm} ratio is similar for these three preparations. *Reproduced with permission from Featherston, E. R., Rose, H. R., McBride, M. J., Taylor, E. M., Boal, A. K., & Cotruvo, J. A., Jr. (2019). Biochemical and structural characterization of XoxG and XoxJ and their roles in lanthanide-dependent methanol dehydrogenase activity.* Chembiochem, 20, *2360–2372, copyright 2019 Wiley-VCH.*

cofactor insertion, thereby reducing Nd incorporation into XoxF in a way not observed with La, if it is imported more efficiently. At endogenous levels of XoxF, less efficient Nd import may be less problematic. Further experiments with XoxFs metalated with different lanthanides will be important to address these questions.

2.2 Notes on dye-linked and XoxG-based activity assays

There are two types of activity assays for MDHs (Scheme 1): dye-linked and cytochrome-based, each of which can be carried out in cuvette or plate reader formats. For assay of chromatography fractions during purifications, activity can be determined by reduction of 2,6-dichlorophenolindolphenol (DCPIP) using phenazine ethosulfate (PES) as an electron acceptor, according to the method of Day and Anthony (Day & Anthony, 1990). Our minor modifications are reported in Cotruvo et al. (2018). Enzyme activity is reported in units (U), where 1U is defined as 1μmol O_2 reduced per min between 15 and 30s after initiating the reaction, which corresponds to 1μmol DCPIP reduced/min. Because the extinction coefficient for DCPIP is pH dependent, we recommend that it be determined in assay

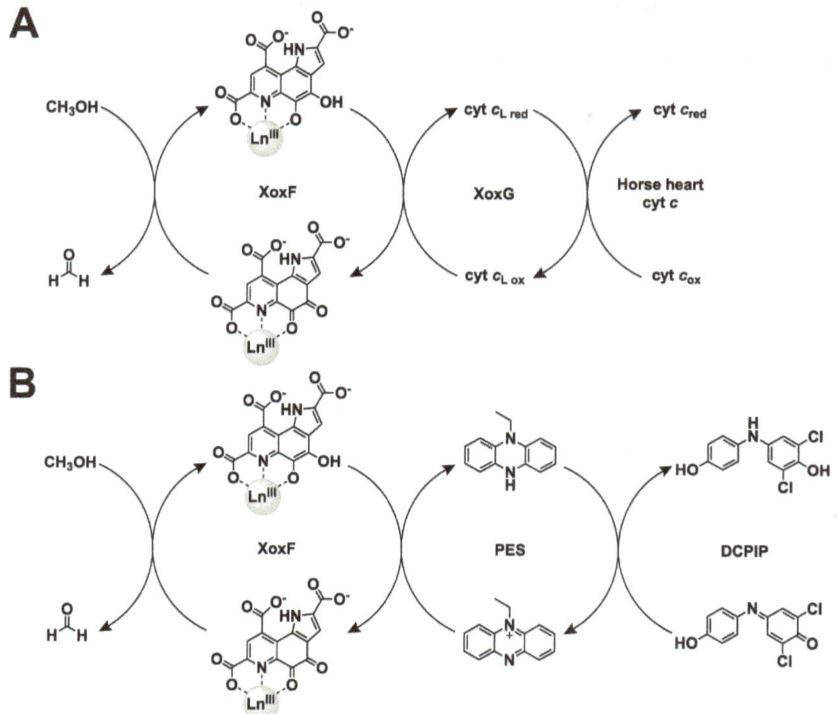

Scheme 1 XoxG-based (A) and dye-linked (B) assay systems for XoxF in *M. extorquens*. In vitro, horse heart cytochrome *c* is a convenient coupling protein because it is readily available. In vivo, another cytochrome (cytochrome c_H) is hypothesized to be the electron acceptor from XoxG. *Reproduced with permission from Featherston, E. R., Rose, H. R., McBride, M. J., Taylor, E. M., Boal, A. K., & Cotruvo, J. A., Jr. (2019). Biochemical and structural characterization of XoxG and XoxJ and their roles in lanthanide-dependent methanol dehydrogenase activity.* Chembiochem, 20, *2360–2372, copyright 2019 Wiley-VCH.*

buffer by comparison with a stock solution prepared in 20 mM MOPS pH 7.0 [$\varepsilon_{600nm} = 20{,}600\,M^{-1}\,cm^{-1}$ at pH 7.0 (Armstrong, 1964)]. In the typical pH 9.0 assay buffer, we have determined $\varepsilon_{600nm} = 22{,}500\,M^{-1}\,cm^{-1}$. Activity assays on purified protein with the artificial, dye-linked system can be carried out on a plate reader by using a modification of these methods, described in detail in Featherston et al. (2019).

We note that chapter "Activity assays of methanol dehydrogenases" by Marie et al. in this volume also describes this method step by step, along with a protocol using an alternative mediator and chromophore that yields similar results. Any of these procedures can also be applied to assay the purified XoxF protein.

Activity assays of the purified protein can also be carried out using the physiological electron acceptor, XoxG (see Section 3 for purification and characterization procedures). Our protocol for XoxG-based assays is adapted from Day and Anthony (Day & Anthony, 1990), which was initially developed to assay the Ca-dependent MDH, MxaFI, with MxaG as electron acceptor, and is described in detail in Featherston et al. (2019). A similar protocol is described in chapter "Activity assays of methanol dehydrogenases" by Marie et al., this volume, based on the reports of Versantvoort et al. (2019) and Featherston et al. (2019).

There are certain points to keep in mind in comparing dye-linked vs. cytochrome-based assays. First, the two assays do not necessarily give the same value for enzyme specific activity. Anthony reported that assays of MxaFI with MxaG yielded lower activities than the dye-linked assay (Cox, Day, & Anthony, 1992; Day & Anthony, 1990), and we observed the same with La- and Ce-metalated XoxF (Featherston et al., 2019). This observation might indicate that interaction between MDH and its electron acceptor is rate-limiting for methanol oxidation in vivo. Second, multiple investigators, including ourselves, have reported lower activities (V_{max}) by the dye-linked assay when Ln-MDHs of the XoxF5 clade (including the *M. extorquens* AM1 XoxF1 discussed here) are metalated with Nd vs. with La (Featherston et al., 2019; Good et al., 2019; Wang et al., 2020). This observation matches slower growth of the native organism in the presence of Nd vs. on La (Huang et al., 2018; Vu et al., 2016; Wang et al., 2020). By contrast, we showed that when La-, Ce-, and Nd-metalated XoxFs were assayed with XoxG as electron acceptor, the V_{max} values were not significantly different for the three enzymes, but the K_m for XoxG increased from 2.5 μM for La-XoxF to 7.6 μM for Nd-XoxF. This result suggests that the slower growth of *M. extorquens* on Nd vs. on La may be due more so to impaired XoxF-XoxG electron transfer—due to impaired redox and/or structural matching of the two proteins (Featherston et al., 2019; Wang et al., 2020)—than on metal uptake and trafficking effects. These considerations emphasize the importance of assaying XoxFs with their native electron acceptors to extract information more relevant to physiological function than the dye-linked assay provides (Huang, Yu, & Chistoserdova, 2018). Although these early results have helped to formulate hypotheses about the effects of different lanthanides on Ln-ADH activity and lanthanide-dependent growth, more studies are needed in a larger number of organisms to understand these phenomena more completely.

3. XoxG

The physiological electron acceptor for XoxF is the *c*-type cytochrome, XoxG (also called a cytochrome c_L). A His-tagged XoxG was first purified by Chistoserdova and co-workers and shown to be able to be reduced by its associated XoxF (Zheng, Huang, Zhao, & Chistoserdova, 2018). We heterologously expressed and purified *M. extorquens* AM1 XoxG from *E. coli*, determined its kinetic parameters as electron acceptor for XoxF, determined its reduction potential, and solved its X-ray crystal structure (Featherston et al., 2019). This study revealed interesting differences relative to the cytochrome c_L associated with the Ca-dependent MDH, MxaG (O'Keeffe & Anthony, 1980; Williams et al., 2006), including an unusually low reduction potential (+172 mV vs. +256 mV), and different structure near the heme-binding pocket. Contemporaneously, the homologous protein from *M. fumariolicum* SolV, which is a fusion of XoxG and XoxJ and therefore called cytochrome c_{GJ}, was reported and biochemically characterized (Versantvoort et al., 2019). Although its structure has not yet been reported, the reduction potential is more similar to that of MxaG (Kalimuthu, Daumann, Pol, Op den Camp, & Bernhardt, 2019); these differences in potential between the different Ln-MDH systems may reflect different metal tolerances of the organisms and their respective MDHs (Featherston et al., 2019). In addition to acting as electron acceptor during catalysis, XoxG may also play other yet to be defined roles in methylotrophy (Zheng et al., 2018).

E. coli is a facile expression system for *c*-type cytochromes like XoxG, but because these bacteria do not normally synthesize *c*-type cytochromes under aerobic conditions, cells must be co-transformed with a plasmid encoding the necessary cytochrome *c* maturation factors, pEC86 (Londer, 2011). As has been previously mentioned by other investigators, His-tags can sometimes interfere with heme loading to the protein (Londer, 2011). Therefore, our purification of XoxG utilized the untagged, wild-type protein. The purity and yield (~6 mg per L culture) are more than sufficient for biochemical and crystallographic studies (Featherston et al., 2019), such that a purification tag is not necessary. The *xoxG* gene was codon-optimized (see Featherston et al. for the sequence) and cloned into pET-24a.

3.1 Materials

For expression and purification:

- Chemically competent or electrocompetent *E. coli* BL21 (DE3) cells
- Plasmids: pEC86 and pET-24a-XoxG
- Luria-Bertani (LB) media
- Terrific broth (TB) media
- Kanamycin
- Chloramphenicol
- Isopropyl β-D-1-thiogalactopyranoside (IPTG)
- 5 mM Magnesium sulfate ($MgSO_4$)
- 1 M Tris, pH 7.4
- DEAE Sepharose Fast Flow column
- Phenyl Sepharose 6 Fast Flow (low substitution) column
- Fast protein liquid chromatography (FPLC) instrument
- HiLoad Superdex 75 pg 16/600 column (120 mL)
- **Buffer G1:** 30 mM Tris, 20% (w/v) sucrose, 1 mM EDTA, pH 7.4
- **Buffer G2:** 50 mM Tris, pH 7.4
- **Buffer G3:** G2 with 35% sat. (194 g/L) ammonium sulfate
- **Buffer G4:** G2 with 20% sat. (106 g/L) ammonium sulfate
- **Buffer G5:** 20 mM MOPS, 100 mM KCl, pH 7.0
- **Buffer G6:** 20 mM MOPS, pH 7.0

For extinction coefficient determination:

- Quartz microcuvette
- Sodium dithionite
- Solution A (0.2 M NaOH, 40% (v/v) pyridine, 500 μM potassium ferricyanide)
- Solution B (0.5 M sodium dithionite, 0.5 M NaOH)

For reduction potential determination:

- Quartz microcuvette with septum cap
- Appropriate redox dye (e.g., DCPIP, 2.5 mM)
- Sodium dithionite
- Xanthine (made fresh, 30 mM in 0.5 M NaOH)
- D-Glucose (1 M)
- Glucose oxidase (8 mg/mL in Buffer G5)
- Bovine liver catalase (made fresh, 1 mg/mL)
- Xanthine oxidase (25 μM in Buffer G5)

  Note: The enzymes used for reduction potential determination are commercially available.

3.2 Expression and purification of XoxG

3.2.1 Protein expression

1. Co-transform *E. coli* BL21 (DE3) cells with pET24a-XoxG and pEC86 (Cm^R) and plate on LB-agar plates containing kanamycin (Km; 50 μg/mL) and chloramphenicol (Cm; 30 μg/mL; antibiotic concentrations in all media), and grow at 30 °C.

   Note: While the co-transformation can be done with chemically competent cells, efficiency is low, so electroporation of electrocompetent cells is recommended.
2. Use a single colony to inoculate a 2-mL Terrific broth (TB)-Km/Cm culture and grow overnight at 30 °C with 200 rpm shaking. Use this culture to inoculate a 100-mL culture at 100 × dilution, grown overnight under the same conditions.
3. Use the 100-mL culture to inoculate 2 × 1.5 L TB-Km/Cm at 100 × dilution, and grow under the same conditions to OD_{600nm} ~0.6 (8–9 h). At this point, induce protein expression by addition of isopropyl β-D-1-thiogalactopyranoside (IPTG) to a concentration of 1 μM. After overnight growth (15 h), harvest cells by centrifugation (7 min, 7000 × *g*, 4 °C). Pre-weigh the centrifuge bottle(s) before harvesting the culture. After centrifugation, pour off the LB supernatant and weigh to determine the mass of the cell pellet. Typical yield is ~8 g cell paste per L culture.

   Note: The IPTG concentration of 1 μM is not a typographical error. Using higher concentrations (even 2–10 μM) yielded noticeably lower XoxG levels in the periplasmic extract (XoxG expression is immediately apparent from the red color of the heme). If this protocol is applied to other cytochromes c_L, we recommend an initial test experiment (200 mL growth) using different IPTG concentrations, then periplasmic extraction and comparison of extract color and bands by gel. This will allow rapid optimization of cofactor incorporation.
4. Immediately use the cell pellet to harvest the periplasmic fraction containing XoxG, according to the procedure below.

3.2.2 Protein purification

Maturation of *c*-type cytochromes occurs in the periplasm; therefore, rather than lysing whole cells, this protocol begins with isolation of the periplasmic fraction, using the osmotic shock method.

Tip: It is advised that you continue through at least to end of step 8 before pausing.

1. Pre-chill a solution of 5 mM $MgSO_4$ on ice.
2. Using a pipet-aid, resuspend the cell paste from above in 40 mL of Buffer G1 per gram of cell paste. After the cell paste is fully resuspended, stir at 150 rpm at room temperature for 10 min. This can be done directly in the centrifuge bottle.
3. Centrifuge for 20 min at 8000 × *g*, 4 °C. Carefully decant the supernatant.
4. Resuspend the cell pellet in 20 mL/g (original wet weight) ice-cold 5 mM $MgSO_4$. Stir at 150 rpm for 10 min at 4 °C (in a cold room) and centrifuge for 10 min at 8000 × *g*, 4 °C. Carefully decant the red supernatant, which is the periplasmic extract, to a beaker and add a stir bar.
5. While stirring the extract, add 0.05 volumes of 1 M Tris, pH 7.4 (e.g., 25 mL for 500 mL extract).

   Note: Extraction efficiency can be visualized by SDS-PAGE. Note that full-length protein, not transported from the cytosol, will be present in the spheroplast, and the transported protein in the extract will be shorter due to signal peptide cleavage.
6. Equilibrate a DEAE Sepharose Fast Flow column (2.5 × 4 cm, 20 mL) in Buffer G2. Apply the periplasmic extract, collecting the flowthrough as a single fraction. Wash the column with 2 column volumes (CV) of Buffer G2, collecting as a single fraction. Combine the light-red flowthrough and wash fractions, which contain XoxG.

   Note: This step removes DNA and some contaminating proteins, which is helpful for subsequent steps.
7. Prepare the combined flowthrough and wash fractions for hydrophobic interaction chromatography by adding solid ammonium sulfate slowly, with stirring, to 35% saturation (194 g/L at 4 °C) (Englard & Seifter, 1990). Meanwhile, equilibrate a Phenyl Sepharose 6 Fast Flow (low substitution) column (2.5 × 4 cm, 20 mL) in Buffer G3.
8. Load the solution to the Phenyl Sepharose column; the column should turn pale red, indicating XoxG binding, and the flowthrough should be colorless.

   Note: This is a convenient pause point for the night.
9. Wash the column with 5 CV of Buffer G3. Elute with 10 CV of Buffer G4 and collect fractions. XoxG-containing fractions can be pooled by inspection of their red color or, more quantitatively, by absorbance at

418 nm. Concentrate the protein to <2 mL using an Amicon Ultra 10-kDa MWCO centrifugal filter device.

10. Load protein solution onto a HiLoad Superdex 75 pg 16/600 FPLC column pre-equilibrated with Buffer G5, and elute at 0.75 mL/min with 1 CV of the same buffer, analyzing by A_{280nm} and A_{418nm}, and collecting 1.5 mL fractions. Pool fractions according to A_{418nm}; XoxG elutes at 67–77 mL. Freeze and store at −80 °C.

3.3 Determination of the extinction coefficients of XoxG

We have determined the absorption spectra and extinction coefficients of the oxidized and reduced forms of XoxG (Fig. 2) by the method of Barr and Guo (Barr & Guo, 2015), which is summarized here.

1. Prepare a 1 mL, ~20 μM solution of XoxG in Buffer G6 (for a novel homolog purified for the first time, use the extinction coefficients below in order to estimate the concentration).
2. Dilute half of this solution 2 × in Buffer G6 and acquire a UV–visible absorption spectrum (240–800 nm). This will be the reference spectrum for oxidized XoxG.

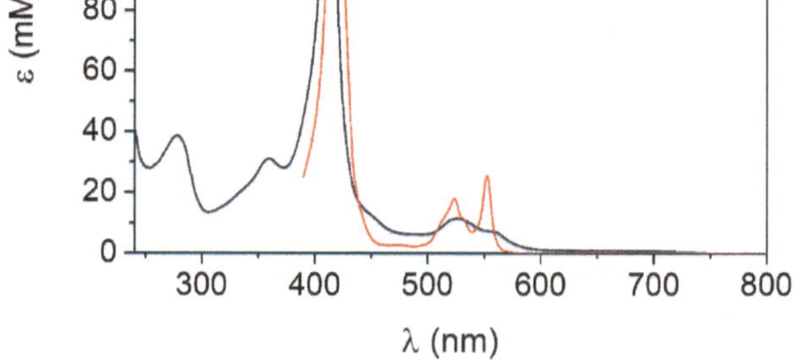

Fig. 2 UV–visible absorption spectra of oxidized (black) and dithionite-reduced (red) XoxG. *Reproduced with permission from Featherston, E. R., Rose, H. R., McBride, M. J., Taylor, E. M., Boal, A. K., & Cotruvo, J. A., Jr. (2019). Biochemical and structural characterization of XoxG and XoxJ and their roles in lanthanide-dependent methanol dehydrogenase activity.* Chembiochem, *20, 2360–2372, copyright 2019 Wiley-VCH.*

3. Use a pipettor to dispense a few grains of sodium dithionite into the cuvette and mix, in order to fully reduce the protein, and acquire another spectrum. This will be the reference spectrum for reduced XoxG.

   Note: Excess dithionite absorbs below ~400 nm, so the spectrum will be unreliable in that region.
4. Mix 250 μL Buffer G6 and 250 μL of Solution A (made fresh) and blank the spectrophotometer on this solution.
5. Mix 250 μL of the ~20 μM solution of XoxG from step 1 with 250 μL of Solution A and take a spectrum.
6. Add 5 μL of Solution B (made fresh) to the cuvette, mix quickly, and acquire a spectrum immediately and every minute until the spectrum no longer changes. Using the absorbance at 550 nm in the final spectrum, determine the concentration of heme [$\varepsilon_{550nm} = 30.27\,mM^{-1}\,cm^{-1}$ for (pyridine)$_2$-heme *c*].
7. Multiply this concentration by 505 μL and divide by 250 μL. This will yield the concentration of heme in the ~20 μM XoxG solution from step 1. Divide this value by 2 to determine the concentration of heme in the oxidized and reduced XoxG samples in steps 2 and 3. From these spectra and the heme concentration, calculate the extinction coefficients of XoxG.

   Maximum absorptions for oxidized *M. extorquens* AM1 XoxG were determined as $\varepsilon_{413nm} = 129\,mM^{-1}\,cm^{-1}$ and $\varepsilon_{527nm} = 11.5\,mM^{-1}\,cm^{-1}$, and for the reduced form were $\varepsilon_{418nm} = 157\,mM^{-1}\,cm^{-1}$ and $\varepsilon_{553nm} = 25.8\,mM^{-1}\,cm^{-1}$. At the isosbestic point (413 nm), $\varepsilon = 129\,mM^{-1}\,cm^{-1}$.

3.4 Determination of the reduction potential of XoxG

The reduction potential of the heme cofactor of XoxG can be determined spectrophotometrically, as described by Raven and co-workers (Efimov et al., 2014) based on a method developed by Massey (Massey, 1991). The method follows spectrophotometrically the reduction of a solution of the protein of interest and a redox-active organic dye standard. At a given time point, the concentrations of the oxidized and reduced forms of all redox-active species can be determined, from which the midpoint reduction potential (E_m) of the protein can be calculated relative to the known potential of the dye. We found DCPIP, for which $E_m = +217\,mV$ at 25 °C, pH 7.0 (Clark, 1960), to be a useful dye for *M. extorquens* XoxG. Other dyes can be used instead, as long as the dye's E_m is within ~50 mV of that of the protein. See Clark (1960) for a compilation of dyes and their reduction potentials.

1. Prepare the oxidized and reduced reference spectra for the dye. For DCPIP, determine the UV–visible spectrum of $DCPIP_{ox}$ in the 350–800 nm range in Buffer G5 by using $\varepsilon_{600nm} = 20.6\,mm^{-1}\,cm^{-1}$, pH 7.0 (Armstrong, 1964), and determine the spectrum of $DCPIP_{red}$ by addition of a slight excess of a freshly prepared solution of sodium dithionite. Convert the absorbance values at each wavelength to extinction coefficients.
2. In a total volume of 500 μL, mix the following in a septum-sealed, capped cuvette: Buffer G5, xanthine (300 μM), D-glucose (5 mM), glucose oxidase (25 μg), and bovine liver catalase (2.5 μg). Incubate the solution for 5 min with periodic inversion to achieve anaerobic conditions.

   Note: D-glucose, glucose oxidase, and catalase act as an O_2-scavenging system; glucose oxidase is a flavoprotein oxidase that reduces O_2 to H_2O_2 as part of its catalytic cycle, and catalase converts H_2O_2 to ½O_2 and H_2O.
3. Add XoxG (final concentration 4 μM, from a stock of 250 μM) and DCPIP (final concentration 25 μM from a 2.5 mM stock) and mix. It is important to keep the volumes added small, to minimally affect anaerobiosis. Acquire an initial spectrum at 350–800 nm. Initiate reduction by addition of xanthine oxidase (20 nM), invert, and acquire spectra every minute for 40–60 min (4800 nm/min scan rate), until spectra are constant for >5 min (Fig. 3A).

   Note: Depending on the activity of the xanthine oxidase preparation, the amount of enzyme may need to be adjusted. It is important that the rate of electron production by xanthine/xanthine oxidase not be too fast, to ensure that the system equilibrates throughout the titration. A total time of at least 40 min for the full reduction of the protein and dye is recommended.
4. Using spreadsheet software, determine the initial concentrations of oxidized protein and oxidized dye by manual fitting of the initial spectrum in step 3 with spectra of DCPIP and XoxG.
5. For each time point, calculate the concentrations of $DCPIP_{ox}$ and $DCPIP_{red}$ from the A_{600nm} value ($DCPIP_{red}$ and XoxG contribute negligibly to the A_{600nm} value and can be ignored); $[DCPIP_{red}]$ is equal to the difference between the initial $[DCPIP_{ox}]$ and $[DCPIP_{ox}]$ at each time point. These values reflect the solution potential at each time point.

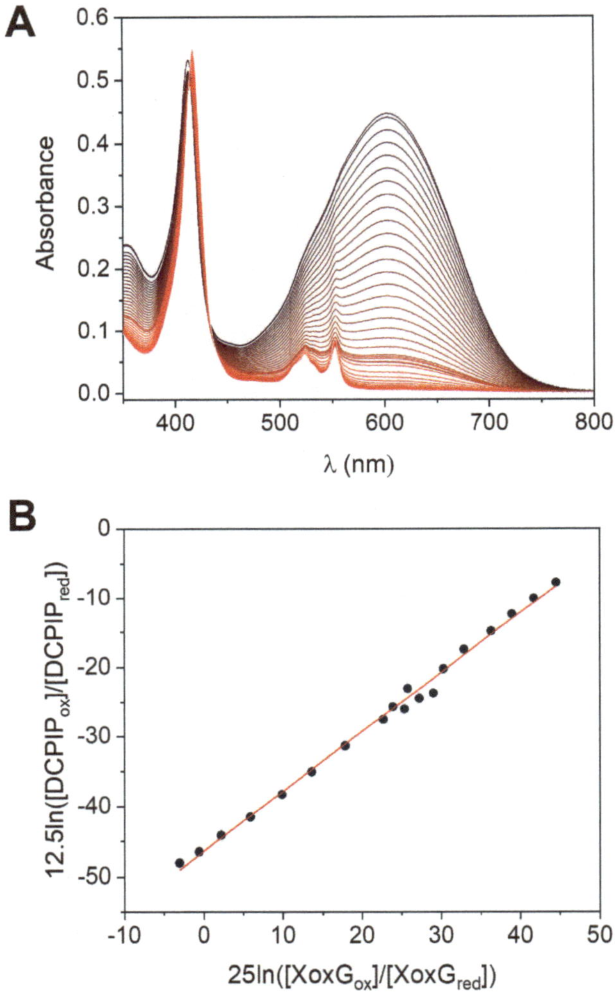

Fig. 3 Spectroscopic determination of XoxG reduction potential. (A) Spectroscopic titration of oxidized XoxG and DCPIP reduced by xanthine/xanthine oxidase over 1 h (black to red lines), with spectra acquired every 1 min. (B) Data analysis for the results in panel (A), according to steps 7 and 8. *Reproduced with permission from Featherston, E. R., Rose, H. R., McBride, M. J., Taylor, E. M., Boal, A. K., & Cotruvo, J. A., Jr. (2019). Biochemical and structural characterization of XoxG and XoxJ and their roles in lanthanide-dependent methanol dehydrogenase activity.* Chembiochem, 20, *2360–2372, copyright 2019 Wiley-VCH.*

6. For each time point, calculate the concentration of reduced XoxG. First, subtract the contribution of $DCPIP_{ox}$ (equivalent to $[DCPIP_{ox}] \times \varepsilon_{418nm(DCPIP)}$) from A_{418nm}. (Note that 418 nm is the λ_{max} for $XoxG_{red}$.) Then, subtract the initial A_{418nm} contribution of $XoxG_{ox}$. Finally, divide by the difference in extinction coefficient between the oxidized and reduced forms of XoxG at 418 nm, $\varepsilon_{418red} - \varepsilon_{418ox} = 56\,mM^{-1}\,cm^{-1}$.

This yields the concentration of $XoxG_{red}$. Subtract this value from the initial concentration of XoxG to yield the concentration of $XoxG_{ox}$ at each time point. The concentrations of each of the redox species at each time point have now been calculated.

7. Determine the midpoint reduction potential (E_m) of XoxG. For each time point, determine (Eqs. 1 and 2):

$$y = 12.5 \ln \frac{[DCPIP_{ox}]}{[DCPIP_{red}]} \tag{1}$$

$$x = 25 \ln \frac{[XoxG_{ox}]}{[XoxG_{red}]} \tag{2}$$

8. Plot y vs. x and determine the regression line. The plot should have a slope of approximately 1, and the midpoint potential of the protein is determined from the y-intercept, which equals $E_{m,protein} - E_{m,dye}$, where $E_{m,dye}$ is the known reduction potential of the dye (Fig. 3B).

   Note: The coefficients in Eq. (1) and (2) reflect the fact that the dye undergoes a two-electron reduction and XoxG undergoes one-electron reduction.

   Note: Points at the beginning and end of the experiment may not be at equilibrium (apparent by deviation from linearity) and can be ignored in the analysis.

4. XoxJ

The third protein encoded by the core Ln-MDH operon is XoxJ. Like its analog in the Ca-MDH system, MxaJ, this protein is a member of the periplasmic (or solute) binding protein (PBP or SBP) family. SBPs are typically associated with recognition of ligands for membrane-bound transporters (Scheepers, Lycklama a Nijeholt, & Poolman, 2016), but XoxJ and MxaJ are not obviously associated with such a system and their functions are unknown. However, the presence of a "J-type" SBP in the operons of PQQ-dependent alcohol dehydrogenases is conserved, and MxaJ (Amaratunga, Goodwin, O'connor, & Anthony, 1997; Van Spanning et al., 1991) and XoxJ (Roszczenko-Jasińska et al., 2020) are essential components of their respective MDH systems. The crystal structures of MxaJ (Choi, Cao, Kim, Lee, & Lee, 2017) and XoxJ (Featherston et al., 2019) reveal general architectures similar to classic SBPs, except they both exhibit exceptionally large cavities, putatively for substrate binding, as well as a β sheet missing several strands. Both of these features suggest that the substrate for XoxJ is likely a protein itself. We refer the reader to Featherston et al. (2019) for structural analysis of XoxJ and a more extended discussion of potential functions.

The protocol described here is expanded from the initial report by Featherston et al., with *M. extorquens* AM1 *xoxJ* having been codon-optimized and cloned into pET-24a. This procedure yields ~3 mg XoxJ per L culture.

4.1 Materials

- Reagents for cell growth and protein expression (see Section 3.1). All solid and liquid growth media contain 50 μg/mL kanamycin (Km)
- Ammonium sulfate
- 5 mM $MgSO_4$
- 1 M Tris, pH 7.4
- DEAE Sepharose Fast Flow column
- Phenyl Sepharose 6 Fast Flow (low substitution) column
- FPLC instrument
- HiPrep Phenyl FF (low sub) 16/10 column
- HiLoad 16/600 Superdex 200 pg column
- **Buffer J1:** 30 mM Tris, 20% (w/v) sucrose, 1 mM EDTA, pH 7.4
- **Buffer J2:** 50 mM Tris, 50 mM NaCl, pH 7.4
- **Buffer J3:** 50 mM Tris, 5% glycerol, pH 7.4, containing 35% sat. $(NH_4)_2SO_4$ (194 g/L)
- **Buffer J4:** 50 mM Tris, 5% glycerol, pH 7.4
- **Buffer J5:** 20 mM MOPS, 100 mM NaCl, pH 7.0

Buffers J1–J4 can be made by adding solids and/or glycerol to a minimal volume of water, adding the appropriate amount of the 1 M Tris stock solution given the final volume desired, and adding water up to volume. No pH adjustment is necessary.

4.2 Expression and purification of XoxJ

4.2.1 Protein expression

1. Transform chemically competent *E. coli* BL21(DE3) cells with pET24a-XoxJ and grow at 37 °C overnight on LB-agar-Km plates. Use a single colony to inoculate 100 mL LB-Km and shake at 200 rpm overnight (~16 h) at 37 °C.
2. Use this culture to inoculate three 2-L cultures of LB-Km (in 6 L flasks) at 100 × dilution (20 mL each). Shake cultures at 200 rpm (37 °C) and monitor the OD_{600nm} periodically. At OD_{600nm} ~ 0.5, induce XoxJ

expression by adding IPTG to a concentration of 0.2 mM. After 2 h of induction, harvest cells by centrifugation (7 min, 7000 × g, 4 °C). Pre-weigh the centrifuge bottle(s) before harvesting the culture. After centrifugation, pour off the LB supernatant and weigh to determine the mass of the cell pellet. Each liter of culture will yield ~3 g cell paste.

Note: Inducing at higher OD_{600nm} or for significantly more than 2 h leads to insoluble material in the periplasmic extract without increasing protein yield. Therefore, inducing at $OD_{600nm} = 0.5$ and for 2 h is recommended.

3. Continue immediately with the periplasmic extraction procedure below.

   Note: Do not freeze the cell pellet. Freezing it will cause cell lysis upon thawing.

4.2.2 Protein purification

This protocol uses the osmotic shock method for periplasmic extraction. Whereas periplasmic proteins like XoxJ can, in principle, be purified from the whole-cell lysate, we recommend extraction of the periplasm rather than whole-cell lysis for two reasons. First, relatively few proteins are present in the periplasm, so your protein of interest begins the purification procedure significantly purer than if cells were lysed. Second, the cytosolic fraction will likely contain your protein of interest that has not been exported, and therefore still contains its signal peptide, and may not have been folded or processed properly (e.g., in the case of XoxJ, formation of a disulfide bond).

Note: It is advised that you continue through at least to end of step 6 before pausing; pausing before running the periplasmic extract through the DEAE column leads to some precipitation overnight, which complicates subsequent steps.

1. Pre-chill a solution of 5 mM $MgSO_4$ on ice.
2. Using a pipet-aid, resuspend the cell paste from above in 40 mL of Buffer J1 per gram of cell paste. After the cell paste is fully resuspended, stir at 150 rpm at room temperature for 10 min. This can be done directly in the centrifuge bottle.
3. Centrifuge for 20 min at 8000 × g, 4 °C. Carefully decant the supernatant.
4. Resuspend the cell pellet in 20 mL/g (original wet weight) ice-cold 5 mM $MgSO_4$. Stir at 150 rpm for 10 min at 4 °C (in a cold room) and centrifuge for 10 min at 8000 × g, 4 °C. Carefully decant the supernatant, which is the periplasmic extract, to a beaker and add a stir bar.

5. While slowly stirring the extract, add 0.05 volumes of 1 M Tris, pH 7.4 (e.g., 25 mL for 500 mL extract), and add solid NaCl to a final concentration of 50 mM.

   Note: Extraction efficiency can be visualized by gel. Note that full-length protein, not transported from the cytosol, will be present in the spheroplast, and the transported protein in the extract will be shorter due to signal peptide cleavage.

   Note: All of the subsequent chromatographic steps can be carried out either at room temperature or 4 °C.

6. Equilibrate a DEAE Sepharose Fast Flow column (2.5 × 4 cm, 20 mL) in Buffer J2. Apply the periplasmic extract, collecting the flowthrough as a single fraction. Wash the column with 5 CV of Buffer J2, collecting as a single fraction. Combine the flowthrough and wash fractions, which contain XoxJ.

   Note: This step removes DNA and some contaminating proteins, which is helpful for subsequent steps.

   Note: This is a convenient pause point for the night. Alternatively, you can pause after loading the gravity-flow Phenyl Sepharose column.

7. Prepare for hydrophobic interaction chromatography by adding solid ammonium sulfate slowly, with stirring, to 35% saturation (194 g/L at 4 °C) (Englard & Seifter, 1990). Equilibrate a Phenyl Sepharose 6 Fast Flow (low substitution) column (2.5 × 2 cm, 10 mL) in Buffer J3.

8. Load the solution to the Phenyl Sepharose column. Wash the column with 15 CV of Buffer J3. Elute with 10 CV Buffer J4 and collect in a single fraction. Concentrate the protein to <5 mL using an Amicon Ultra 10-kDa MWCO centrifugal filter device.

   Note: The primary purpose of this step is to rapidly concentrate the protein so that it can be loaded more easily onto the FPLC. If you have a loop larger than 5 mL, the protein does not need to be concentrated as far.

9. Weigh out $(NH_4)_2SO_4$ (194 mg per mL of XoxJ, giving ~35% saturation) and dissolve in the concentrated XoxJ solution.

10. Purify XoxJ by FPLC using a hydrophobic column, e.g., HiPrep Phenyl FF (low sub) 16/10 column (GE Healthcare). Pre-equilibrate the column with Buffer J3, and load the protein using a 5 mL capillary loop. Wash the column at 2 mL/min with 4 CV Buffer J3, and elute with a 200 mL linear gradient of 0–100% Buffer J4 at 2 mL/min. Collect 4 mL fractions in peak fractionation mode (10 mAU threshold, detect A_{280nm} and A_{260nm}). XoxJ elutes as the major A_{280nm} peak,

spanning 25–15% sat. $(NH_4)_2SO_4$. Pool these fractions and concentrate by centrifugal filtration (Amicon Ultra 10kDa MWCO) to <2mL.

11. In order to remove ammonium sulfate and a contaminating protein at ~15kDa, purify XoxJ further by size-exclusion chromatography, e.g., on a HiLoad 16/600 Superdex 200pg column. Pre-equilibrate the column with Buffer J5, and load the protein using a 2mL capillary loop. Rinse the loop with 4mL Buffer J5, and elute the column with 1 CV Buffer J5 at 0.75mL/min. Collect 2mL fractions.

12. Analyze fractions by SDS-PAGE, pool XoxJ-containing fractions, and concentrate by centrifugal filtration to the desired volume. XoxJ can be concentrated at least to 25mg/mL, which was the protein concentration used for crystallography (Featherston et al., 2019). Determine the protein concentration using $\varepsilon_{280nm} = 34.0\,mM^{-1}\,cm^{-1}$, estimated by ExPASy (Gasteiger et al., 2005).

Note: Purified XoxJ shows two bands on an SDS-PAGE gel. Mass spectrometry analysis suggests that the larger (minor) band corresponds to the protein after cleavage of the signal peptide before Gln27 (28.6kDa), and the smaller (major) band corresponds to signal peptide cleavage before Leu32 (28.0kDa) (Featherston et al., 2019). Because both bands are present throughout the purification, including in the whole-cell post-induction sample run on the gel, these results are consistent with alternative *E. coli* signal peptidase recognition sites, rather than proteolysis during purification. We have not been successful in separating these two forms, but the protein still crystallizes well, and the first residue visible in the structure is Val37.

5. Lanmodulin

Lanmodulin (LanM) was initially identified during a purification of XoxF from *M. extorquens* AM1, and it is the first natural, selective macrochelator for lanthanides to be discovered and characterized (Cotruvo et al., 2018). Its discovery also revealed the gene cluster responsible for uptake of lanthanides and delivery into the cytosol in this bacterium (Mattocks et al., 2019; Ochsner et al., 2019; Roszczenko-Jasińska et al., 2020). Although its function is not yet fully understood, it is one of a relatively small number of proteins upregulated in vitro and in planta in *M. extorquens* PA1 (along with XoxF, XoxG, XoxJ, PqqT, and three uncharacterized proteins) in response to lanthanum (Ochsner et al., 2019), and gene expression studies have also shown it to be upregulated in response to lanthanides in several other bacteria (Masuda et al., 2018; Wegner, Gorniak, Riedel, Westermann, & Küsel, 2019). However, it is not essential for growth of *M. extorquens* in the presence of lanthanides, at least under the conditions studied to date (Ochsner et al., 2019; Roszczenko-Jasińska et al., 2020). In addition, although it is found in many alphaproteobacterial methylotrophs, it is not conserved in all lanthanide-utilizing bacteria (Chistoserdova, 2019; Cotruvo et al., 2018).

LanM consists of four 12-residue EF-hand motifs, each separated by 12–13 amino acid residues. EF-hand motifs are common recognition sites for Ca^{II} ions in biology (Gifford, Walsh, & Vogel, 2007), and, while these sites also non-specifically bind Ln^{III} ions (Drake, Lee, & Falke, 1996; Edington et al., 2018), LanM stands out for its tight and highly selective lanthanide binding. Biochemical studies revealed that the protein binds 3 equiv. of Ln^{III} ions with picomolar affinity (one in each of the first three EF hands), whereas the fourth EF hand does not bind metal ions tightly. LanM is largely disordered in its apo state, but in the presence of any rare earth, it adopts a compact, ordered form, stabilized by hydrophobic packing interactions as revealed by the NMR solution structure of the Y^{III}-bound form (Cook et al., 2019). This conformational change is extremely (10^8-fold) selective for lanthanides over Ca^{II} (Cotruvo et al., 2018) as well as for all non-rare earths tested to date (Deblonde et al., 2020). Although all lanthanides induce the conformational change at picomolar concentrations, the protein is biased toward the lighter Ln^{III} ions, albeit in an intriguing way: even though the biologically relevant early lanthanides bind more weakly to the protein than the later Ln^{III} ions (Deblonde et al., 2020), they more efficiently cause the conformational change of the protein (Cotruvo et al., 2018; Mattocks et al., 2019). For recent discussions of what this protein is teaching us about the principles of selective lanthanide recognition, and what is still yet to be learned, we refer the reader to Cotruvo (2019) and Featherston and Cotruvo (2021).

Initial expression studies in *E. coli* demonstrated that the protein was exported to the periplasm, and we verified the predicted signal peptide cleavage site. Yields of LanM purified from the periplasm were low. Because the protein contains no Cys residues (which could be oxidized to disulfides in the periplasm), we have found that higher protein yields are achievable via cytosolic expression of a construct with Ala22 as the first amino acid coded after the N-terminal Met. In our initial reports of LanM we used constructs that had a His_6-tag appended to either the N-terminus or C-terminus of the protein, which both yielded ~20 mg/L culture. Protocols for purification of those constructs can be found in Cotruvo et al. (2018).

However, we noticed that our yields were being limited by poor binding to the Ni-NTA column. Furthermore, and more importantly, we noted that the Ln^{III}-LanM K_d values for the C-terminally tagged protein were twofold higher than for the N-terminally tagged or untagged [produced by cleavage of the tag by tobacco etch virus (TEV) protease] proteins, an observation explained by the NMR structure (Cook et al., 2019). Therefore, we sought to develop a purification protocol for the wild-type protein, with no affinity tag.

Here, we introduce a purification protocol for wild-type LanM, with markedly improved yield (70–85 mg/L culture) relative to our initial reports. We also describe a simple method to determine metal-binding stoichiometry for lanthanide-binding proteins.

5.1 Materials

For purification:

- Reagents for cell growth and protein expression (see Section 3.1). All solid and liquid growth media contain 50 μg/mL kanamycin (Km)
- Protease inhibitor cocktail (e.g., Roche Complete Mini tablets)
- DNase I
- Phenylmethanesulfonyl fluoride (PMSF, 100 mM stock in isopropanol)
- Q-Sepharose Fast Flow
- FPLC instrument with S75 column (120 mL)
- **Buffer M1:** 50 mM Tris, 10 mM NaCl, 1 mM EDTA, 5% glycerol, pH 8.0
- **Buffer M2:** 50 mM Tris, 1 M NaCl, 1 mM EDTA, 5% glycerol, pH 8.0
- **Buffer M3:** 30 mM MOPS, 100 mM KCl, 5% glycerol, pH 7.0

For xylenol orange competition assay:

- **Buffer M4:** 20 mM MES, 20 mM acetate, 100 mM KCl, pH 6.1, stirred with 10 g/L Chelex-100 for 1 h followed by re-adjustment to pH 6.1
- Stock solution of 1.25 mM Ln^{III} of interest in Buffer M4
- Xylenol orange (XO), ~5 mM in water
- Quartz microcuvette

5.2 Expression and purification of LanM

5.2.1 Protein expression

1. Transform chemically competent *E. coli* BL21(DE3) cells with pET24a-LanM [this plasmid is analogous to those constructed to express the cytosolic LanM proteins initially described by Cotruvo et al., but without the His_6-tag] and grow at 37 °C overnight on LB-agar-Km plates. Use a

single colony to inoculate 100 mL LB-Km and shake at 200 rpm overnight (~16 h) at 37 °C.

2. Use this culture to inoculate one 2-L culture (in a 6-L flask) at 50 × dilution. Shake cultures at 200 rpm (37 °C) and monitor the OD_{600nm} periodically. At OD_{600nm} ~ 0.6, induce LanM expression by adding IPTG to a concentration of 0.2 mM.
3. After 3 h of induction, harvest cells by centrifugation (7 min, 7000 × *g*, 4 °C) and transfer to pre-weighed 50-mL conical tubes. Typical yield is ~2–3 g per L culture. Flash freeze cell paste in liquid nitrogen and store at −80 °C.

5.2.2 Protein purification

1. On ice, resuspend the cell pellet in 5 mL of Buffer M1 per g of cell paste. Buffer M1 should be supplemented with 1 protease tablet per 10 mL, DNase (2 U/mL), and PMSF (0.25 mM).
2. Lyse cells by sonication and pellet insoluble material by centrifugation at 40,000 × *g*, 45 min, 4 °C.
3. The following step can be done at room temperature or in a cold room at 4 °C, as desired. Pre-equilibrate a 20 mL (2.5 × 4 cm) Q-Sepharose Fast Flow column in Buffer M1. Load the lysate to the column slowly, at ~2 mL/min. Wash column slowly with 1 CV Buffer M1.

   Note: We find it easiest to run the Q-Sepharose step using a gravity-flow column, but the protocol can, of course, be adapted to an FPLC instrument.
4. Fill a gradient mixer (e.g., CBS Scientific GM-200) with 80 mL solutions of Buffer M1 and Buffer M2. Start the gradient and elute the protein from the Q-Sepharose column at the same flow rate as above. Collect 2 mL fractions in microcentrifuge tubes. It is only necessary to collect fractions for the first half (~80 mL) of the gradient, as LanM elutes with relatively high purity close to the beginning, at ~100–200 mM NaCl. Fractions can be screened by SDS-PAGE (16% Tris-Glycine). See Fig. 4A for a representative gel.

   Note: LanM runs on the gel as a major band just above 10 kDa, in addition to several lower molecular weight bands [see fig. S7 in Cotruvo et al. (2018)]. Based on mass spectrometry analysis, these additional bands do not correspond to proteolysis; instead, they seem to be the result of multiple protein conformations, perhaps induced by adventitious trace metal binding.

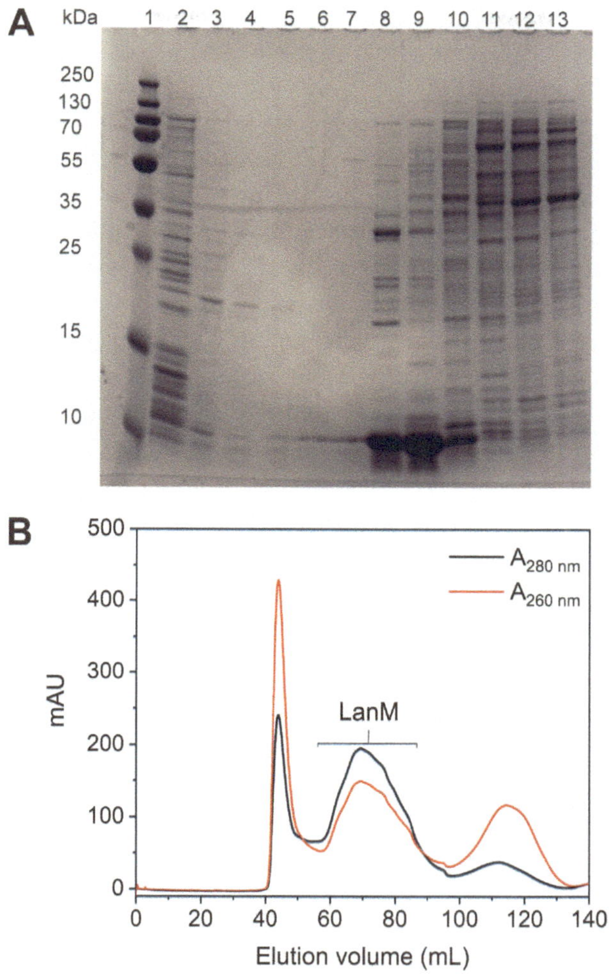

Fig. 4 Purification of wild-type LanM. (A) SDS-PAGE analysis of the Q-Sepharose column step, as follows: Lane 1, MW ladder; Lane 2, flowthrough; Lanes 3–5, wash fractions 1–3 (5–6 mL each); Lanes 6–13, elution fractions 1, 5, 10, 15, 20, 30, 35 (2 mL each). LanM is the strong band just above the 10 kDa marker, in lanes 8–10. (B) Chromatogram for the Superdex 75 pg step, displaying the broad elution of LanM. Fractions containing pure LanM are indicated.

5. Pool desired fractions and concentrate by centrifugal filtration (Amicon Ultra 10 kDa MWCO) to <2 mL.
6. In order to remove DNA and minor contaminating proteins, load protein solution onto a HiLoad Superdex 75 pg 16/600 column pre-equilibrated with Buffer M3 and load the protein using a 2 mL capillary loop.

Rinse the loop with 2 mL Buffer M3 and elute with 1 CV of the same buffer at 0.75 mL/min. Follow elution at 280 and 260 nm. Collect 2 mL fractions.

Note: The elution profile will be a broad peak between ~60 and 85 mL on the 120 mL Superdex 75 pg column (Fig. 4B). This result is presumably associated with intrinsic disorder of the apoprotein as well as, perhaps, weak binding of trace metals in the buffer.

7. Pool LanM-containing fractions based on SDS-PAGE analysis and concentrate to the desired volume (e.g., 2 mL) by centrifugal filtration (10 kDa MWCO). Add 5 g Chelex-100 to 500 mL Buffer M3 and stir in a beaker at room temperature. Add LanM to a dialysis cassette (e.g., Slide-A-Lyzer MWCO 3500) or dialysis tubing and dialyze overnight at room temperature.
8. Determine the protein concentration using $\varepsilon_{275\text{nm}} = 1400\,M^{-1}\,cm^{-1}$ (Cotruvo et al., 2018). *M. extorquens* LanM possesses one tyrosine, no tryptophans, and three phenylalanines; the fine structure below 275 nm associated with Phe absorption should be visible if the protein preparation has high purity [fig. S8 in Cotruvo et al. (2018)]. Freeze protein in liquid nitrogen at ~1 mM.

5.3 Determination of lanthanide-binding stoichiometry using xylenol orange

A fast and simple method to determine the lanthanide-binding stoichiometry of a newly isolated protein is an assay using xylenol orange (XO) as a competitive colorimetric reagent. A solution (pH 6.1) contains the protein of interest and XO, into which a solution of a lanthanide ion of interest is titrated. The affinity of XO for Ln^{III} ions under these conditions is in the 1–10 μM range (Munshi & Dey, 1968; Tonosaki & Otomo, 1962); therefore, if the protein has high-nanomolar affinity for Ln^{III} ions, or better, it will preferentially bind the metal ions, and the spectrum of XO will remain unchanged. Once the protein's metal-binding site(s) is (are) saturated, the metal will instead bind to the dye, leading to an increase in the absorbance at ~575 nm. From the volume of titrant added to reach the first increase of XO absorption, the metal-binding stoichiometry of the protein can be calculated. This method is a useful preliminary experiment before more time- and protein-intensive methods to determine binding affinities are pursued, such as isothermal titration calorimetry (Quinn, Carpenter, Croteau, &

Wilcox, 2016) or the spectrophotometric methods described in chapters "Spectrophotometric methods to probe the solution chemistry of lanthanide complexes with macromolecules" by Deblonde and "Determination of affinities of lanthanide-binding proteins using chelator-buffered titrations" by Mattocks et al. of the accompanying volume, Volume 651.

For best results, we suggest ~10–20 μM metal-binding sites in solution. In the method below, we use 5 μM LanM, because it binds 3 equiv. of Ln^{III} ions tightly—the average stoichiometry across all of the lanthanides determined using this method is 2.9 ± 0.3 [see fig. S11 in Cotruvo et al. (2018]. When we applied this method to demonstrate that LanD (META1p1781 or LutD) binds 1 equiv. of Ln^{III} ions, we used 10 μM (Mattocks et al., 2019). In the latter case, interestingly, the relative K_ds of XO and LanD were such that the protein outcompeted XO for light Ln^{III} ions (La-Gd) but not heavy Ln^{III} ions (roughly Ho-Lu) [see fig. S17 in Mattocks et al. (2019)].

We note that there is a relatively narrow range of pH in which xylenol orange is an effective colorimetric reagent for Ln^{III} ions (Lyle & Rahman, 1963); therefore, control of the pH is important during the experiment.

1. Blank the UV–visible spectrophotometer (240–800 nm) with a microcuvette containing 500 μL of Buffer M4.
2. Add your protein of interest to the microcuvette to a concentration of ~5 μM, and collect a spectrum over the 240–800 nm region. Calculate the protein concentration using a known extinction coefficient.

   Note: An accurate starting protein concentration is important for quantitative analysis of metal-binding stoichiometry.

   Note: Because of the pH sensitivity of the assay, if the protein is stored in a buffer at a different pH, it should either be quite concentrated (e.g., for LanM, ~1 mM) or the protein should be exchanged into Buffer M4 before dilution for the assay.
3. Add XO to a final concentration of ~5 μM and collect a spectrum in the 240–800 nm region.
4. Titrate 1.25 mM $LnCl_3$ solution in 0.5 μL aliquots (0.25 equiv. if 5 μM protein is present), mix, and acquire spectra between 240 and 800 nm after each addition.

   Note: To mix, either use a P200 pipettor or invert the cuvette with parafilm on top. If you do the latter, use a different spot on the parafilm each time mixing in order to avoid a color change before the actual endpoint of the titration.

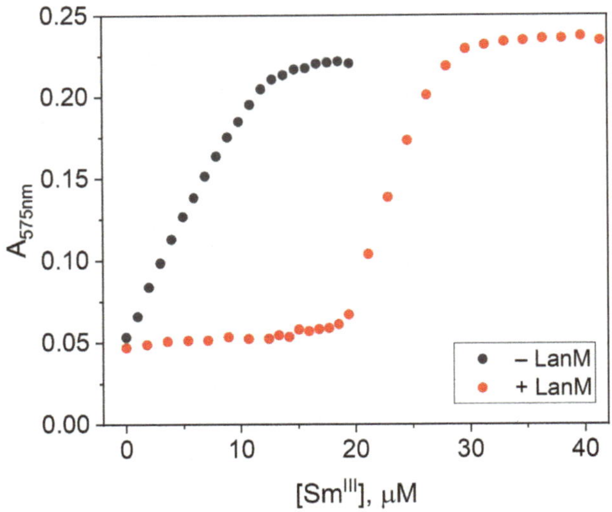

Fig. 5 Titration of xylenol orange with $SmCl_3$ in the presence and absence of LanM, monitored at 575 nm. The concentration of LanM in the experiment was 6.0 μM. The endpoint calculated by the method described in the text was at 19.4 μM Sm^{III}, which corresponds to 3.2 equiv. *Adapted with permission from Cotruvo, J. A., Jr., Featherston, E. R., Mattocks, J. A., Ho, J. V., & Laremore, T. N. (2018). Lanmodulin: A highly selective lanthanide-binding protein from a lanthanide-utilizing bacterium.* Journal of the American Chemical Society, 140, *15056–15061. Copyright 2018 American Chemical Society.*

5. Continue the titration until there is no further change in the Ln^{III}-XO absorption feature at ~575 nm (λ_{max} is dependent on Ln^{III} ion in the experiment).

   Note: Commercially available XO is impure, as evidenced by lack of precise isosbestic points during titrations. This fact does not affect the result of the titrations.

6. Correct the raw spectra for baseline drift and for volume change of the solution upon each addition of metal stock. Plot the corrected absorption at the λ_{max} of ~575 nm for each titration point against the metal ion concentration (Fig. 5). The metal ion concentration at which $A = A_{initial} + 0.1 (A_{final} - A_{initial})$, the point of 10% saturation of XO, is taken as an estimate of the point at which the tight binding (≪μM) sites of the protein of interest are saturated.

   Note: XO adheres slightly to the microcuvette; after each experiment, we suggest filling the cuvette with 6 M HCl and letting it sit for 5–10 min, followed by your typical cuvette washing procedure (e.g., rinsing with water and acetone).

6. PqqT

Decades of work, especially by Klinman and co-workers, has recently resulted in elucidation of the mechanism of biosynthesis of the organic cofactor of the ADHs, pyrroloquinoline quinone (PQQ) (Duine, 1999; Koehn et al., 2019; Zhu & Klinman, 2020), which occurs in the cytosol. However, less is known about PQQ trafficking once synthesized. The enzymes that use the cofactor are periplasmic, and it is also known that the cofactor is released into growth media by some of the bacteria that synthesize it (van Kleef & Duine, 1989). Furthermore, not all organisms that encode PQQ-dependent enzymes are able to make the cofactor, suggesting that PQQ transfer between organisms is important in the biosphere, as is the case for other cofactors such as cobalamins (Shelton et al., 2019).

Interested in investigating this exchange of PQQ between organisms, we hypothesized that a mechanism for PQQ uptake might exist. *M. extorquens* AM1 META1p1737 was a putative SBP encoded, along with putative ABC transporter subunits, by genes adjacent to *xoxF*. Like LanM, this protein is one of the handful of proteins upregulated by the presence of lanthanides in the growth medium of *M. extorquens* (Good et al., 2019; Ochsner et al., 2019). Using spectrophotometry, spectrofluorometry, and isothermal titration calorimetry (ITC), we demonstrated that this SBP, which we named PqqT, binds PQQ tightly (Ho & Cotruvo, 2019). We note that PQQ transport activity of the full system (a putative "PqqTUV" system) has not yet been demonstrated in cells, and the structure of protein has not yet been reported. We proposed that the function of cytosolic PQQ uptake might be to supplement the pool of the cofactor endogenously biosynthesized. It may also suggest that, even though the cofactor is used in the periplasm, that exogenous PQQ cannot equilibrate with the periplasmic pool used for ADH activation, for reasons of regulation and/or proper cofactor chaperoning.

As is its mechanism of biosynthesis, the solution chemistry and spectroscopy of PQQ are complex. In aqueous solution, PQQ exists in an equilibrium between its hydrated and unhydrated forms (Scheme 2). This equilibrium is also altered by the presence of nucleophiles and metal ions (Dekker, Duine, Frank, Verweil, & Westerling, 1982; Lumpe & Daumann, 2019). However, because these forms have different absorption and fluorescence properties, they can be used in certain cases to study PQQ binding to proteins, as has been done with the PQQ biosynthetic protein PqqC (Magnusson, RoseFigura,

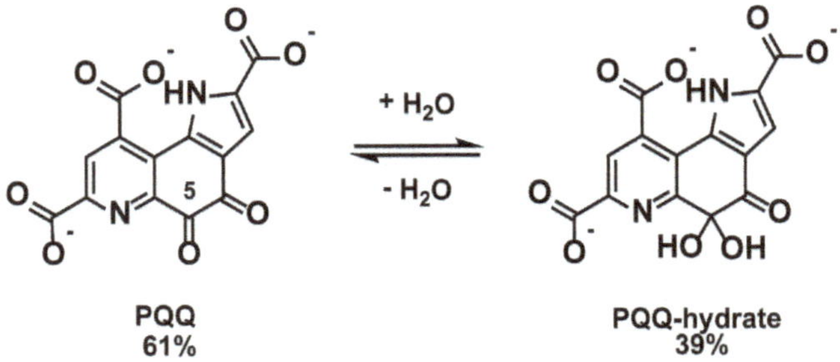

Scheme 2 Hydration equilibrium speciation for PQQ. *Reproduced with permission from Ho, J. V., & Cotruvo, J. A., Jr. (2019). A periplasmic binding protein for pyrroloquinoline quinone.* Biochemistry, *58, 2665–2669. Copyright 2019 American Chemical Society.*

Toyama, Schwarzenbacher, & Klinman, 2007) and with PqqT (Ho & Cotruvo, 2019). The affinities of these proteins for PQQ are 2nM for the former and 50nM for the latter.

Here we describe a new purification method for PqqT as well as provide helpful information for investigation of PQQ binding. As discussed in our original report, native PqqT possesses a Cys residue (C192) that is desulfurated to Ala when exported to the periplasm in *E. coli*, a modification that does not affect PQQ binding. Therefore, we will use a C192A variant, cloned into pET24a, that can be expressed in the cytosol (with Ala28 as the first residue), which yields 5mg PqqT per L culture. A form with the Cys can be studied, but it requires an anaerobic spectroscopy setup (see our paper for a description). Further work is necessary to determine whether the Cys, while dispensable for PQQ binding in vitro, is playing a functional role in vivo. By discussing the case of PqqT here, we aim to provide guidance for researchers who wish to use these methods to investigate PQQ-binding proteins. Of course, PQQ will not necessarily undergo these same changes upon protein binding in all cases, as these changes depend on affinity and the structure of the binding site.

6.1 Materials

- Reagents for cell growth and protein expression (see Section 3.1). All solid and liquid growth media contain 50μg/mL kanamycin (Km)
- Protease inhibitor cocktail (e.g., Roche Complete Mini tablets)
- DNase I
- Phenylmethanesulfonyl fluoride (PMSF, 100mM stock in isopropanol)
- Q-Sepharose Fast Flow
- FPLC instrument

- HiLoad Superdex 75 pg 16/600
- **Buffer T1:** 50 mM Tris, 10 mM NaCl, 1 mM EDTA, pH 8.0
- **Buffer T2:** 50 mM Tris, 300 mM NaCl, 1 mM EDTA, pH 8.0
- **Buffer T3:** 30 mM MOPS, 100 mM KCl, pH 7.0

6.2 Expression and purification of PqqT

6.2.1 Protein expression

1. Transform chemically competent *E. coli* BL21(DE3) cells with pET24a-PqqT(C192A) and grow at 37 °C overnight on LB-agar-Km plates. Use a single colony to inoculate 100 mL LB-Km and shake at 200 rpm overnight (~16 h) at 37 °C.
2. Use this culture to inoculate one 2-L culture (in a 6-L flask) at 50 × dilution. Shake cultures at 200 rpm (37 °C) and monitor the OD_{600nm} periodically. At OD_{600nm} ~0.6, cool the flasks in a 4 °C cold room for 15 min and lower the temperature of the shaker to 18 °C. Return flask to the shaker and induce PqqT expression by adding IPTG to a concentration of 0.2 mM.
3. After induction overnight (~18 h), harvest cells by centrifugation (7 min, 7000 × *g*, 4 °C) and transfer to pre-weighed 50-mL conical tubes. Typical yield is ~2.5 g per L culture. Flash freeze cell paste in liquid nitrogen and store at −80 °C.

6.2.2 Protein purification

1. On ice, resuspend the cell pellet in 5 mL of Buffer T1 per g of cell paste. Buffer T1 should be supplemented with 1 protease tablet per 10 mL, DNase (2 U/mL), and PMSF (0.25 mM).
2. Lyse cells by sonication and pellet insoluble material by centrifugation at 40,000 × *g*, 20 min, 4 °C.
3. Pre-equilibrate a 15 mL (2.5 × 3 cm) Q-Sepharose Fast Flow column (gravity-flow) in Buffer T1. Load the lysate to the column slowly. Wash column with 2 CV Buffer T1.
4. Fill a gradient mixer (e.g., CBS Scientific GM-200) with 70 mL solutions of Buffer T1 and Buffer T2. Start the gradient and elute the protein from the Q-Sepharose column. Collect 1 mL fractions in microcentrifuge tubes and screen by SDS-PAGE. PqqT (32 kDa) elutes at ~100–200 mM NaCl.
5. Pool desired fractions and concentrate by centrifugal filtration (Amicon Ultra 10 kDa MWCO) to <2 mL.

6. Load protein solution onto a HiLoad Superdex 75 pg 16/600 column pre-equilibrated with Buffer T3 and load the protein using a 2 mL capillary loop. Rinse the loop with 2 mL Buffer M3 and elute with 1 CV of the same buffer at 0.75 mL/min. Follow elution at 280 and 260 nm. Collect 1.5 mL fractions. PqqT elutes at ~60–70 mL.
7. Pool fractions based on SDS-PAGE analysis and concentrate to the desired volume by centrifugal filtration (10 kDa MWCO). Determine the protein concentration using $\varepsilon_{280\text{nm}} = 32.4\,\text{mM}^{-1}\,\text{cm}^{-1}$, estimated using ExPASy (Gasteiger et al., 2005), and store at −80 °C.

6.3 Spectroscopic methods to probe PQQ binding

As noted above, binding of PQQ to proteins may be amenable to several spectroscopic approaches, taking advantage of spectral shifts in the PQQ/hydrate equilibrium (Scheme 2) and possibly other interactions accompanying protein binding. If observed, these shifts can be used to determine stoichiometry and even affinity. Titrations may be carried out in either direction; because it is the change in the PQQ spectrum associated with protein binding that is being monitored in these experiments, however, it is easier to start with PQQ in the cuvette and titrate protein into it.

For absorbance measurements, PQQ concentration is determined spectrophotometrically using $\varepsilon_{322\text{nm}} = 8963\,\text{M}^{-1}\,\text{cm}^{-1}$, which is an isosbestic point between the two forms (Dekker et al., 1982). In our experiments, we focused on binding stoichiometry, so the PQQ and PqqT concentrations were >100-fold higher than the K_d to ensure full binding. Fig. 6A shows that the UV–visible spectrum of PQQ is increasingly perturbed as PqqT is titrated in; at the endpoint of the titration, the spectrum resembles that of the unhydrated cofactor—suggesting that the protein specifically recognizes it over the PQQ-hydrate.

The fluorescence spectrum of PQQ in aqueous solution ($\lambda_{ex} = 375$ nm) exhibits a broad emission centered at 488 nm. This spectrum was suggested by early studies to be attributable largely to the PQQ-hydrate, with PQQ itself having little or no fluorescence (Dekker et al., 1982). Prior titrations of PQQ with PqqC showed a bathochromic shift of this emission to $\lambda_{max} = 442$ nm, accompanied by an enhancement of the fluorescence, upon binding of PQQ to the enzyme (Magnusson et al., 2007). In the case of PqqT, the fluorescence emission also exhibits a bathochromic shift in the presence of the protein, but only to 470 nm, and the intensity of the resulting signal decreases significantly compared to that of the cofactor in solution

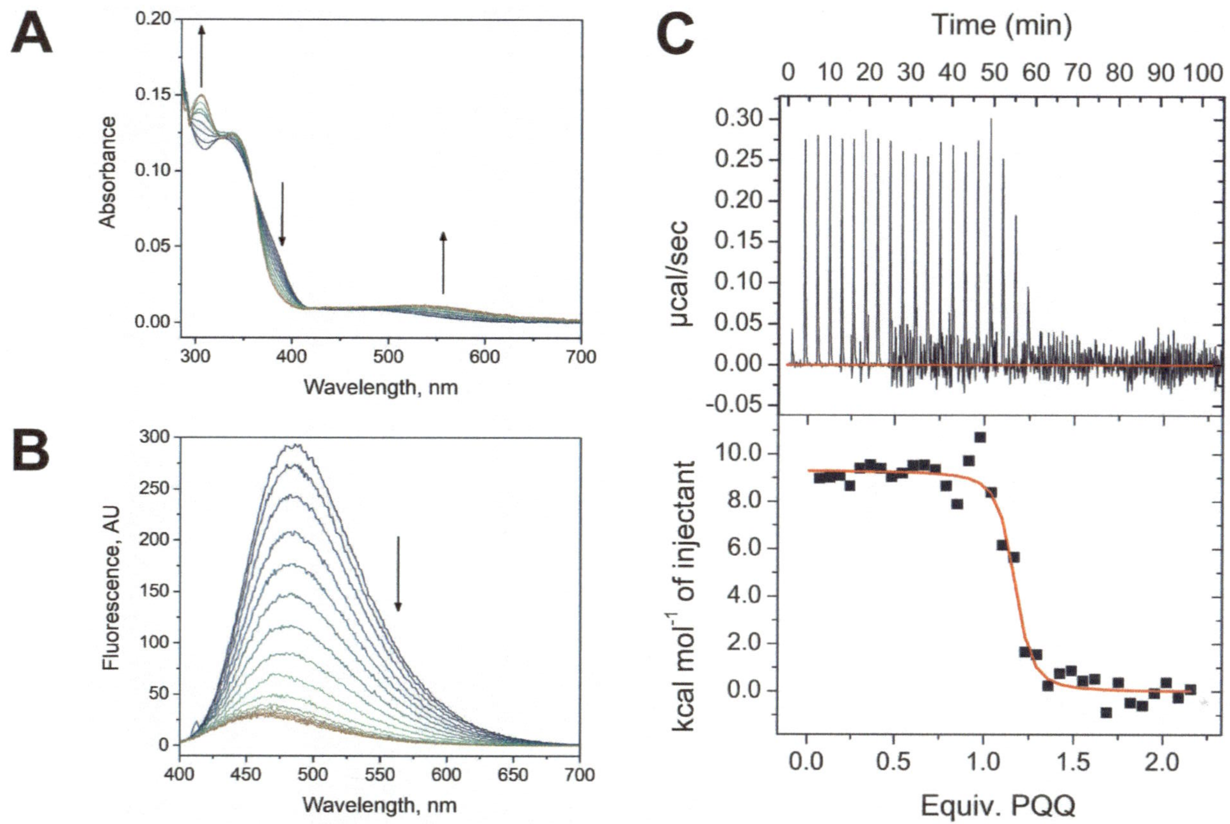

Fig. 6 Biophysical methods to study PQQ binding to PqqT. (A) Spectrophotometric titration of 13 μM PQQ with 0–18 μM PqqT in 1.5 μM increments (dark blue to gray). Binding stoichiometry (1.1 PQQ/PqqT) can be estimated once PqqT addition no longer alters the spectrum of PQQ. (B) Spectrofluorometric titration of 13 μM PQQ with 0–20 μM PqqT in 1 μM increments (dark blue to gray). Binding stoichiometry (1.3 PQQ/PqqT) can be estimated once PqqT addition no longer alters the spectrum of PQQ. (C) ITC analysis of PQQ binding to PqqT, yielding a $K_d = 50$ nM and $n = 1.2$. *Adapted with permission from Ho, J. V., & Cotruvo, J. A., Jr. (2019). A periplasmic binding protein for pyrroloquinoline quinone.* Biochemistry, *58, 2665–2669. Copyright 2019 American Chemical Society.*

(Fig. 6B) (Ho & Cotruvo, 2019). We suggested that this might represent the fluorescence spectrum of the unhydrated PQQ (which to the best of our knowledge has not been definitively reported); however, the spectral shifts of PQQ accompanying binding to PqqT are not identical to those with PqqC. Structural characterization of the PqqT-PQQ complex will help to resolve this point.

Spectrofluorometric titrations can also be used to determine the protein-PQQ K_d if the concentration of PQQ in the experiment is on a similar order to the K_d. We refer the reader to Magnusson et al. (2007) for the application of this approach to PqqC. Binding affinity can also be determined by ITC, which also provides other thermodynamic parameters such as stoichiometry, enthalpy, and entropy change (Fig. 6C). For a practical summary of the ITC technique, see Pierce, Raman, and Nall (1999); for a more involved discussion of the theory, see Indyk and Fisher (1998). A protocol for ITC study of PqqT is provided in Ho and Cotruvo (2019).

7. Summary

We have provided detailed protocols for expression, purification, and characterization of proteins involved in lanthanide and PQQ utilization in *M. extorquens* AM1. These proteins, the first representatives of the lanthanome to be characterized beyond the ADHs, illustrate the ability to use the facile and ubiquitous *E. coli* expression platform to yield wild-type proteins (without affinity tags) from either the cytosol or periplasm. With defined protein expression and purification methods available and assay protocols developed, these methods can be easily applied to proteins from other lanthanide-utilizing organisms to expand our knowledge of the biochemistry of lanthanide and PQQ acquisition, trafficking, and utilization.

Acknowledgments

This work was supported by the National Science Foundation (CHE-1945015 to J.A.C.). J.A.C. also acknowledges the Pennsylvania State University Department of Chemistry and a Louis Martarano Career Development Professorship for support.

References

Akberdin, I. R., Collins, D. A., Hamilton, R., Oshchepkov, D. Y., Shukla, A. K., Nicora, C. D., et al. (2018). Rare earth elements alter redox balance in Methylomicrobium alcaliphilum 20Z(R). *Frontiers in Microbiology*, *9*, 2735. https://doi.org/10.3389/fmicb.2018.02735.

Amaratunga, K., Goodwin, P. M., O'connor, C. D., & Anthony, C. (1997). The methanol oxidation genes *mxaFJGIR(S)ACKLD* in *Methylobacterium extorquens*. *FEMS Microbiology Letters*, *146*, 31–38. https://doi.org/10.1111/j.1574-6968.1997.tb10367.x.
Anthony, C. (1982). *The biochemistry of methylotrophs*. London: Academic.
Anthony, C., & Williams, P. (2003). The structure and mechanism of methanol dehydrogenase. *Biochimica et Biophysica Acta*, *1647*, 18–23. https://doi.org/10.1016/s1570-9639(03)00042-6.
Armstrong, J. D. (1964). The molar extinction coefficient of 2,6-dichlorophenol indophenol. *Biochimica et Biophysica Acta*, *86*, 194–197. https://doi.org/10.1016/0304-4165(64)90180-1.
Barr, I., & Guo, F. (2015). Pyridine hemochromagen assay for determining the concentration of heme in purified protein solutions. *Bio-Protocol*, *5*, e1594. https://doi.org/10.21769/BioProtoc.1594.
Chistoserdova, L. (2011). Modularity of methylotrophy, revisited. *Environmental Microbiology*, *13*, 2603–2622. https://doi.org/10.1111/j.1462-2920.2011.02464.x.
Chistoserdova, L. (2019). New pieces to the lanthanide puzzle. *Molecular Microbiology*, *111*, 1127–1131. https://doi.org/10.1111/mmi.14210.
Chistoserdova, L., & Kalyuzhnaya, M. G. (2018). Current trends in methylotrophy. *Trends in Microbiology*, *26*, 703–714. https://doi.org/10.1016/j.tim.2018.01.011.
Chistoserdova, L., Kalyuzhnaya, M. G., & Lidstrom, M. E. (2009). The expanding world of methylotrophic metabolism. *Annual Review of Microbiology*, *63*, 477–499. https://doi.org/10.1146/annurev.micro.091208.073600.
Choi, J. M., Cao, T.-P., Kim, S. W., Lee, K. H., & Lee, S. H. (2017). MxaJ structure reveals a periplasmic binding protein-like architecture with unique secondary structural elements. *Proteins*, *85*, 1379–1386. https://doi.org/10.1002/prot.25283.
Chu, F., Beck, D. A. C., & Lidstrom, M. E. (2016). MxaY regulates the lanthanide-mediated methanol dehydrogenase switch in Methylomicrobium buryatense. *PeerJ*, *4*, e2435. https://doi.org/10.7717/peerj.2435.
Clark, W. M. (1960). *Oxidation-reduction potentials of organic systems*. Baltimore: The Williams and Wilkins Company.
Cook, E. C., Featherston, E. R., Showalter, S. A., & Cotruvo, J. A., Jr. (2019). Structural basis for rare earth element recognition by *Methylobacterium extorquens* lanmodulin. *Biochemistry*, *58*, 120–125. https://doi.org/10.1021/acs.biochem.8b01019.
Cotruvo, J. A., Jr. (2019). The chemistry of lanthanides in biology: Recent discoveries, emerging principles, and technological applications. *ACS Central Science*, *5*, 1496–1506. https://doi.org/10.1021/acscentsci.9b00642.
Cotruvo, J. A., Jr., Featherston, E. R., Mattocks, J. A., Ho, J. V., & Laremore, T. N. (2018). Lanmodulin: A highly selective lanthanide-binding protein from a lanthanide-utilizing bacterium. *Journal of the American Chemical Society*, *140*, 15056–15061. https://doi.org/10.1021/jacs.8b09842.
Cox, J. M., Day, D. J., & Anthony, C. (1992). The interaction of methanol dehydrogenase and its electron acceptor, cytochrome c_L in methylotrophic bacteria. *Biochimica et Biophysica Acta*, *1119*, 97–106. https://doi.org/10.1016/0167-4838(92)90240-e.
Day, D. J., & Anthony, C. (1990). Methanol dehydrogenase from *Methylobacterium extorquens* AM1. *Methods in Enzymology*, *188*, 210–216. https://doi.org/10.1016/0076-6879(90)88035-9.
Deblonde, G. J.-P., Mattocks, J. A., Park, D. M., Reed, D. W., Cotruvo, J. A., Jr., & Jiao, Y. (2020). Selective and efficient biomacromolecular extraction of rare-earth elements using lanmodulin. *Inorganic Chemistry*, *59*, 11855–11867. https://doi.org/10.1021/acs.inorgchem.0c01303.
Dekker, R. H., Duine, J. A., Frank, J., Verweil, P. E. J., & Westerling, J. (1982). Covalent addition of H_2O, enzyme substrates and activators to pyrrolo-quinoline quinone, the

coenzyme of quinoproteins. *European Journal of Biochemistry*, *125*, 69–73. https://doi.org/10.1111/j.1432-1033.1982.tb06652.x.

Delaney, N. F., Kaczmarek, M. E., Ward, L. M., Swanson, P. K., Lee, M.-C., & Marx, C. J. (2013). Development of an optimized medium, strain and high-throughput culturing methods for *Methylobacterium extorquens*. *PLoS One*, *8*, e62957. https://doi.org/10.1371/journal.pone.0062957.

Drake, S. K., Lee, K. L., & Falke, J. J. (1996). Tuning the equilibrium ion affinity and selectivity of the EF-hand calcium binding motif: Substitutions at the gateway position. *Biochemistry*, *35*, 6697–6705. https://doi.org/10.1021/bi952430l.

Duine, J. A. (1999). The PQQ story. *Journal of Bioscience and Bioengineering*, *88*, 231–236. https://doi.org/10.1016/S1389-1723(00)80002-X.

Edington, S. C., Gonzalez, A., Middendorf, T. R., Halling, D. B., Aldrich, R. W., & Baiz, C. R. (2018). Coordination to lanthanide ions distorts binding site conformation in calmodulin. *Proceedings. National Academy of Sciences. United States of America*, *115*, E3126–E3134. https://doi.org/10.1073/pnas.1722042115.

Efimov, I., Parkin, G., Millett, E. S., Glenday, J., Chan, C. K., Weedon, H., et al. (2014). A simple method for the determination of reduction potentials in heme proteins. *FEBS Letters*, *588*, 701–704. https://doi.org/10.1016/j.febslet.2013.12.030.

Englard, S., & Seifter, S. (1990). Precipitation techniques. *Methods in Enzymology*, *182*, 287–300. https://doi.org/10.1016/0076-6879(90)82024-v.

Farhan Ul Haque, M., Kalidass, B., Bandow, N., Turpin, E. A., DiSpirito, A. A., & Semrau, J. D. (2015). Cerium regulates expression of alternative methanol dehydrogenases in Methylosinus trichosporium OB3b. *Applied and Environmental Microbiology*, *81*, 7546–7552. https://doi.org/10.1128/AEM.02542-15.

Featherston, E. R., & Cotruvo, J. A., Jr. (2021). The biochemistry of lanthanide acquisition, trafficking, and utilization. *Biochimica et Biophysica Acta, Molecular Cell Research*, *1868*, 118864. https://doi.org/10.1016/j.bbamcr.2020.118864.

Featherston, E. R., Rose, H. R., McBride, M. J., Taylor, E. M., Boal, A. K., & Cotruvo, J. A., Jr. (2019). Biochemical and structural characterization of XoxG and XoxJ and their roles in lanthanide-dependent methanol dehydrogenase activity. *Chembiochem*, *20*, 2360–2372. https://doi.org/10.1002/cbic.201900184.

Fitriyanto, N. A., Fushimi, M., Matsunaga, M., Pertiwiningrum, A., Iwama, T., & Kawai, K. (2011). Molecular structure and gene analysis of Ce^{3+}-induced methanol dehydrogenase of *Bradyrhizobium* sp. MAFF211645. *Journal of Bioscience and Bioengineering*, *111*, 613–617. https://doi.org/10.1016/j.jbiosc.2011.01.015.

Gasteiger, E., Hoogland, C., Gattiker, A., Duvaud, S., Wilkins, M. R., Appel, R. D., et al. (2005). Protein identification and analysis tools on the ExPASy server. In J. M. Walker (Ed.), *The proteomics protocols handbook* (pp. 571–607). Totowa, NJ: Humana Press. https://doi.org/10.1385/1-59259-890-0:571.

Gifford, J. L., Walsh, M. P., & Vogel, H. J. (2007). Structures and metal-ion-binding properties of the Ca^{2+}-binding helix-loop-helix EF-hand motifs. *The Biochemical Journal*, *405*, 199–221. https://doi.org/10.1042/BJ20070255.

Good, N. M., Moore, R. S., Suriano, C. J., & Martinez-Gomez, N. C. (2019). Contrasting in vitro and in vivo methanol oxidation activities of lanthanide-dependent alcohol dehydrogenases XoxF1 and ExaF from *Methylobacterium extorquens* AM1. *Scientific Reports*, *9*, 4248. https://doi.org/10.1038/s41598-019-41043-1.

Good, N. M., Vu, H. N., Suriano, C. J., Subuyuj, G. A., Skovran, E., & Martinez-Gomez, N. C. (2016). Pyrroloquinoline quinone ethanol dehydrogenase in *Methylobacterium extorquens* AM1 extends lanthanide-dependent metabolism to multicarbon substrates. *Journal of Bacteriology*, *198*, 3109–3118. https://doi.org/10.1128/JB.00478-16.

Gu, W., Haque, M. F. U., Dispirito, A. A., & Semrau, J. D. (2016). Uptake and effect of rare earth elements on gene expression in *Methylosinus trichosporium* OB3b. *FEMS Microbiology Letters*, *363*, fnw129. https://doi.org/10.1093/femsle/fnw129.

Gu, W., & Semrau, J. D. (2017). Copper and cerium-regulated gene expression in Methylosinus trichosporium OB3b. *Applied Microbiology and Biotechnology, 101*, 8499–8516. https://doi.org/10.1007/s00253-017-8572-2.

Hibi, Y., Asai, K., Arafuka, H., Hamajima, M., Iwama, T., & Kawai, K. (2011). Molecular structure of La^{3+}-induced methanol dehydrogenase-like protein in *Methylobacterium radiotolerans*. *Journal of Bioscience and Bioengineering, 111*, 547–549. https://doi.org/10.1016/j.jbiosc.2010.12.017.

Ho, J. V., & Cotruvo, J. A., Jr. (2019). A periplasmic binding protein for pyrroloquinoline quinone. *Biochemistry, 58*, 2665–2669. https://doi.org/10.1021/acs.biochem.9b00358.

Huang, J., Yu, Z., & Chistoserdova, L. (2018). Lanthanide-dependent methanol dehydrogenases of XoxF4 and XoxF5 clades are differentially distributed among methylotrophic bacteria and they reveal different biochemical properties. *Frontiers in Microbiology, 9*, 1366. https://doi.org/10.3389/fmicb.2018.01366.

Huang, J., Yu, Z., Groom, J., Cheng, J. F., Tarver, A., Yoshikuni, Y., et al. (2019). Rare earth element alcohol dehydrogenases widely occur among globally distributed, numerically abundant and environmentally important microbes. *The ISME Journal, 13*, 2005–2017. https://doi.org/10.1038/s41396-019-0414-z.

Indyk, L., & Fisher, H. F. (1998). Theoretical aspects of isothermal titration calorimetry. *Methods in Enzymology, 295*, 350–364. https://doi.org/10.1016/s0076-6879(98)95048-0.

Kalimuthu, P., Daumann, L. J., Pol, A., Op den Camp, H. J. M., & Bernhardt, P. V. (2019). Electrocatalysis of a europium-dependent bacterial methanol dehydrogenase with its physiological electron acceptor cytochrome c_{GJ}. *Chemistry—A European Journal, 25*, 8760–8768. https://doi.org/10.1002/chem.201900525.

Keltjens, J. T., Pol, A., Reimann, J., & Op den Camp, H. J. M. (2014). PQQ-dependent methanol dehydrogenases: Rare-earth elements make a difference. *Applied Microbiology and Biotechnology, 98*, 6163–6183. https://doi.org/10.1007/s00253-014-5766-8.

Koehn, E. M., Latham, J. A., Armand, T., Evans, R. L., Tu, X., Wilmot, C. M., et al. (2019). Discovery of hydroxylase activity for PqqB provides a missing link in the pyrroloquinoline quinone biosynthetic pathway. *Journal of the American Chemical Society, 141*, 4398–4405. https://doi.org/10.1021/jacs.8b13453.

Krause, S. M. B., Johnson, T., Samadhi Karunaratne, Y., Fu, Y., Beck, D. A. C., Chistoserdova, L., et al. (2017). Lanthanide-dependent cross-feeding of methane-derived carbon is linked by microbial community interactions. *Proceedings of the National Academy of Sciences of the United States of America, 114*, 358–363. https://doi.org/10.1073/pnas.1619871114.

Londer, Y. Y. (2011). Expression of recombinant cytochromes c in *E. coli*. In J. T. C. Evans, & M.-Q. Xu (Eds.), *Heterologous gene expression in E. coli: Methods and protocols* (pp. 123–150). Totowa, NJ: Humana Press. https://doi.org/10.1007/978-1-61737-967-3_8.

Lumpe, H., & Daumann, L. J. (2019). Studies of redox cofactor pyrroloquinoline quinone and its interaction with lanthanides(III) and calcium(II). *Inorganic Chemistry, 58*, 8432–8441. https://doi.org/10.1021/acs.inorgchem.9b00568.

Lumpe, H., Pol, A., Op den Camp, H. J. M., & Daumann, L. J. (2018). Impact of the lanthanide contraction on the activity of a lanthanide-dependent methanol dehydrogenase—A kinetic and DFT study. *Dalton Transactions, 47*, 10463–10472. https://doi.org/10.1039/c8dt01238e.

Lyle, S. J., & Rahman, M. M. (1963). Complexometric titration of yttrium and the lanthanons—I: A comparison of direct methods. *Talanta, 10*, 1177–1182. https://doi.org/10.1016/0039-9140(63)80170-8.

Magnusson, O. T., RoseFigura, J. M., Toyama, H., Schwarzenbacher, R., & Klinman, J. P. (2007). Pyrroloquinoline quinone biogenesis: Characterization of PqqC and its H84N and H84A active site variants. *Biochemistry, 46*, 7174–7186. https://doi.org/10.1021/bi700162n.

Massey, V. (1991). A simple method for determination of redox potentials. In B. Curtis, S. Roncho, & G. Zanetti (Eds.), *Flavins and flavoproteins 1990* (pp. 59–66). Berlin: Walter de Gruyter & Co.

Masuda, S., Suzuki, Y., Fujitani, Y., Mitsui, R., Nakagawa, T., Shintani, M., et al. (2018). Lanthanide-dependent regulation of methylotrophy in *Methylobacterium aquaticum* strain 22A. *mSphere*, *3*, e00462-00417. https://doi.org/10.1128/mSphere.00462-17.

Mattocks, J. A., Ho, J. V., & Cotruvo, J. A., Jr. (2019). A selective, protein-based fluorescent sensor with picomolar affinity for rare earth elements. *Journal of the American Chemical Society*, *141*, 2857–2861. https://doi.org/10.1021/jacs.8b12155.

Munshi, K. N., & Dey, A. K. (1968). Absorptiometric study of the chelates formed between the lanthanoids and xylenol orange. *Microchimica Acta*, *56*, 1059–1065. https://doi.org/10.1007/BF01221178.

Nakagawa, T., Mitsui, R., Tani, A., Sasa, K., Tashiro, S., Iwama, T., et al. (2012). A catalytic role of XoxF1 as La^{3+}-dependent methanol dehydrogenase in *Methylobacterium extorquens* strain AM1. *PLoS One*, *7*, e50480. https://doi.org/10.1371/journal.pone.0050480.

Ochsner, A. M., Hemmerle, L., Vonderach, T., Nüssli, R., Bortfeld-Miller, M., Hattendorf, B., et al. (2019). Use of rare earth elements in the phyllosphere colonizer *Methylobacterium extorquens* PA1. *Molecular Microbiology*, *111*, 1152–1166. https://doi.org/10.1111/mmi.14208.

O'Keeffe, D. T., & Anthony, C. (1980). The two cytochromes *c* in the facultative methylotroph *Pseudomonas* AM1. *The Biochemical Journal*, *192*, 411–419. https://doi.org/10.1042/bj1920411.

Picone, N., & Op den Camp, H. J. M. (2019). Role of rare earth elements in methanol oxidation. *Current Opinion in Chemical Biology*, *49*, 39–44. https://doi.org/10.1016/j.cbpa.2018.09.019.

Pierce, M. M., Raman, C. S., & Nall, B. T. (1999). Isothermal titration calorimetry of protein–protein interactions. *Methods*, *19*, 213–221. https://doi.org/10.1006/meth.1999.0852.

Pol, A., Barends, T. R. M., Dietl, A., Khadem, A. F., Eygensteyn, J., Jetten, M. S. M., et al. (2014). Rare earth metals are essential for methanotrophic life in volcanic mudpots. *Environmental Microbiology*, *16*, 255–264. https://doi.org/10.1111/1462-2920.12249.

Quinn, C. F., Carpenter, M. C., Croteau, M. L., & Wilcox, D. E. (2016). Isothermal titration calorimetry measurements of metal ions binding to proteins. *Methods in Enzymology*, *567*, 3–21. https://doi.org/10.1016/bs.mie.2015.08.021.

Roszczenko-Jasińska, P., Vu, H. N., Subuyuj, G. A., Crisostomo, R. V., Cai, J., Lien, N. F., et al. (2020). Gene products and processes contributing to lanthanide homeostasis and methanol metabolism in *Methylorubrum extorquens* AM1. *Scientific Reports*, *10*, 12663. https://doi.org/10.1038/s41598-020-69401-4.

Scheepers, G. H., Lycklama a Nijeholt, J. A., & Poolman, B. (2016). An updated structural classification of substrate-binding proteins. *FEBS Letters*, *590*, 4393–4401. https://doi.org/10.1002/1873-3468.12445.

Shelton, A. N., Seth, E. C., Mok, K. C., Han, A. W., Jackson, S. N., Haft, D. R., et al. (2019). Uneven distribution of cobamide biosynthesis and dependence in bacteria predicted by comparative genomics. *The ISME Journal*, *13*, 789–804. https://doi.org/10.1038/s41396-018-0304-9.

Skovran, E., Raghuraman, C., & Martinez-Gomez, N. C. (2019). Lanthanides in methylotrophy. *Current Issues in Molecular Biology*, *33*, 101–116. https://doi.org/10.21775/cimb.033.101.

Taubert, M., Grob, C., Howat, A. M., Burns, O. J., Dixon, J. L., Chen, Y., et al. (2015). XoxF encoding an alternative methanol dehydrogenase is widespread in coastal marine environments. *Environmental Microbiology*, *17*, 3937–3948. https://doi.org/10.1111/1462-2920.12896.

Tonosaki, K., & Otomo, M. (1962). Spectrophotometric determination of cerium(III) and some rare earths with xylenol orange. *Bulletin of the Chemical Society of Japan*, *35*, 1683–1686. https://doi.org/10.1246/bcsj.35.1683.

van Kleef, M. A., & Duine, J. A. (1989). Factors relevant in bacterial pyrroloquinoline quinone production. *Applied and Environmental Microbiology*, *55*, 1209–1213. https://doi.org/10.1128/aem.55.5.1209-1213.1989.

Van Spanning, R. J., Wansell, C. W., De Boer, T., Hazelaar, M. J., Anazawa, H., Harms, N., et al. (1991). Isolation and characterization of the moxJ, moxG, moxI, and moxR genes of Paracoccus denitrificans: Inactivation of moxJ, moxG, and moxR and the resultant effect on methylotrophic growth. *Journal of Bacteriology*, *173*, 6948–6961. https://doi.org/10.1128/jb.173.21.6948-6961.1991.

Versantvoort, W., Pol, A., Daumann, L. J., Larrabee, J. A., Strayer, A. H., Jetten, M. S. M., et al. (2019). Characterization of a novel cytochrome c_{GJ} as the electron acceptor of XoxF-MDH in the thermoacidophilic methanotroph *Methylacidiphilum fumariolicum* SolV. *Biochimica et Biophysica Acta, Proteins and Proteomics*, *1867*, 595–603. https://doi.org/10.1016/j.bbapap.2019.04.001.

Vu, H. N., Subuyuj, G. A., Vijayakumar, S., Good, N. M., Martinez-Gomez, N. C., & Skovran, E. (2016). Lanthanide-dependent regulation of methanol oxidation systems in *Methylobacterium extorquens* AM1 and their contribution to methanol growth. *Journal of Bacteriology*, *198*, 1250–1259. https://doi.org/10.1128/JB.00937-15.

Wang, L., Hibino, A., Suganuma, S., Ebihara, A., Iwamoto, S., Mitsui, R., et al. (2020). Preference for particular lanthanide species and thermal stability of XoxFs in Methylorubrum extorquens strain AM1. *Enzyme and Microbial Technology*, *136*, 109518. https://doi.org/10.1016/j.enzmictec.2020.109518.

Wegner, C.-E., Gorniak, L., Riedel, S., Westermann, M., & Küsel, K. (2019). Lanthanide-dependent methylotrophs of the family Beijerinckiaceae: Physiological and genomic insights. *Applied and Environmental Microbiology*, *86*(1), e01830-01819. https://doi.org/10.1128/AEM.01830-19.

Wehrmann, M., Berthelot, C., Billard, P., & Klebensberger, J. (2018). The PedS2/PedR2 two-component system is crucial for the rare earth element switch in *Pseudomonas putida* KT2440. *mSphere*, *3*, e00376-00318. https://doi.org/10.1128/mSphere.00376-18.

Wehrmann, M., Billard, P., Martin-Meriadec, A., Zegeye, A., & Klebensberger, J. (2017). Functional role of lanthanides in enzymatic activity and transcriptional regulation of pyrroloquinoline quinone-dependent alcohol dehydrogenases in *Pseudomonas putida* KT2440. *mBio*, *8*, e00570-00517. https://doi.org/10.1128/mBio.00570-17.

Williams, P., Coates, L., Mohammed, F., Gill, R., Erskine, P., Bourgeois, D., et al. (2006). The 1.6 Å x-ray structure of the unusual *c*-type cytochrome, cytochrome c_L, from the methylotrophic bacterium *Methylobacterium extorquens*. *Journal of Molecular Biology*, *357*, 151–162. https://doi.org/10.1016/j.jmb.2005.12.055.

Zheng, Y., Huang, J., Zhao, F., & Chistoserdova, L. (2018). Physiological effect of XoxG(4) on lanthanide-dependent methanotrophy. *mBio*, *9*, e02430-02417. https://doi.org/10.1128/mBio.02430-17.

Zhu, W., & Klinman, J. P. (2020). Biogenesis of the peptide-derived redox cofactor pyrroloquinoline quinone. *Current Opinion in Chemical Biology*, *59*, 93–103. https://doi.org/10.1016/j.cbpa.2020.05.001.

CHAPTER SEVEN

Transposon mutagenesis for methylotrophic bacteria using *Methylorubrum extorquens* AM1 as a model system

Huong N. Vu[†], Gabriel A. Subuyuj[‡], Ralph Valentine Crisostomo[§], and Elizabeth Skovran*

Department of Biological Sciences, San José State University, San José, CA, United States
*Corresponding author: e-mail address: elizabeth.skovran@sjsu.edu

Contents

[†] Present address: Department of Microbiology, University of Georgia, Athens, GA, United States
[‡] Present address: Department of Microbiology and Molecular Genetics, University of California At Davis, Davis, CA, United States
[§] Present address: Molecular Biology Institute, University of California At Los Angeles, Los Angeles, CA, United States

Methods in Enzymology, Volume 650
ISSN 0076-6879
https://doi.org/10.1016/bs.mie.2021.01.015

Abstract

Transposon mutagenesis utilizes transposable genetic elements that integrate into a recipient genome to generate random insertion mutations which are easily identified. This forward genetic approach has proven powerful in elucidating complex processes, such as various pathways in methylotrophy. In the past decade, many methylotrophic bacteria have been shown to possess alcohol dehydrogenase enzymes that use lanthanides (Lns) as cofactors. Using *Methylorubrum extorquens* AM1 as a model organism, we discuss the experimental designs, protocols, and results of three transposon mutagenesis studies to identify genes involved in different aspects of Ln-dependent methanol oxidation. These studies include a selection for transposon insertions that prevent toxic intracellular formaldehyde accumulation, a fluorescence-imaging screen to identify regulatory processes for a primary Ln-dependent methanol dehydrogenase, and a phenotypic screen for genes necessary for function of a Ln-dependent ethanol dehydrogenase. We anticipate that the methods described in this chapter can be applied to understand other metabolic systems in diverse bacteria.

1. Introduction

Methylotrophy is defined as the ability to use reduced single carbon (C_1) substrates, such as methanol and methylamine, as a sole carbon and energy source (Anthony, 1982). Over the last decade, it was discovered that many Gram-negative methylotrophic bacteria encode alcohol dehydrogenase (ADH) enzymes that use a subset of lanthanides (Lns), or rare earth elements, as a cofactor for methanol oxidation (Chistoserdova & Kalyuzhnaya, 2018; Fitriyanto et al., 2011; Hibi et al., 2011; Nakagawa et al., 2012; Pol et al., 2014). This exciting discovery has opened new areas of study in biology and has allowed the cultivation of new microorganisms in the laboratory that were previously uncultivable (Kato, Takashino, Igarashi, & Kitagawa, 2020; Lv, Sahin, & Tani, 2018, 2020; Pol et al., 2014; Skovran & Martinez-Gomez, 2015).

Since its isolation in 1961, *Methylorubrum extorquens* AM1 has served as a model organism for the understanding of aerobic methylotrophic metabolism (Peel & Quayle, 1961). In *M. extorquens* AM1, methanol metabolism begins with the oxidation of methanol to formaldehyde or formate in the periplasmic space (Fig. 1). *M. extorquens* AM1 encodes three distinct types of ADH enzymes capable of oxidizing methanol. Each of these ADHs contains pyrroloquinoline quinone (PQQ) in the active site and a cation, which polarizes PQQ for electron extraction (Lumpe, Pol, den Camp, & Daumann, 2018; McSkimming, Cheisson, Carroll, & Schelter, 2018; Prejanò, Marino, & Russo, 2017). MxaFI is a Ca-dependent methanol dehydrogenase

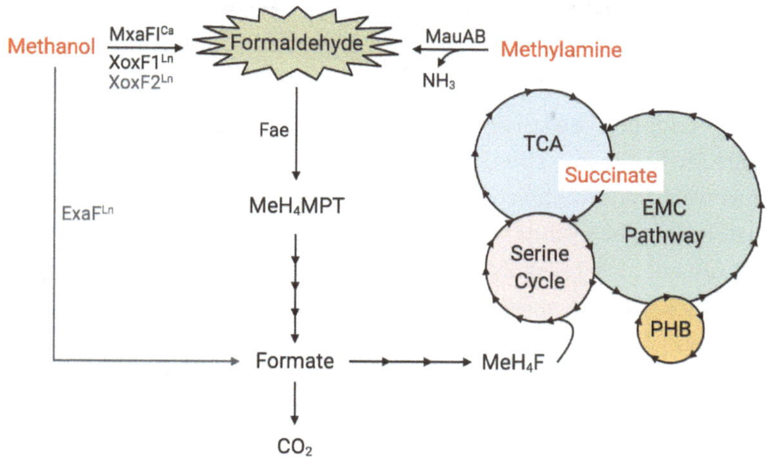

Fig. 1 Ln-dependent and -independent growth of *M. extorquens* AM1 on different carbon sources. XoxF2 and ExaF (gray) play a minor role in methanol oxidation under tested laboratory growth conditions. Formaldehyde is highlighted green to indicate a toxic metabolite. Each arrow represents a reaction in a pathway or cycle. Abbreviations: Ca, calcium; Ln, lanthanide; MeDH, methanol dehydrogenase; EtDH, ethanol dehydrogenase; MaDH, methylamine dehydrogenase; Fae, formaldehyde-activating enzyme; MeH_4MPT, methylene tetrahydromethanopterin; MeH_4F, methylene tetrahydrofolate; TCA, tricarboxylic acid cycle; EMC, ethylmalonyl-CoA; PHB, polyhydroxybutyrate cycle.

(MeDH) which oxidizes methanol to formaldehyde (Anthony & Zatman, 1967), whereas the two XoxF-type ADHs encoded in the *M. extorquens* AM1 genome use light Lns (La-Sm) to catalyze the same reaction. XoxF1 is the dominant XoxF enzyme produced during laboratory growth conditions, while the contribution of XoxF2 is only observed in the absence of XoxF1 (Skovran, Palmer, Rountree, Good, & Lidstrom, 2011; Vu et al., 2016). *M. extorquens* AM1 also encodes a Ln-dependent ExaF ethanol dehydrogenase (EtDH), which contributes to methanol oxidation but converts methanol directly to formate (Fig. 1) (Good, Moore, Suriano, & Martinez-Gomez, 2019; Good et al., 2016). Which ADHs are produced by *M. extorquens* AM1 in part depends on Ln availability. When Lns are present, transcriptional repression of the *mxa* genes occurs, while expression of the *xoxF* and *exaF* genes are upregulated (Skovran, Raghuraman, & Martinez-Gomez, 2019; Vu et al., 2016). This regulatory event is commonly referred to as the Ln- or rare earth-switch (Chu, Beck, & Lidstrom, 2016; Gu, Farhan Ul Haque, DiSpirito, & Semrau, 2016; Masuda et al., 2018; Skovran et al., 2019; Vu et al., 2016). Expression of *exaF* is further upregulated by ethanol, its preferred substrate (Skovran et al., 2019).

In addition to methanol, *M. extorquens* AM1 can use methylamine as a C_1 substrate for growth. Methylamine is also oxidized to formaldehyde in the periplasm, but this reaction is catalyzed by methylamine dehydrogenase (MaDH) encoded by the *mauAB* genes (Chistoserdov, Chistoserdova, McIntire, & Lidstrom, 1994). Once in the cytoplasm, formaldehyde is oxidized to formate, which can be further oxidized to CO_2 or assimilated into cell material via the serine cycle (Crowther, Kosály, & Lidstrom, 2008; Large, Peel, & Quayle, 1961, 1962a, 1962b; Large & Quayle, 1963) (Fig. 1). Besides C_1 substrates, *M. extorquens* AM1 is a facultative methylotroph capable of growth with C_2, C_3, and C_4 substrates. These capabilities have facilitated genetic studies to identify genes required for methylotrophy, since strains that have mutations in genes essential for methylotrophy often still grow with multi-carbon substrates (Marx, O'Brien, Breezee, & Lidstrom, 2003; Van Dien et al., 2003).

Many of the gene clusters that are required for methylotrophic growth in *M. extorquens* AM1 were first identified and categorized using biochemical or complementation analysis of mutants generated by chemical and UV mutagenesis (Anthony, 1982; Heptinstall & Quayle, 1970; Nunn & Lidstrom, 1986; Springer, Chou, Fan, Lee, & Lidstrom, 1995). Additional genes essential for methylotrophy were later pinpointed using chromosomal walking from identified gene clusters and were confirmed by constructing insertions in the identified genes (Chistoserdova, Chen, Lapidus, & Lidstrom, 2003). Though these early studies provided significant insights into methylotrophy, the mutagens used failed to facilitate the identification of key genes found outside of the known methylotrophic islands. Early efforts to use Tn5 transposon mutagenesis were successful, but only identified previously discovered methylotrophic genes. However, only ~4000 transposon mutants were screened for defects in methanol oxidation (Lee, Stone, Goodwin, & Holloway, 1991; Whitta, Sinclair, & Holloway, 1985). The addition of an IS*phoA*/hah-Tc mini-Tn5 transposon (D'Argenio, Gallagher, Berg, & Manoil, 2001) to the toolkit used to study methylotrophs via random mutagenesis has allowed further progress in the identification of methylotrophy genes. These studies discovered dihydromethanopterin reductase, the QscR transcriptional regulator of the serine cycle genes, and two genes involved in the ethylmalonyl-CoA pathway, among others (Kalyuzhnaya & Lidstrom, 2003; Korotkova, Lidstrom, & Chistoserdova, 2005; Marx et al., 2003). More recently, mutagenesis strategies using the same transposon were used to identify novel genes that contribute to Ln and methanol metabolism (Roszczenko-Jasińska et al., 2020).

In this chapter, we first describe the transposon mutagenesis study published by Roszczenko-Jasinka et al. (Roszczenko-Jasińska et al., 2020) that selected for transposon mutants defective in Ln-dependent methanol oxidation. We then introduce a transcriptional reporter fusion vector, pHV4, and show how this reporter can be used in combination with transposon mutagenesis to identify cellular processes required for *xoxF1* expression. Lastly, we present a pilot study that uses replica printing to screen for genes required for ExaF function. While this chapter focuses on the use of transposon mutagenesis in *M. extorquens* AM1 to identify processes required for Ln metabolism, these methods can be extended to diverse studies in other bacteria.

2. General scheme for IS*phoA*/hah-Tc mutagenesis

All experiments described in this chapter use the IS*phoA*/hah-Tc transposon located on pCM639 (Jacobs et al., 2003) (Fig. 2A). This suicide vector lacks an origin of replication recognized by *M. extorquens* AM1 but can be propagated in *Escherichia coli*. The Tn5 transposase, which facilitates the mobilization of the IS*phoA*/hah transposon, is encoded by *tnpA* located upstream of an ampicillin resistance cassette. The inability of pCM639 to

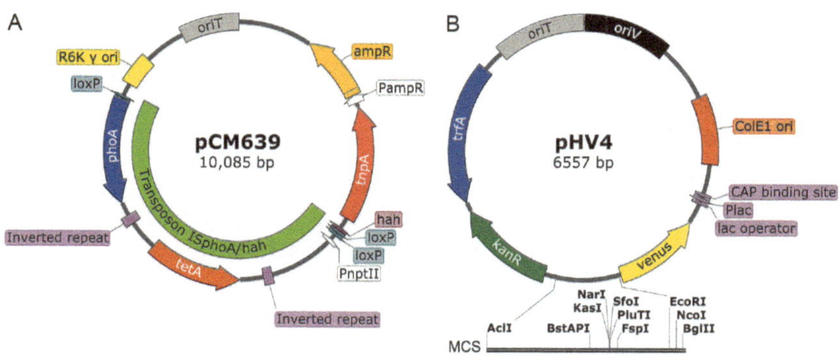

Fig. 2 Vector maps of (A) pCM639 and (B) pHV4. Abbreviations: oriT, origin of transfer; R6K γ ori, R6K γ origin of replication; *loxP*, Cre recombinase recognition site; *phoA*, alkaline phosphatase gene; *tetA*, tetracycline efflux pump gene; P*nptII*, neomycin resistance gene promoter; hah, fused human influenza hemagglutinin epitope tag and His_6-tag; *tnpA*, Tn5 transposase gene; P*ampR*, ampicillin resistance gene promoter; *ampR*, β-lactamase gene; oriV, IncP origin of replication; ColE1 ori, ColE1 origin of replication; *trfA*, oriV trans-acting activator gene; *kanR*, aminoglycoside phosphotransferase gene; *venus*, modified YFP gene; P*lac*, *lac* promoter; MCS, multiple cloning site.

replicate in *M. extorquens* AM1 and the presence of a *tetA* gene located within IS*phoA*/hah provide the ability to select for successful incorporation of the transposon into the recipient genome. The *loxP* sites flanking the transposon are recognized by Cre-Lox recombinase encoded on pCM158 and can be used to remove the insertion from the genome if desired (Jacobs et al., 2003; Manoil, 2000; Marx & Lidstrom, 2002).

We constructed a new transcriptional reporter fusion vector, pHV4, which can be used in combination with pCM639 for fluorescence-based expression screening (Fig. 2B). A kanamycin resistance cassette was cloned into the *Acl*I and *Afe*I sites of pAP5 (Skovran et al., 2011) in the opposite orientation of *venus*, which encodes a modified yellow fluorescence protein (YFP) (Nagai et al., 2002). This plasmid carries the origins of replication recognized by *M. extorquens* AM1 and *E. coli*, and thus can replicate in both species. Upstream of *venus*, promoter regions of interest can be cloned into a multiple cloning site. Plasmid pHV4 can be conjugated into *M. extorquens* AM1 via triparental mating (Section 4.3). When used in combination with pCM639, insertions of IS*phoA*/hah that disrupt genes encoding regulatory components of the cloned promoter will result in induction or repression of *venus* under appropriate conditions. The isolates carrying corresponding insertions will appear bright or dark when excited with green *epi* light. A general workflow for a transposon mutagenesis experiment is shown in Fig. 3 and detailed protocols are described in Sections 4 and 5.

3. Transposon mutagenesis studies used to identify gene products involved in Ln and methanol metabolism

3.1 Selection-based transposon mutagenesis: Identification of gene products required for XoxF-mediated methanol oxidation

In *M. extorquens* AM1, methanol is oxidized to formaldehyde by the Ca-dependent MxaFI-MeDH when Lns are absent from the growth medium (Anthony & Zatman, 1967), and by XoxF1 and XoxF2 when Lns are present (Good et al., 2019; Vu et al., 2016) (Fig. 1). If methanol is available and is oxidized to formaldehyde, a mutation in *fae* results in an intracellular accumulation of formaldehyde, which inhibits growth of *M. extorquens* AM1 (Vorholt, Marx, Lidstrom, & Thauer, 2000). Taking advantage of the methanol-sensitive phenotype of the *fae* mutant strain, a transposon mutant hunt was designed to select for insertions that prevent

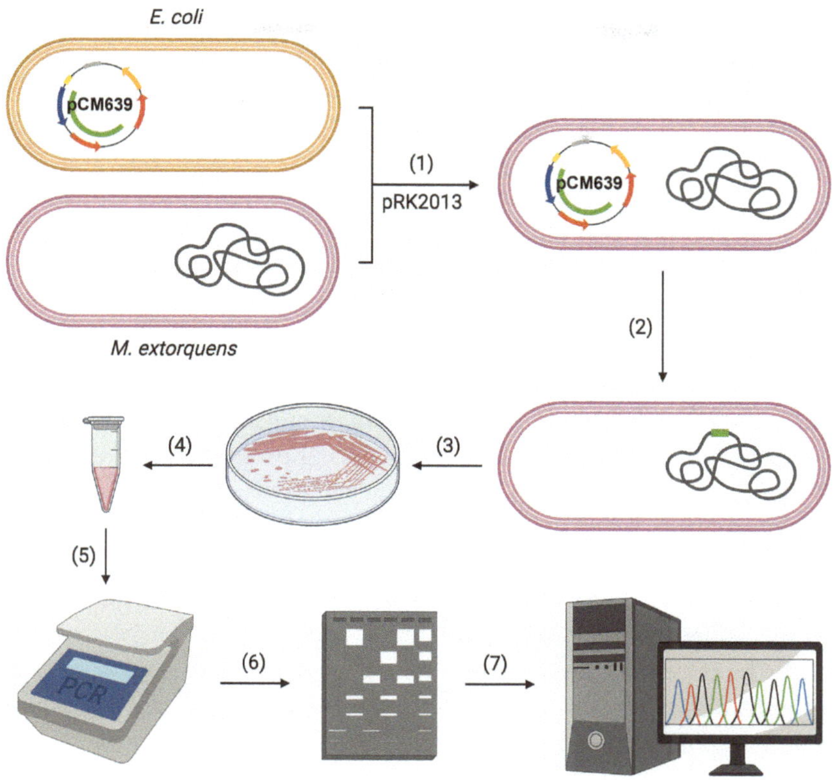

Fig. 3 Schematic workflow of IS*phoA*/hah-Tc transposon mutagenesis. (1) pCM639 is conjugated into *M. extorquens* via triparental mating. (2) Desired IS*phoA*/hah insertions are selected or screened under appropriate conditions. (3) Phenotypes of putative mutants are confirmed. (4) gDNA carrying IS*phoA*/hah insertions is purified. (5) Nested PCR using degenerate primers is used to amplify the DNA region adjacent to the transposon. (6) PCR products are visualized on 1% agarose gel. (7) After Sanger sequencing, the location of the transposon is determined using a sequence alignment tool. Image was created with BioRender.com.

La-dependent methanol oxidation and therefore formaldehyde accumulation (Roszczenko-Jasińska et al., 2020). An additional mutation in *mxaF* was used to ensure that methanol oxidation to formaldehyde was facilitated by XoxF1 and XoxF2. The IS*phoA*/hah-Tc transposon was introduced into the *mxaF fae* strain and growth was selected on minimal medium containing methanol and succinate. Strains that were disrupted in a gene required for methanol oxidation could grow using succinate as a carbon source. More than 600 transposon mutants were isolated and the insertion locations mapped to the genome. Notable insertions that prevented or reduced methanol oxidation in the *mxaF fae* mutant strain mapped to the *lut* (lanthanide utilization and transport) gene cluster, genes involved in PQQ and heme biosynthesis, and genes of unknown function. Further characterization of these mutant strains is discussed in detail by Roszczenko-Jasińska et al. (Roszczenko-Jasińska et al., 2020).

3.2 Screen-based transposon mutagenesis using a transcriptional reporter fusion: Mutant hunt design to identify proteins and processes that affect *xoxF1* expression

This section describes a fluorescence-based transposon mutagenesis screen designed to identify potential gene products affecting the transcriptional regulation of *xoxF1*. Previous experiments showed that Ln availability influences which type of MeDH is produced (XoxF or MxaFI) in different methylotrophs (Chu et al., 2016; Gu et al., 2016; Masuda et al., 2018; Vu et al., 2016). When La is provided as a Ln source, the *xox1* operon is upregulated in *M. extorquens* AM1; when La is not available, expression of the *xox1* operon is repressed. To take advantage of this Ln-switch in the mutant hunt design, the *xox1* promoter was cloned into pHV4 (Fig. 2B) and regulation of expression by La was validated in the wild-type strain background (Fig. 4C).

Transposon mutants can be selected on MP methanol succinate La Tc^{10} Km^{50} Rif^{50} medium then patched onto fresh media to screen for changes in fluorescence (Fig. 4A). Screening for fluorescence in individual colonies instead of patches has proven unreliable and the phenotypes are less clear. While use of this reporter in the wild-type strain background would be unbiased in the types of processes that can be identified (Fig. 4B), ~60,000 colonies would need to be screened to ensure the mutant hunt is saturated. Use of the *mxaF fae* strain background and the approach described in Section 3.1 allows for an enrichment of strains with altered *xox1* expression. However, this mutant hunt is biased to only identify genes required for XoxF function.

In our pilot study, MP methanol succinate La Tc^{10} Km^{50} Rif^{50} medium was used to select for *mxaF fae* transposon mutants that were defective in methanol oxidation. These colonies were patched onto MP methanol succinate La media and allowed to grow for 2 days before screening for changes in fluorescence. Both increases and decreases in *xox1* expression were detected using this method (Fig. 4B and D). Sequences from two transposon insertions were obtained. The dark patch shown in Fig. 4B contains an

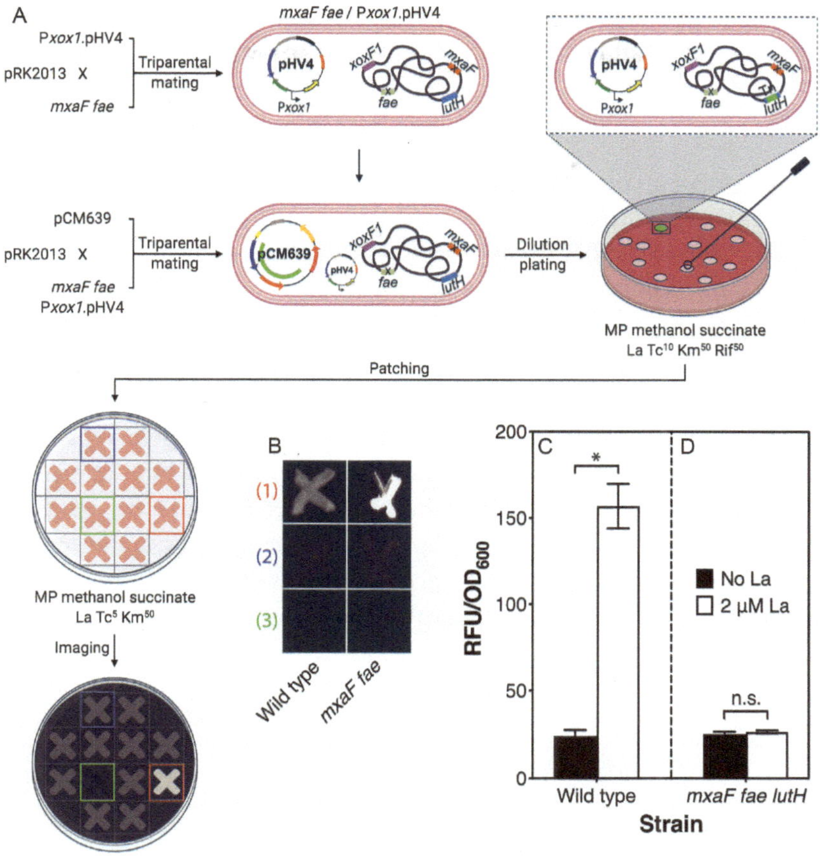

Fig. 4 Fluorescence screening for changes in *xox1* expression in the *mxaF fae* background. (A) Schematic overview of the experiment. (B) When La is available, expression from the *xox1* promoter occurs and *venus* is induced resulting in a fluorescent patch (2). Insertion of a transposon in the *lutH* La transporter gene results in repression of *xox1* regardless of La availability (3). Disruption of a gene that prevents *xox1* expression, such as *xoxF1* or *mxbM*, results in increased fluorescence (1). Expression from the *xox1* promoter was measured in (C) wild type and (D) *mxaF fae lutH* strains grown in methanol succinate media in the absence and presence of La. An asterisk indicates $P < 0.0001$ as determined by two-tailed unpaired Student's *t*-test. Abbreviation: P*xox1*, *xox1* operon promoter, RFU/OD_{600}, relative fluorescence units normalized to optical density at 600nm; n.s., not significant ($P > 0.05$, $n = 3$). Image was created with BioRender.com.

insertion in the *lutH* TonB-dependent La transporter gene, while the bright patch contains an insertion in the *xoxF1* gene. We have previously shown that mutations in *xoxF1* result in increased expression of the *xox1* operon (Vu et al., 2016), and mutations in *lutH* result in decreased expression of the *xox1* operon (Roszczenko-Jasińska et al., 2020). This pilot study can be scaled-up to determine if additional requirements for *xox1* expression can be identified.

3.3 Screen-based transposon mutagenesis using replica printing: Mutant hunt design to identify proteins and processes required for ExaF function

M. extorquens AM1 is capable of oxidizing both methanol and methylamine to formaldehyde (Fig. 1). In the absence of *mxaF, xoxF1,* and *xoxF2*, ExaF can facilitate methanol growth but at a reduced rate (Good et al., 2016; Vu et al., 2016). As ExaF does not participate in methylamine utilization, a *mxaF xoxF1 xoxF2* deletion strain that obtains a transposon in a gene required for ExaF function will grow on MP methylamine La medium but not methanol La medium (Fig. 5). This mutant hunt is expected to identify genes that encode proteins required for Ln uptake and homeostasis, genes essential for all Ln-dependent ADHs such as those involved in PQQ biosynthesis, and genes specific for ExaF function including *exaF* itself.

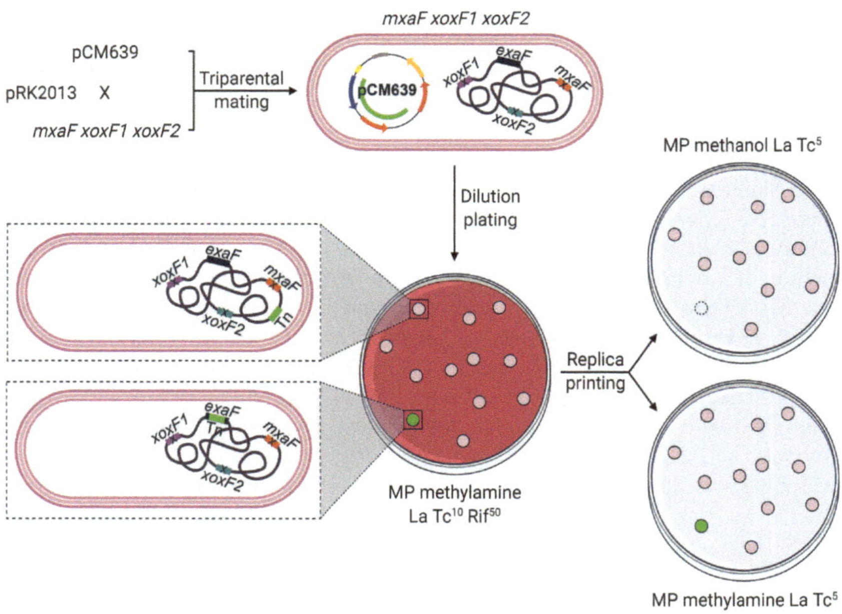

Fig. 5 Diagram of replica-printing screen to identify gene products required for ExaF function. A colony that can grow on methylamine La medium but not methanol La medium is highlighted. Image was created with BioRender.com.

In a pilot mutant hunt, approximately 5000 colonies carrying independent insertions of IS*phoA*/hah-Tc were screened for the ability to grow on MP La-supplemented methylamine medium with the concurrent inability to grow on MP methanol La medium (Fig. 5). A total of 264 putative mutants showed varying degrees of growth reduction on methanol. After phenotypic confirmation, 11 isolates lacking the ability to utilize methanol in the presence of La were further pursued and sequenced to identify the insertion locations (Table 1). Insertions in four genes (*MexAM1_META1p1292*, *MexAM1_META1p1750*, *MexAM1_META1p1785*, and *MexAM1_META1p3908*) were also identified in the selection-based experiment (see Section 3.1) and published elsewhere (Roszczenko-Jasińska et al., 2020). This protocol serves as a proof of concept for future replica printing screen-based transposon

Table 1 Transposon insertion locations and putative function of gene products identified in the *mxaF xoxF1 xoxF2* strain background.

Strain[a]	Growth[b]	Transposon location	Putative function
Wild type	++		
mxaF xoxF1 xoxF2	+		
Tn0	−	*MexAM1_META1p1785*	LutH, TonB-dependent receptor
Tn2 and Tn5	−	*MexAM1_META1p3908*	Aminopeptidase
Tn7	+/−	*MexAM1_META1p2478*	XerD, tyrosine recombinase
Tn13	−	*MexAM1_META1p1292*	CycL, cytochrome biogenesis
Tn146	+/−	*MexAM1_META1p2906*	Putative glycosyl hydrolase
Tn168	−	*MexAM1_META1p1799*	Conserved hypothetical protein
Tn170	−	*MexAM1_META1p1135*	Conserved hypothetical protein
Tn260	−	*MexAM1_META1p4613*	Two-component response regulator
Tn261	−	*MexAM1_META1p1800*	Putative sensor kinase
Tn262	−	*MexAM1_META1p1750*	PqqB, PQQ biosynthesis

[a]Tn transposon strains contain a transposon insertion in the *mxaF xoxF1 xoxF2* strain background.
[b]Growth was assessed on MP methanol La plates. ++, wild-type growth; +, reduced growth with colony size ~1/2 of wild type; +/−, poor growth with colony size ~¼ of wild type; −, no growth.

mutagenesis studies. Novel genes identified from this approach include genes encoding transcriptional regulators, hypothetical proteins, a tyrosine recombinase and a glycosyl hydrolase (Table 1). It is likely that some insertions identified here are not responsible for the loss of ExaF activity, but the inability to grow on MP methanol La medium is rather due to a spontaneous mutation. Strain reconstruction followed by phenotypic testing is needed to confirm growth requirements. Additional gene products required for ExaF function are expected to be identified upon scale-up of this mutant hunt since insertions in *exaF* were not found in the pilot study.

4. Materials and methods

4.1 Media

Table 2 lists media used in the transposon mutagenesis studies.

4.2 Equipment and materials

Table 3 lists the media, materials, and equipment required for each transposon mutagenesis experiment described in this chapter. "General" refers to components required for all transposon mutagenesis experiments.

Table 2 Common media used in IS*phoA*/hah-Tc transposon mutagenesis.

Media	Abbreviation	Component	Final concentration
Lysogeny broth	LB	Tryptone	10 g/L
		Yeast extract	5 g/L
		Sodium chloride	5 g/L
Nutrient broth	NB	Nutrient broth mix	8 g/L
Methylobacterium PIPES	MP	(Delaney et al., 2013)	(Delaney et al., 2013)
Carbon source[a]		Methanol	125 mM
		Methylamine	35 mM
		Succinate	15 mM
Cofactor[a]	Ca	Calcium	20 μM
	La	Lanthanum	2 μM
Antibiotics	Tc^{10}	Tetracycline	10 μg/mL
	Tc^{5}	Tetracycline	5 μg/mL
	Km^{50}	Kanamycin	50 μg/mL
	Rif^{50}	Rifamycin	50 μg/mL
Additive		Agar	1.5% wt/vol

[a]Addition to MP media.

Table 3 Media, materials and equipment.

Experiment	**Media/Materials/Equipment**	
General	Media	Nutrient agar (NA) plates Lysogeny agar (LA) Tc^{10} plates LA Km^{50} plates MP broth without carbon source/trace metals (MP blank) MP succinate broth MP succinate plates
	Materials/ Equipment	70%–95% (vol/vol) ethanol Micropipettors Sterile micropippette tips Sterile 1.5 mL microfuge tubes Sterile applicator sticks/loops Glass/metal cell spreader Standing incubator Vortexer Sharpie
Selection	Media	MP methanol + succinate La Tc^{10} Rif^{50} plates MP methanol + succinate La Tc^{5} plates
Fluorescence screen	Media	MP succinate Km^{50} Rif^{50} plates MP succinate Km^{50} plates MP methanol + succinate La Tc^{10} Km^{50} Rif^{50} plates MP methanol + succinate La Tc^{5} Km^{50} plates MP methanol + succinate Tc^{5} Km^{50} plates
	Equipment	Fluorescence imager
Screen	Media	MP methylamine La Tc^{10} Rif^{50} plates MP methylamine La Tc^{5} plates MP methanol La Tc^{5} plates
	Materials/ Equipment	Replica-printing tools (cylindrical plastic or metal block and metal ring) Sterile 16 cm × 16 cm velvetine cloths
DNA purification and nested PCR	Materials/ Equipment	Sterile deionized water 1 × Tris-borate-EDTA (TBE) buffer (Cold Spring Harbor Protocols, 2006) 100 mM dNTPs 100% dimethyl sulfoxide (DMSO) 10 × DreamTaq buffer DreamTaq DNA polymerase

Continued

Table 3 Media, materials and equipment.—cont'd

	100 μM primers (see Table 4)
	1% agarose gels
	Sterile culture tubes
	0.2 mL thin-wall PCR tubes
	Qiagen's DNeasy UltraClean Microbial kit
	Heat block
	Vortex-Genie 2 vortexer
	Vortex adapter for 24 (1.5–2 mL) tubes
	PCR purification kit
	Thermal cycler
	Gel electrophoresis apparatus
	Gel imaging system or transilluminator
	Spectrophotometer for DNA quantification
	Shaking incubator

4.3 Procedures

4.3.1 General steps for all mutant hunts

Day 1: Streak *M. extorquens* AM1 strains for isolated colonies.

1. Using a sterile applicator stick (or loop), streak *M. extorquens* AM1 recipient strain for isolated colonies onto an MP succinate plate from a cryostock. Incubate at 29 °C for 4 days. If the recipient strain contains pHV4, streak onto an MP succinate Km^{50} plate.

Day 5: Preparation of strains for triparental mating

2. Streak the recipient strain from step 1 for biomass growth (thick growth over the entire plate, no isolated colonies) onto an MP succinate plate (with Km^{50} if strain with pHV4 is used). Approximately 5 colonies are picked using a sterile applicator stick and streaked in a single vertical line over half an MP succinate plate. The same stick is used to create a lawn of growth by streaking horizontally back and forth over the cells deposited in the vertical line. This procedure is repeated on the second half of the plate. Incubate the plate at 29 °C for 2 days. One biomass plate is sufficient for two matings.
3. Streak the *E. coli* strain carrying pCM639 for isolated colonies onto an LA Tc^{10} plate from a cryostock. Incubate at 37 °C for 1 day or at 29 °C for 2 days.
4. Streak the *E. coli* strain carrying pRK2013 for isolated colonies onto an LA Km^{50} plate from a cryostock. Incubate at 37 °C for 1 day or at 29 °C for 2 days.

Day 7: Triparental mating

5. Using a sterile applicator stick (or loop), gently remove the biomass of the recipient strain from one half of the MP succinate plate and transfer it to the center of an NA plate by rolling the stick onto the agar surface until no visible cells remain.
6. Using a sterile applicator stick, take a visible mass of *E. coli* pCM639 (~3–5 colonies) from the LA Tc^{10} plate and roll the stick into the *M. extorquens* biomass on the NA plate. Spread the cells around in a zig-zag pattern to create a circle of approximately 2 cm in diameter.
7. Using a sterile applicator stick, repeat step 6 with *E. coli* pRK2013 cells from the LA Km^{50} plate.
8. Incubate the NA mating plate at 29 °C for 1–2 days.

   Note: If the plate is mistakenly incubated at 37 °C for several hours and then placed at 29 °C, the mutant hunt will still be successful. If the mating plate is incubated at 37 °C for the full 1–2 days, *M. extorquens* AM1 will often survive. However, native insertion elements encoded in the *M. extorquens* AM1 chromosome will actively insert into other regions of the DNA and complicate identification of the mutation responsible for the selected or screened phenotype.

Day 8: Plating of transposon mutants

9. Pipette 1 mL MP blank onto the NA mating plate.
10. Sterilize a glass cell spreader by submerging it in 70%–95% ethanol and passing it through a flame several times to ensure all ethanol has evaporated.
11. Cool the sterilized spreader by touching the agar surface around the edge of the NA mating plate.
12. Using the sterilized spreader, dislodge cells from the NA mating plate.
13. Tilt the NA mating plate and use the spreader to push cells toward the bottom edge of the plate.
14. Gently pipette the dislodged cells up and down to break up any visible cell clumps before transferring the cell culture into a sterile 1.5 mL microfuge tube. Gently vortex to mix.
15. For a selection-based mutant hunt, serially dilute the mating mixture 1:10 and 1:100 in MP blank contained in sterile 1.5 mL microfuge tubes to yield a final volume of 1 mL. For a screen-based mutant hunt, serially dilute the mating mixture 1:50 and 1:500 using MP blank to yield a final volume of 1 mL.
16. If selecting for a phenotype in addition to a transposon insertion (e.g., methanol resistance), use a sterilized cell spreader to spread 100 μL of

the 1:100 mating mixture onto an MP La Tc^{10} Rif^{50} plate containing the appropriate carbon source(s). Repeat for the 1:10 and undiluted mating mixtures. When scaling up, spread 100 µL of each dilution on multiple plates.

Note: Rif^{50} is included in the medium to counter-select against *E. coli* strains and Tc^{10} is added to select for transposon insertion mutants.

17. As a control, plate 100 µL of each dilution onto an MP succinate Tc^{10} Rif^{50} plate, which will depict the frequency of transposition without further selection.

18. For a screen-based mutant hunt where all cells containing a transposon will grow on the initial Tc^{10} Rif^{50} plate (or Tc^{10} Km^{50} Rif^{50} if the recipient contains pHV4), 50–100 µL of the 1:50 and 1:500 dilutions are plated instead to obtain better separation of colonies.

Note: A small-scale pilot experiment using the same conditions should be performed to determine the appropriate dilution factor and amount of cell culture to be plated. Approximately 100–200 colonies per plate are desired.

19. Incubate the spread plates for 5–7 days at 29 °C. Very small colonies will appear after >7 days, but these mutants typically do not contain a transposon insertion and can be discarded.

4.3.2 Steps specific for each mutant hunt

4.3.2.1 Selection-based mutant hunt

1. To confirm the selected phenotype, pick each colony from the MP Tc^{10} Rif^{50} plate and streak for isolation onto the same media, except reduce the concentration of Tc from 10 to 5 µg/mL and omit Rif.

4.3.2.2 Screen-based mutant hunt using pHV4

1. Using a sharpie, draw a grid of approximately 1 cm × 1 cm squares on the bottom portion of the MP Tc^{5} Km^{50} agar plate(s) to be patched and imaged.

Note: Number the top right corner of each square in the grid for easy identification of a patch of interest during the imaging step.

2. To make a patch, use a sterile applicator stick and pick an individual isolated colony from the appropriate dilution plate and draw an "X" in the first square of the plate to be imaged. Use the same stick with cells to patch on another plate if comparison between different conditions is needed (e.g., no La versus La condition).

3. Repeat step 2 until each square in the grid is patched with an individual colony.

   Note: Multiple plates will be needed depending on the number of well-separated colonies obtained from the dilution plates.
4. Incubate the patched plates at 29 °C for 1–2 days.
5. Remove the lid from the incubated plate and follow the user manual instructions for fluorescence imaging as imaging methods differ between instruments. For the studies reported herein, green *epi* light was used to detect Venus fluorescence.

   Note: Image the patch plates once per day the first time the experiment is performed to determine the optimal time for imaging. 2 days was optimal when screening for *xox1* expression changes. To set up an optimized imaging method specific for the fluorescent reporter protein being used, it is necessary to troubleshoot the exposure time, brightness level, fluorescence filter type, and distance of the imaging tray from the camera.
6. Examine the image and identify patches containing expression levels of interest.
7. Use a sterile applicator stick to swab the edge of a patch of interest and streak for isolation onto an MP succinate Tc^{5} Km^{50} plate.
8. Incubate the plate at 29 °C for 4 days.
9. Generate a cryostock of each strain exhibiting the phenotype of interest for downstream analysis. For *M. extorquens* AM1, grow a 2 mL culture to late exponential phase. Add 100 μL of DMSO to a sterile cryovial. Add 900 μL of culture and mix by inverting several times. Freeze at −80 °C.

4.3.2.3 Screen-based mutant hunt using replica printing

1. For easy alignment, use a sharpie to mark a 12 o'clock position on the metal ring, the MP Tc^{10} Rif^{50} dilution plate, and the two MP Tc^{5} plates containing alternative carbon source to be printed on.

   Note: Prior to replica-printing, plates should be inspected to ensure there is no condensation on the agar surface, which could potentially smear the printed cells and interfere with screening procedure. It is important that the plates to be printed are freshly pulled from a 29 °C incubator. If plates are left out at room temperature or placed at 4 °C, the colonies become sticky and are difficult to transfer via replica plating. *M. extorquens* AM1 prints well to two plates; a test condition plate (methanol La Tc in our pilot study) and a control plate where all colonies should grow (methylamine La Tc in our pilot study). Replica printing to more than two plates is not always successful when working with *M. extorquens* AM1.

2. Briefly wipe the replica-printing tools with 70%–95% ethanol.
3. Attach a sterile velveteen cloth (do not use velvet) onto the cylindrical replica printing block such that the sterile velveteen side is facing up. Use the metal or plastic ring to secure the cloth.

   Note: Velveteen is used for the short length of the cloth fibers. Use of velvet will result in smeared colony prints. Replica printing cloths can be purchased or cut (16 cm × 16 cm) from fabric purchased at a fabric store. Clean cloths can be stacked with the printing side facing down, wrapped in bulk with aluminum foil, and autoclaved. Contaminated cloths can be autoclaved, washed with water (no detergent), dried, and re-autoclaved for future use.
4. Ensuring the marks on the MP Tc^{10} Rif^{50} dilution plate and metal ring are aligned, take off the lid and gently press the plate onto the velveteen fabric; press evenly around all areas of the plate. To avoid having the agar detach from the petri dish, tilt the replica printing block with the plate attached, and gently peel the plate off of the velveteen beginning at the top. Set the MP Tc^{10} Rif^{50} dilution plate aside.
5. To transfer the cells onto the screening plates, repeat step 4. Print colonies onto the MP Tc^{5} control plate last to assess how well the colonies transferred.
6. Incubate the screening plates at 29 °C for 3–4 days.
7. Aligning the marks on the two screening plates, look for colonies that can grow on the control plate but not on the experimental test plate.
8. To confirm the phenotype, use sterile applicator sticks to patch the colonies of interest onto fresh media (the experimental test media and control media). Incubate at 29 °C for 2 days.
9. Streak isolates that show the expected phenotype onto MP succinate Tc^{5} plates for downstream analysis.

4.3.3 Determination of transposon insertion sites

This section describes an optimized degenerate nested PCR protocol (Manoil, 2000; Marx et al., 2003; Roszczenko-Jasińska et al., 2020) used to identify the transposon insertion sites. Briefly, two pairs of primers (Table 4) are used in two successive PCR steps to amplify the genomic DNA (gDNA) flanking the transposon insertion site. In the first step, the Tn*phoA*-II primer binds specifically to the transposon and the CEKG 2B primer binds randomly to multiple locations in the genome (Manoil, 2000) (Fig. 6). The product of the first PCR reaction is used as template

Table 4 PCR primer sequences.

Step	Primer #	Primer	Sequence	References
PCR 1	1	Tn*phoA*-II	GTGCAGTAATATCGCCCTGAGCA	Manoil (2000)
	2	CEKG 2B	GGCCACGCGTCGACTAGTAC$(N)_{10}$ACGCC	Manoil (2000)
PCR 2	3	hah-1	ATCCCCCTGGATGGAAAACGG	Manoil (2000)
	4	CEKG 4	GGCCACGCGTCGACTAGTAC	Manoil (2000)
Sequencing	5	CM-Tn seq	AAACGGGAAAGGTTCCGTCCA	Marx et al. (2003)

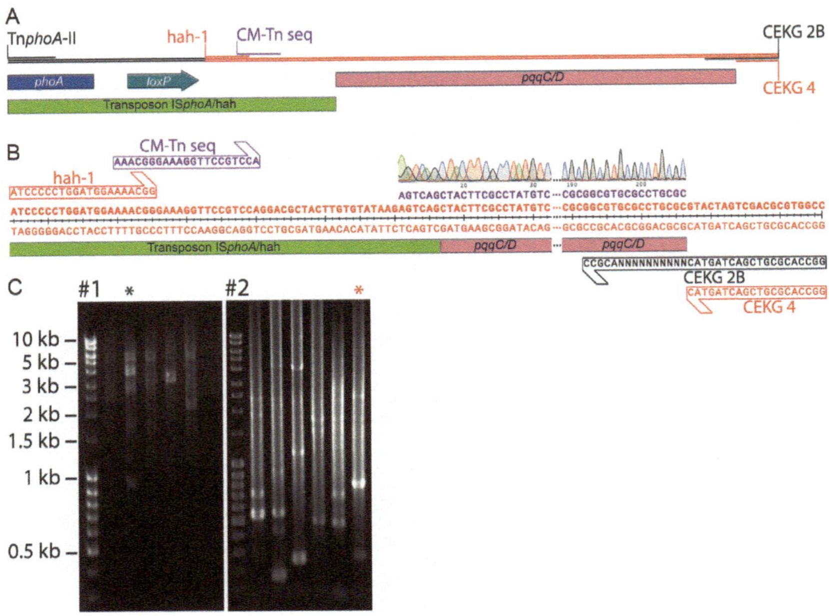

Fig. 6 Nested PCR using degenerate primers to identify the location of the transposon insertion site. Example of an insertion in *pqqC/D* is shown. (A) Binding sites of primers to the transposon and adjacent DNA region. (B) Alignment of a Sanger sequencing result to the *M. extorquens* AM1 genome. (C) Visualization of PCR products on a 1% agarose gel. Black, red, and purple highlight represents primers or DNA sequence from PCR#1, PCR#2, and sequencing, respectively. Asterisks indicate the amplification products obtained from the *pqqC/D* mutant shown in (A) and (B).

DNA for the second PCR reaction. Here, the hah-1 primer binds to a transposon region amplified during the first PCR step and the CEKG 4 primer binds to a sequence incorporated by the CEKG 2B primer. The PCR products are purified, sequenced, and analyzed using DNA alignment tools.

4.3.3.1 gDNA purification

1. Inoculate one colony from each transposon mutant isolate into 2 mL MP succinate (Delaney et al., 2013). Incubate at 29 °C in a shaking incubator for 1–2 days
2. Isolate genomic DNA using a Qiagen DNeasy UltraClean Microbial kit according to manufacturer instructions

   Note: Include the optional 65 °C 10-min heating step for better lysis of *M. extorquens* AM1.
3. Quantify the purified gDNA, which will be the template DNA for the first PCR step

Note: A minimum of 10ng gDNA is needed in the first round of PCR for the product to be detectable (Chun, Edenberg, Kelley, & Goebl, 1997). Instead of isolating gDNA, a bacterial colony can be used as a DNA template source, but the result may not be as robust compared to using purified gDNA.

4.3.3.2 First round of PCR

1. Set up the first PCR reaction according to Table 5
   Note: Use of DMSO instead of 0.1 mg/mL BSA (Marx et al., 2003) in the first and second PCR reaction mixes gives more consistent results.
2. Run samples in a thermal cycler using the settings in Table 6
3. Make a 1% (wt/vol) agarose gel using 1 × TBE buffer and run 5 μL of each PCR reaction to visualize
4. Image the agarose gel after staining with ethidium bromide or other DNA stain
   Note: A smear or faint bands are commonly observed (Fig. 6C).

4.3.3.3 Second round of PCR

1. Use the product from the first PCR reaction as the template DNA for the second PCR reaction and set up the reaction according to Table 7
2. Run samples in a thermal cycler using the settings in Table 8
3. Run 5 μL of each reaction on a 1% agarose gel
4. Image the agarose gel after staining
   Note: A smear is still expected but one or more distinct bands should now be visible (Fig. 6C).

Table 5 First PCR reaction mix.

Reagent	Stock concentration	Final concentration	Volume per reaction
Sterile deionized water			17.5 μL
DreamTaq buffer	10 ×	1 ×	2.5 μL
dNTPs	100 mM	1 mM	0.25 μL
DMSO	100%	6%	1.5 μL
Tn*phoA*-II primer	100 μM	1 μM	0.25 μL
CEKG 2B primer	100 μM	1 μM	0.25 μL
DreamTaq DNA polymerase	5 U/μL	0.05 U/μL	0.25 μL
Purified gDNA template	Varied	Varied	2.5 μL
Total reaction volume			25 μL

Table 6 First PCR thermal cycler settings.

Step	Temperature	Duration	Number of cycles
1	95 °C	2 min	
2	95 °C	30 s	Repeat step 2–4 for 6 cycles
3	40 °C	30 s	
4	72 °C	3 min	
5	94 °C	30 s	Repeat step 5–7 for 25 cycles
6	65 °C	30 s	
7	72 °C	3 min	
8	72 °C	6 min	
9	4 °C	Hold	

Table 7 Second PCR reaction mix.

Reagent	Stock concentration	Final concentration	Volume per reaction
Sterile deionized water			39.5 μL
DreamTaq buffer	10 ×	1 ×	5 μL
dNTPs	100 mM	1 mM	0.5 μL
DMSO	100%	5%	2.5 μL
hah-1 primer	100 μM	1 μM	0.5 μL
CEKG 4 primer	100 μM	1 μM	0.5 μL
DreamTaq DNA polymerase	5 U/μL	0.05 U/μL	0.5 μL
DNA template (PCR 1 product)	Varied	Varied	1 μL
Total reaction volume			50 μL

Table 8 Second PCR thermal cycler settings.

Step	Temperature	Duration	Number of cycles
1	95 °C	2 min	
2	94 °C	30 s	Repeat step 2–4 for 30 cycles
3	65 °C	30 s	
4	72 °C	3 min	
5	72 °C	6 min	
6	4 °C	Hold	

4.3.3.4 Sequencing

1. Use a PCR purification kit to clean the samples from the second PCR reaction for sequencing
2. Use a spectrophotometer to quantify the sample concentrations
3. Prepare the samples for sequencing according to the sequencing facility specifications using primer CM-Tn seq as the sequencing primer
4. Analyze sequencing data using NCBI Blast or other DNA alignment tools

   Note: Expect that ~1%–2% of transposon mutants will have pCM639 integrated into the genome rather than an insertion due to a transposition event.

5. Concluding remarks

Transposon mutagenesis is a powerful genetic tool that can facilitate the identification of gene products required for a process of interest. Experimental cost is low such that this technique can be applied in a wide variety of research institutions and be incorporated into student research experiments in the classroom (Roszczenko-Jasińska et al., 2020). This chapter describes a published study that identified novel genes required for Ln-dependent methanol oxidation along with two additional unpublished pilot studies that have the potential to uncover additional gene products involved in Ln metabolism when scaled up.

Acknowledgments

This work was funded by a California State University Program for Education and Research in Biotechnology grant. Special thanks to Timothy Andriese who helped modify the PCR reaction conditions to obtain more robust PCR amplification and sequencing results, and to Richard Ngo for developing software to rapidly map transposon insertion sequences to the *M. extorquens* AM1 genome.

References

Anthony, C. (1982). *The biochemistry of Methylotrophs. Vol. 439*. London: Academic press.

Anthony, C., & Zatman, L. (1967). The microbial oxidation of methanol. The prosthetic group of the alcohol dehydrogenase of *Pseudomonas* sp. M27: A new oxidoreductase prosthetic group. *The Biochemical Journal*, *104*(3), 960–969.

Chistoserdov, A. Y., Chistoserdova, L. V., McIntire, W. S., & Lidstrom, M. E. (1994). Genetic organization of the mau gene cluster in *Methylobacterium extorquens* AM1: Complete nucleotide sequence and generation and characteristics of mau mutants. *Journal of Bacteriology*, *176*(13), 4052–4065.

Chistoserdova, L., Chen, S.-W., Lapidus, A., & Lidstrom, M. E. (2003). Methylotrophy in *Methylobacterium extorquens* AM1 from a genomic point of view. *Journal of Bacteriology, 185*(10), 2980–2987.

Chistoserdova, L., & Kalyuzhnaya, M. G. (2018). Current trends in methylotrophy. *Trends in Microbiology, 26*(8), 703–714.

Chu, F., Beck, D. A., & Lidstrom, M. E. (2016). MxaY regulates the lanthanide-mediated methanol dehydrogenase switch in *Methylomicrobium buryatense*. *PeerJ, 4*, e2435.

Chun, K. T., Edenberg, H. J., Kelley, M. R., & Goebl, M. G. (1997). Rapid amplification of uncharacterized transposon-tagged DNA sequences from genomic DNA. *Yeast, 13*(3), 233–240.

Cold Spring Harbor Protocols. (2006). TBE buffer. *Cold Spring Harbor Protocols, 2006*(1). pdb.rec8458.

Crowther, G. J., Kosály, G., & Lidstrom, M. E. (2008). Formate as the main branch point for methylotrophic metabolism in *Methylobacterium extorquens* AM1. *Journal of Bacteriology, 190*(14), 5057–5062.

D'Argenio, D. A., Gallagher, L. A., Berg, C. A., & Manoil, C. (2001). *Drosophila* as a model host for *Pseudomonas aeruginosa* infection. *Journal of Bacteriology, 183*(4), 1466–1471.

Delaney, N. F., Kaczmarek, M. E., Ward, L. M., Swanson, P. K., Lee, M.-C., & Marx, C. J. (2013). Development of an optimized medium, strain and high-throughput culturing methods for *Methylobacterium extorquens*. *PLoS One, 8*(4), e62957.

Fitriyanto, N. A., Fushimi, M., Matsunaga, M., Pertiwiningrum, A., Iwama, T., & Kawai, K. (2011). Molecular structure and gene analysis of Ce^{3+}-induced methanol dehydrogenase of *Bradyrhizobium* sp. MAFF211645. *Journal of Bioscience and Bioengineering, 111*(6), 613–617.

Good, N. M., Moore, R. S., Suriano, C. J., & Martinez-Gomez, N. C. (2019). Contrasting *in vitro* and *in vivo* methanol oxidation activities of lanthanide-dependent alcohol dehydrogenases XoxF1 and ExaF from *Methylobacterium extorquens* AM1. *Scientific Reports, 9*(1), 1–12.

Good, N. M., Vu, H. N., Suriano, C. J., Subuyuj, G. A., Skovran, E., & Martinez-Gomez, N. C. (2016). Pyrroloquinoline quinone ethanol dehydrogenase in *Methylobacterium extorquens* AM1 extends lanthanide-dependent metabolism to multicarbon substrates. *Journal of Bacteriology, 198*(22), 3109–3118.

Gu, W., Farhan Ul Haque, M., DiSpirito, A. A., & Semrau, J. D. (2016). Uptake and effect of rare earth elements on gene expression in *Methylosinus trichosporium* OB3b. *FEMS Microbiology Letters, 363*(13).

Heptinstall, J., & Quayle, J. (1970). Pathways leading to and from serine during growth of *Pseudomonas* AM1 in C_1 compounds or succinate. *The Biochemical Journal, 117*(3), 563–572.

Hibi, Y., Asai, K., Arafuka, H., Hamajima, M., Iwama, T., & Kawai, K. (2011). Molecular structure of La^{3+}-induced methanol dehydrogenase-like protein in *Methylobacterium radiotolerans*. *Journal of Bioscience and Bioengineering, 111*(5), 547–549.

Jacobs, M. A., Alwood, A., Thaipisuttikul, I., Spencer, D., Haugen, E., Ernst, S., et al. (2003). Comprehensive transposon mutant library of *Pseudomonas aeruginosa*. *Proceedings of the National Academy of Sciences, 100*(24), 14339–14344.

Kalyuzhnaya, M. G., & Lidstrom, M. E. (2003). QscR, a LysR-type transcriptional regulator and CbbR homolog, is involved in regulation of the serine cycle genes in *Methylobacterium extorquens* AM1. *Journal of Bacteriology, 185*(4), 1229–1235.

Kato, S., Takashino, M., Igarashi, K., & Kitagawa, W. (2020). Isolation and genomic characterization of a proteobacterial methanotroph requiring lanthanides. *Microbes and Environments, 35*(1), ME19128.

Korotkova, N., Lidstrom, M. E., & Chistoserdova, L. (2005). Identification of genes involved in the glyoxylate regeneration cycle in *Methylobacterium extorquens* AM1, including two new genes, meaC and meaD. *Journal of Bacteriology, 187*(4), 1523–1526.

Large, P., Peel, D., & Quayle, J. (1961). Synthesis of cell constituents by methanol-and formate-grown *Pseudomonas* AM1, and methanol-grown *Hyphomicrobium vulgare*. *The Biochemical Journal*, *81*, 470–480.

Large, P., Peel, D., & Quayle, J. (1962a). Microbial growth on C_1 compounds. 3. Distribution of radioactivity in metabolites of methanol-grown *Pseudomonas* AM1 after incubation with [^{14}C] methanol and [^{14}C] bicarbonate. *The Biochemical Journal*, *82*(3), 483–488.

Large, P., Peel, D., & Quayle, J. (1962b). Microbial growth on C_1 compounds. 4. Carboxylation of phosphoenolpyruvate in methanol-grown *Pseudomonas* AM1. *The Biochemical Journal*, *85*(1), 243–250.

Large, P., & Quayle, J. (1963). Microbial growth on C_1 compounds. 5. Enzyme activities in extracts of *Pseudomonas* AM1. *The Biochemical Journal*, *87*(2), 386–396.

Lee, K. E., Stone, S., Goodwin, P., & Holloway, B. (1991). Characterization of transposon insertion mutants of *Methylobacterium extorquens* AM1 (*Methylobacterium* strain AM1) which are defective in methanol oxidation. *Microbiology*, *137*(4), 895–904.

Lumpe, H., Pol, A., den Camp, H. J. O., & Daumann, L. J. (2018). Impact of the lanthanide contraction on the activity of a lanthanide-dependent methanol dehydrogenase–a kinetic and DFT study. *Dalton Transactions*, *47*(31), 10463–10472.

Lv, H., Sahin, N., & Tani, A. (2018). Isolation and genomic characterization of *Novimethylophilus kurashikiensis* gen. Nov. sp. nov., a new lanthanide-dependent methylotrophic species of *Methylophilaceae*. *Environmental Microbiology*, *20*(3), 1204–1223.

Lv, H., Sahin, N., & Tani, A. (2020). *Methylotenera oryzisoli* sp. nov., a lanthanide-dependent methylotrophic bacteria isolated from rice field soil. *International Journal of Systematic and Evolutionary Microbiology*, *70*(4), 2713–2718.

Manoil, C. (2000). [3] tagging exported proteins using *Escherichia coli* alkaline phosphatase gene fusions. *Methods in Enzymology*, *326*, 35–47.

Marx, C. J., & Lidstrom, M. E. (2002). Broad-host-range *cre-lox* system for antibiotic marker recycling in gram-negative bacteria. *BioTechniques*, *33*(5), 1062–1067.

Marx, C. J., O'Brien, B. N., Breezee, J., & Lidstrom, M. E. (2003). Novel methylotrophy genes of *Methylobacterium extorquens* AM1 identified by using transposon mutagenesis including a putative dihydromethanopterin reductase. *Journal of Bacteriology*, *185*(2), 669–673.

Masuda, S., Suzuki, Y., Fujitani, Y., Mitsui, R., Nakagawa, T., Shintani, M., et al. (2018). Lanthanide-dependent regulation of methylotrophy in *Methylobacterium aquaticum* strain 22A. *mSphere*, *3*(1).

McSkimming, A., Cheisson, T., Carroll, P. J., & Schelter, E. J. (2018). Functional synthetic model for the lanthanide-dependent quinoid alcohol dehydrogenase active site. *Journal of the American Chemical Society*, *140*(4), 1223–1226.

Nagai, T., Ibata, K., Park, E. S., Kubota, M., Mikoshiba, K., & Miyawaki, A. (2002). A variant of yellow fluorescent protein with fast and efficient maturation for cell-biological applications. *Nature Biotechnology*, *20*(1), 87–90.

Nakagawa, T., Mitsui, R., Tani, A., Sasa, K., Tashiro, S., Iwama, T., et al. (2012). A catalytic role of XoxF1 as La^{3+}-dependent methanol dehydrogenase in *Methylobacterium extorquens* strain AM1. *PLoS One*, 7(11), e50480.

Nunn, D., & Lidstrom, M. (1986). Isolation and complementation analysis of 10 methanol oxidation mutant classes and identification of the methanol dehydrogenase structural gene of *Methylobacterium* sp. strain AM1. *Journal of Bacteriology*, *166*(2), 581–590.

Peel, D., & Quayle, J. (1961). Microbial growth on C_1 compounds. 1. Isolation and characterization of *Pseudomonas* AM1. *The Biochemical Journal*, *81*(3), 465–469.

Pol, A., Barends, T. R., Dietl, A., Khadem, A. F., Eygensteyn, J., Jetten, M. S., et al. (2014). Rare earth metals are essential for methanotrophic life in volcanic mudpots. *Environmental Microbiology*, *16*(1), 255–264.

Prejanò, M., Marino, T., & Russo, N. (2017). How can methanol dehydrogenase from *Methylacidiphilum fumariolicum* work with the alien CeIII ion in the active center? A theoretical study. *Chemistry - A European Journal, 23*(36), 8652–8657.

Roszczenko-Jasińska, P., Vu, H. N., Subuyuj, G. A., Crisostomo, R. V., Cai, J., Lien, N. F., et al. (2020). Gene products and processes contributing to lanthanide homeostasis and methanol metabolism in *Methylorubrum extorquens* AM1. *Scientific Reports, 10*(1), 1–15.

Skovran, E., & Martinez-Gomez, N. C. (2015). Just add lanthanides. *Science, 348*(6237), 862–863.

Skovran, E., Palmer, A. D., Rountree, A. M., Good, N. M., & Lidstrom, M. E. (2011). XoxF is required for expression of methanol dehydrogenase in *Methylobacterium extorquens* AM1. *Journal of Bacteriology, 193*(21), 6032–6038.

Skovran, E., Raghuraman, C., & Martinez-Gomez, N. C. (2019). Lanthanides in methylotrophy, methylotrophs and methylotroph communities. *Current Issues in Molecular Biology, 33*, 101–116.

Springer, A. L., Chou, H.-H., Fan, W.-H., Lee, E., & Lidstrom, M. E. (1995). Methanol oxidation mutants in *Methylobacterium extorquens* AM1: Identification of new genetic complementation groups. *Microbiology, 141*(11), 2985–2993.

Van Dien, S. J., Okubo, Y., Hough, M. T., Korotkova, N., Taitano, T., & Lidstrom, M. E. (2003). Reconstruction of C_3 and C_4 metabolism in *Methylobacterium extorquens* AM1 using transposon mutagenesis. *Microbiology, 149*(3), 601–609.

Vorholt, J. A., Marx, C. J., Lidstrom, M. E., & Thauer, R. K. (2000). Novel formaldehyde-activating enzyme in *Methylobacterium extorquens* AM1 required for growth on methanol. *Journal of Bacteriology, 182*(23), 6645–6650.

Vu, H. N., Subuyuj, G. A., Vijayakumar, S., Good, N. M., Martinez-Gomez, N. C., & Skovran, E. (2016). Lanthanide-dependent regulation of methanol oxidation systems in *Methylobacterium extorquens* AM1 and their contribution to methanol growth. *Journal of Bacteriology, 198*(8), 1250–1259.

Whitta, S., Sinclair, M., & Holloway, B. (1985). Transposon mutagenesis in *Methylobacterium* AM1 (*Pseudomonas* AM1). *Microbiology, 131*(6), 1547–1549.

CHAPTER EIGHT

Metals and methylotrophy: Via global gene expression studies

Zachary J. Johnson[a], Dennis D. Krutkin[a], Pavlo Bohutskyi[b], and Marina G. Kalyuzhnaya[a,*]
[a]Department of Biology, San Diego State University, San Diego, CA, United States
[b]Pacific Northwest National Laboratory, Richland, WA, United States
*Corresponding author: e-mail address: mkalyuzhnaya@sdsu.edu

Contents

Abstract

A number of minerals, such as copper, cobalt, and rare earth elements (REE), are essential modulators of microbial one-carbon metabolism. This chapter provides an overview of the gene expression study design and analysis protocols for uncovering REE-induced changes in methylotrophic bacteria. By interrogating relationships and differences in total gene expression induced by mineral micronutrients, a deeper understanding of gene regulation at a systems scale can be gained. With careful design and execution of RNA-sequencing experiments, thorough processing and assessment of read quality can be utilized to assess and adjust for possible biases. By ensuring only quality data are utilized in downstream processes, differential gene expression, overrepresented analyses, and gene-set enrichment analyses provide reliable and reproducible representation of pathways and functions which are being affected by changes in environmental conditions.

1. Introduction: Metals and methylotrophy

Methylotrophy can be described as the ability to utilize single carbon (C1) compounds, such as methane, methanol, and methylamine, as a sole source of carbon and energy (Anthony, 1982). While most methylotrophic

Methods in Enzymology, Volume 650
ISSN 0076-6879
https://doi.org/10.1016/bs.mie.2021.01.046

organisms can use multi-carbon compounds, methylotrophy requires a highly specialized metabolic network for reduced C1-carbon oxidation and assimilation (Chistoserdova, 2011). The methylotrophy metabolic network is surprisingly redundant (Fig. 1). Two different methane-monooxygenase (MMO) enzymes drive methane oxidation to methanol (Ross & Rosenzweig, 2017; Semrau, DiSpirito, Gu, & Yoon, 2018). Two to four pyrroloquinoline quinone (PQQ)-linked methanol dehydrogenases (MDH) contribute to methanol oxidation (Chistoserdova, 2019). Several metabolic routes exist for formaldehyde oxidation (FAox), and up to four formate dehydrogenases (FDH) contribute to the final step of the primary oxidation (Chistoserdova, Laukel, Portais, Vorholt, & Lidstrom, 2004). The C1 carbon assimilation pathways are often redundant, with up to three routes predicted for some microbial species. This metabolic redundancy enables C1 conversion at different environmental settings and is often controlled by the availability

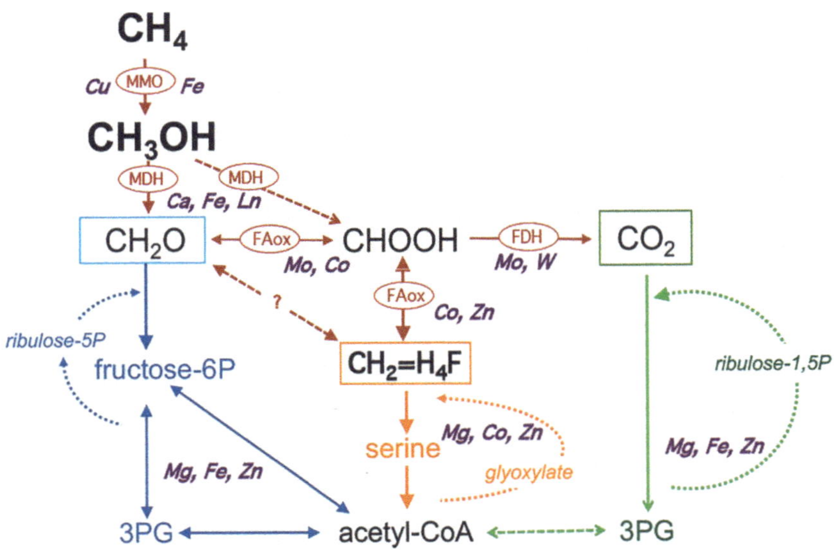

Fig. 1 Metal requirements for the core C_1-metabolism. MMO, methane-monooxygenase; MDH, methanol dehydrogenases; FAox, formaldehyde oxidation pathways; FDH, formate dehydrogenases. *Blue*: The ribulose monophosphate (RuMP) pathway for formaldehyde assimilation; *gold*, serine cycle and tetrahydrofolate pathway for formate assimilation, *green*, Calvin cycle for CO_2 assimilation.

of micronutrients, such as copper, calcium, molybdenum, tungsten, cobalt, and nickel (summarized in Fig. 1).

In addition to enzyme activity modulation, some minerals induce changes in gene expression which result in alternative metabolic pathway utilization, a phenomenon described as a switch. Several switches have been described for methanotrophs: copper switch (Semrau, DiSpirito, & Yoon, 2010), lanthanum switch (Semrau et al., 2018) and molybdenum switch (Akberdin et al., 2018; Chistoserdova et al., 2004). Some studies have highlighted the large complexity of metabolic responses and suggested metabolic cross-talk between different metals (Gu & Semrau, 2017; Semrau et al., 2018). In most cases, -omics studies were essential steps for uncovering the molecular mechanisms underlying the system performances and outputs.

While some lanthanide (Ln)-induced changes could be easily detected by routine gene expression studies, uncovering redox-driven metabolic alterations required more diligent data digging. In our original study, we combined an array of -omics approaches, including proteomics, metabolomics, and enzymatic studies to establish the Ln-induced redox shift (Akberdin et al., 2018). Here we present computational strategies that provide the same outcome based on gene expression data alone.

2. Gene expression studies: Experimental design

A strong experimental design is required to maximize statistical significance and minimize the cost of the RNA-sequencing (RNA-seq) studies. A successful RNA-seq experimental design requires an explicit question for a specific goal. While there is a broad spectrum of scientific questions that can be addressed through RNA-seq experiments, quantitative differential gene expression (DGE) profiling between samples is arguably the most prevalent. Major RNA-seq experimental design considerations include the number and type of replicates, library construction, library size, and sequencing depth. Each experimental design feature may introduce specific biases or changes in relative gene expression level, these changes must be modeled during processing to reduce false discovery rate during DGE analysis.

The DGE protocols described here were designed and tested with the RNA-seq data generated for *Methylomicrobium alcaliphilum* 20Z cells grown in a continuous culture (turbidostatic) on mineral nitrate medium with or without lanthanides and with methane as the sole source of carbon (Akberdin et al., 2018). The discovery of the lanthanide-dependent methanol dehydrogenase (Ln-Mdh) transformed our vision of the diversity and distribution of methylotrophy in various ecosystems. The Ln-XoxF enzymes became an example of unique biochemistry which relies on rare earth elements (Picone & Op den Camp, 2019). In addition to the activation of the methanol-converting enzymes, lanthanides mediate gene expression of calcium-dependent methanol dehydrogenase, induce a Ln-transporter system, and alter redox-pathways in methylotrophs (Akberdin et al., 2018; Groom, Ford, Pesesky, & Lidstrom, 2019). The experimental set-ups for uncovering micronutrient-induced changes require chemically clean labware (flasks) without metal parts, as any contamination or leaching from metals can impact data reproducibility. We found that single-use plastic labware, such as disposable culture vessels (i.e., Eppendorf BioBLU Single Use Vessels) provide reproducible data. Protocols for setting up a bioreactor culture with C1 compounds as carbon feedstocks are provided in Collins and Kalyuzhnaya (2018). The protocol for RNA extraction is presented below.

2.1 cDNA library preparation and sequencing

cDNA library preparation is conducted through RNA molecule extraction, fragmentation, adapter ligation, and PCR amplification followed by sequencing. The various library preparations available can affect relative expression levels and specific biases can be introduced during PCR amplification and sequencing.

Illumina short-read RNA-seq, which generates 50–300 bp reads (Oshlack, Robinson, & Young, 2010) is still the most commonly used RNA-sequencing technology due to its high accuracy, experimental throughput, and relatively affordable price (Stark, Grzelak, & Hadfield, 2019). Still, there are inherent biases associated with PCR amplification and sequencing these reads. There are two categories of inherent biases which arise: (1) between-lane distributional differences (sequencing depth) and (2) within-lane gene-specific effects (GC biases, expression level, gene-length) (Risso, Schwartz, Sherlock, & Dudoit, 2011). To prevent high numbers of false-positive results, it becomes important to model these biases when performing DGE analysis (Love, Hogenesch, & Irizarry, 2016).

2.1.1 PROTOCOL 1: RNA extraction

2.1.1.1 Reagents

Mini-BeadBeater Zirconia-Silicate Beads, 0.1–0.15 mm (BioSpec Products 11079101Z).

Stop solution: 5% equilibrated phenol in ethanol: take 5 mL of Water Saturated Phenol (Ambion, cat# AM9710) add 45 mL of ethanol (200 proof). Store at 4 °C. Use within 1–2 weeks.

RNeasy Mini Kit (Qiagen, ID 74104).

β-Mercaptoethanol (β-ME).

RNase-Free DNase Set (Qiagen).

PureLink DNase I (ThermoFisher Scientific).

Note: The protocol below is based on the Qiagen RNA extraction protocol. PureLink RNA mini kit (Thermofisher) or QuickRNA (ZymoReserach) can be also used with appropriate modifications (see manufacturer instructions). However, we do not recommend the TRIzol reagent, which reduces the mRNA yield.

Equipment: A refrigerated, general-purpose centrifuge with rotors/adapters for 1–50 mL tubes (5702R Eppendorf or similar), a high-speed benchtop homogenizer (such as MPBio beat-beater or Qiagen TissueLyser).

2.1.1.2 Before you start

1. Prepare 2 mL screw-cap tubes (1 tube per sample): Add 0.5 g of 0.1 mm zirconia-silicate beads.
2. Prepare buffers and solutions for RNA extraction, including RLT solution (provided with the RNeasy Mini Kit; add 10 μL β-ME per 1 mL Buffer RLT); RW1 solution (provided with the RNeasy Mini Kit; add ethanol).

2.1.1.3 Steps

3. Transfer samples of batch or bioreactor cultures (45 mL, with OD = 0.3–1) into tubes containing 5 mL of the stop solution (5% water-equilibrated phenol in ethanol). The stop solution helps to stabilize cellular RNA and eliminates the stress–response induced by centrifugation at cold (typically 4 °C) conditions.
4. Collect cells by centrifugation at 2500 × *g* for 15 min at 4 °C. Remove and discard supernatant. Proceed to step 5 immediately.

5. Resuspend collected cell pellets in 350 μL of RLT buffer and transfer into prepared 2 mL screw-cap tubes with 0.5 g of 0.1 mm zirconia-silicate beads.
6. Homogenize samples using a mini-beater or TissueLyser for 1 min, cool on ice for 2 min. Repeat this step 2–3 times.
7. Centrifuge for 2 min and transfer supernatant into a new tube.
8. Add one volume (about 300–350 μL) ethanol, mix and transfer onto an RNeasy spin column. Centrifuge for 15–30 s and transfer the RNeasy spin column into a new collection tube.
9. Perform on-column DNase digestion (follow the manufacturer's protocol).
10. Add 500 μL Buffer RPE (provided with the RNeasy Mini Kit; add ethanol) to the RNeasy spin column and centrifuge. Discard the flow-through. Repeat the step two times.
11. Transfer the RNeasy spin column into 1.5 mL tube. Add 50 μL RNase-free water and centrifuge for 1 min at 8000 × *g* to elute the RNA.
12. Measure RNA concentration using nanodrop or similar spectrophotometer. Expected concentrations are in the range of 200–500 ng/μL for cell samples collected at OD = 0.3–1. The purity of RNA preparations should be confirmed by negative PCR and positive RT-PCR.

Most gene expression experiments in microbial systems require additional ribosomal RNA depletion steps (described in Kalyuzhnaya, Beck, & Chistoserdova, 2011). High-resolution automated electrophoresis is typically used for validating the integrity of RNA sample preparations (Mueller, Lightfoot, & Schroeder, 2004). It should be noted that both the RNA integrity check and rRNA depletion services are often provided by the sequencing center or facility as a standardized step in library construction. The RNA-samples described here were sequenced on an Illumina HiSeq2500 with ~50 million/sample SR50 reads by IGM Genomics Center, University of California, San Diego. The raw Fastq files could be downloaded from http://sci.sdsu.edu/kalyuzhlab/omics-datasets/.

2.1.2 PROTOCOL 2: Processing RNA-seq data

The main steps in RNA-seq processing workflow for DGE analysis are shown in Fig. 2 and include the following steps.

1. Set-up Bioinformatic pipeline.
2. Preprocessing and Trimming Low-Quality Reads. A preprocessing report is generated using MultiQC to cumulate individual FastQC reports. Adapter content and sequence quality inform whether reads need to be trimmed further using Trimmomatic.

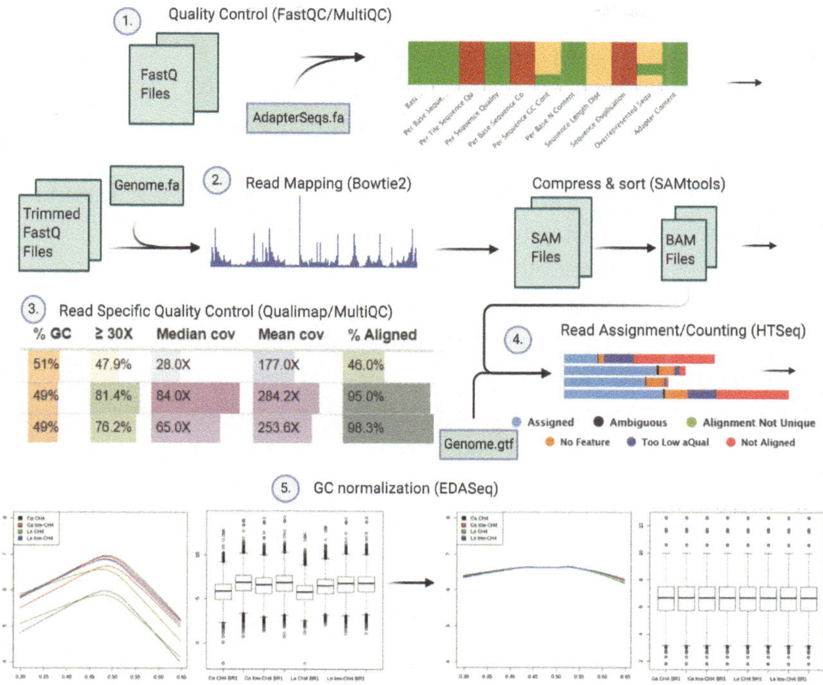

Fig. 2 Overview of RNA-seq data processing pipeline. (1) Preprocessing quality control using FastQC and MultiQC. (2) Read mapping to the reference genome using Bowtie2. (3) Read-specific quality control using Qualimap and MultiQC. (4) Assigning mapped reads to genes and counting the number of reads using HTSeq. (5) GC normalization of read counts using EDASeq.

3. High-quality reads are passed to Bowtie2 to map them to an indexed genome, Sequence Alignment/Map (SAM) files are sorted and converted to binary alignment (BAM) files using Samtools. Alignment-specific QC is generated using Qualimaps.
4. Aligned reads are counted using feature locations obtained from a gene transfer file (GTF) or general feature format file (GFF3); raw counts obtained from HTSeq are output as gene-count tables. At this time, MultiQC can be used again to obtain a cumulative report of all preprocessing steps, informing on sample mapping quality and biases.
5. Raw-reads are normalized for differences in sequencing depth and gene-length or GC-content using EDASeq.
6. Normalized counts are passed to DESeq2 where differential gene expression analysis is used to determine how gene expression varies based on experimental conditions.

2.1.2.1 Bioinformatics pipeline set-up

A typical DGE RNA-seq bioinformatics pipeline will process and assess the quality of reads, normalize the raw read counts, perform DGE analysis, and enrich for the function of differentially expressed genes. There are a multitude of downloadable and web-based applications which can be utilized to conduct RNA-seq analysis. To ensure reproducible results, it is important to retain data sources, package versions, and processing scripts.

Package managers are often employed to create environments and install packages for the analysis. Creating unique environments allows the user to isolate packages and dependencies for easily reproducible and shareable code. The Conda package management system (Anaconda Software Distribution, 2016) is one of the most versatile and easy-to-use package managers. It can be installed through Anaconda, a full-stack scientific Python development suite which is installed along with a graphical user interface (gui); or as miniconda, a light-weight alternative that does not include packages upon downloading. Miniconda is appropriate when accessing remote servers via a secure shell (ssh), but either can be installed on all operating systems.

Once an environment is set-up for the RNA-seq pipeline, bioinformatics packages can be downloaded from the bioconda channel (Grüning et al., 2018). Setting up the appropriate channels and installing packages from the bioconda channel is further described on the bioconda github page. The packages used in this analysis include: Python 3.6.1 (Van Rossum & Drake, 2009), R 3.6.1 (R Core Team, 2013), FastQC 0.11.9 (Andrews, 2010), MultiQC 1.9 (Ewels, Magnusson, Lundin, & Käller, 2016), Trimmomatic 0.39 (Bolger, Lohse, & Usadel, 2014), Bowtie2 2.4.1 (Langmead & Salzberg, 2012), Samtools 1.7 (Li et al., 2009), Qualimap 2.2.2a (García-Alcalde et al., 2012), HTSeq 0.12.4 (Anders, Pyl, & Huber, 2014), EDAseq 2.20 (Risso et al., 2011), DESeq2 1.26 (Love, Huber, & Anders, 2014), pheatmap 1.0.8, dendextend 1.14.0, AnnotationForge 1.32.0 (Carlson & Pagès, 2020), clusterProfiler 3.18.0 (Yu, Wang, Han, & He, 2012), and Pathview 3.12 (Luo & Brouwer, 2013).

2.1.2.2 Preprocessing and trimming low-quality reads

Raw RNA-seq data are often returned from sequencing centers as compressed fastq files containing the sequence reads and a phred quality score encoded as ASCII characters. The first step of processing RNA-seq data requires checking the read quality to identify and remove poor-quality reads that can arise from library preparation flaws and sequencing errors.

FastQC can be used to assess multiple aspects of RNA-seq quality including sequence quality and GC-content; it can also check for the presence of adapter sequences. Once FastQC is executed on each sample, a cumulative report can be generated using multiQC.

MultiQC returns an HTML document containing all statistics and graphs generated by FastQC for each sample including numerous quality-control tests shown in the MultiQC heatmap (Fig. 2, step 1). Importantly, the GC-content and overrepresented sequences test can help flag biases in sequenced reads that can be accounted for using a normalization procedure if they persist throughout mapping. Some other FastQC failure flags are often harmless to the analysis. For example, per-base sequence content may be raised due to an unequal distribution of bases in early reads caused by biased selection of random primers. Another common failure is sequence duplication when non-unique sequences make up more than 50% of the total sequences (Andrews, 2010). This occurs in many RNA-seq libraries which have been over-sequenced to identify low-expressed transcripts. The presence of adapter sequences can also cause warnings when any sequence is present in more than 5% of reads and failure when it is over 10%. If adapter sequences are present, they can be removed using Trimmomatic. Given specified single-end or paired-end library-prep adapter sequences in a fasta format, Trimmomatic will identify and remove them from the reads. Trimmomatic can also trim low-quality reads that may occur toward the ends of sequences by removing bases off the 3′ and 5′ ends of reads that fall below a given threshold while ensuring all reads are over a given length (Bolger et al., 2014). Read trimming may be particularly useful when reanalyzing data from early RNA-seq studies where datasets had low average base calling quality and susceptibility to adapter contamination. If used, read trimming should be done with caution; Macmanes (2014) suggests trimming nucleotides which fall below a PHRED score of 2 or 5, with other studies similarly finding aggressive trimming can negatively affect accuracy of expression estimates (Williams, Baccarella, Parrish, & Kim, 2016).

2.1.2.3 Mapping RNA-seq reads and feature counting

Once RNA-seq reads are checked for quality and adapter sequences are removed, the reads can then be mapped against a reference genome or assembled into a transcriptome and annotated (de novo mapping). In this study, we used a reference-based mapping to align sequenced reads to the *M. alcaliphilum 20Z* genome obtained from the Microbial Genome Annotation and Analysis (MaGE) platform (Vallenet, 2006). A splice-aware aligner is not required when aligning to a reference transcriptome or prokaryotic genome, and Bowtie2 (Langmead & Salzberg, 2012) is a frequently used tool for mapping to a reference transcriptome or bacterial genome. Before mapping, the reference genome is indexed to increase the mapping speed. Mapped reads are then output in the Sequence Alignment/Map (SAM) format and converted into a compressed binary (BAM) file before being sorted with Samtools (Li et al., 2009) for mapping quality assessment and feature counting. These steps are shown in Fig. 2 (Step 2).

Alignment-specific quality control can be checked using the BAM QC function from the Qualimap (García-Alcalde et al., 2012) package. The overall summary statistics and figures showing genome mapping coverage and GC-content over mapped reads are shown in Fig. 2 (Step 3). From these figures and statistics, we can understand the relative quality and quantity of mapped reads. Read coverage is an accurate determinant of sample quality and GC statistics can further extend the study of GC bias in mapped reads.

Several programs have been developed for counting genomic features; FeatureCounts and HTSeq are both popular for processing RNA-seq data (Anders et al., 2014; Liao, Smyth, & Shi, 2013). Both programs utilize a feature annotation file, strand protocol, feature count type, and counting protocol to assign and count mapped reads to genomic features (Fig. 2, step 4). While these programs employ slightly different algorithms, they produce comparable results. The final report generated by MultiQC provides a summary and visualizes statistics from all previous steps of the pipeline; it can be customized to include or exclude specific elements. At this step, the data can be checked for any significant flaws including contaminated samples or incorrect mapping. Samples with low percentage of aligned reads (<10%) can be further assessed using clustering techniques such as PCA.

2.1.2.4 Normalization

Count normalization (Fig. 2, step 5) is necessary to make accurate comparisons of gene expression between different genes within a sample and genes between different samples. Normalization procedures fall into two categories: (1) within-lane gene-specific effects (GC biases, expression level, gene-length) and (2) between-lane distributional differences (sequencing depth).

2.1.2.4.1 RPKM, FPKM, and TPM RPKM (reads per kilobase million) is used to represent single-end data, while FPKM (fragments per kilobase million) represents paired-end data and takes into consideration that there are two reads per mapping feature. Both methods account for sequencing depth and gene-length, but not RNA composition. These methods can be used for comparisons between genes within a sample, but not for between-sample comparisons or differential expression analysis. RPKM and FPKM are calculated by dividing by 1,000,000, normalizing for sequencing depth, then dividing by the length of the gene in kilobases. With TPM (transcripts per kilobase million), gene-length is normalized first, followed by sequencing depth. TPM can be used for gene-count comparisons within or between samples in the same sequencing order, but should not be used for differential expression analyses.

2.1.2.4.2 DESeq2 Normalization DESeq2 has its own internal normalization method in which counts are divided by sample-specific size factors calculated by a median ratio of gene counts relative to the geometric mean of the gene. This method is specifically meant for gene-count comparisons between samples, normalizing for varying sequencing depths, and allowing for differential expression analysis. The median of ratios method does not consider GC or gene-length biases.

2.1.2.4.3 GC bias normalization If GC bias is observed throughout the RNA-seq processing stages, further exploratory analysis can be conducted to visualize the biases which may have been introduced into the data and correct for them. To accomplish this analysis, EDAseq (Risso et al., 2011) was implemented to further explore the count data. EDAseq can be used to normalize sequenced reads while accounting for GC bias as a covariate. Full-quantile within- and between-lane normalization were implemented to normalize for GC bias and sequencing depth differences between samples respectively as described by Risso et al. (2011). These normalizing procedures are modeled as offsets (Robinson, Mccarthy, & Smyth, 2009), leaving the count data unchanged. This model can then be passed to DESeq2 in place of the function estimateSizeFactors as described in section 6.2 of the EDASeq vignette (Risso, 2018).

Once counts have been normalized, the count matrix, normalized offsets, and additional metadata are passed to DESeq2 to construct a DESeqDataSet. The variance stabilizing transformation (vst) or regularized log transformation (rlog) from DESeq2 can then be used to extract transformed, normalized counts for visualizing expression level changes.

2.2 Differential gene expression analysis

Batch effects can be modeled during DESeqDataSet construction prior to differential gene expression analysis to account for covariates in balanced experimental designs. In this example, two concentrations of methane, optimal (CH_4) and low (low-CH_4) were used to grow samples in media containing either calcium (Ca) or lanthanum (La). Methane concentrations were modeled as a covariate in the expanded design matrix; however, any number of covariates or interactions between variables can be modeled in the design matrix, including differences in library preparation, as long as samples are balanced between different conditions.

Differential gene expression analysis is then used to detect genes which are significantly differentially expressed between two different conditions. DESeq2 uses a generalized linear model with a negative binomial distribution to model count data. The results of the differential expression analysis are accessed using the results() function. Differentially expressed genes can then be extracted. Here, genes with known function were selected with a Log2FoldChange (LFC) $|LFC| > 2$ and adjusted p-value (p_{adj}) < 0.01. Differentially expressed genes were then hierarchically clustered using dendextend, annotated by function assigned from MaGE, and plotted using pheatmap.

Comparing the differential gene expression results for both the DESeq2-normalized (Fig. 3A) and GC-normalized data (Fig. 3B), we see similar upregulation of the La-dependent methanol dehydrogenase (*xoxF*) in the presence of La (Fig. 3, cluster 2) while the *mxa*-operon, encoding the Ca-dependent MDH, associate cytochrome and accessory chaperons, is downregulated in the presence of La (Fig. 3A, cluster 3/4; Fig. 3B, 4/5). Importantly, the NAD-dependent formate dehydrogenase was found to be highly differentially expressed in the GC-normalized data; the same conclusion was made based on a combination of metabolomic and proteomics analyses, but was overlooked with DESeq2's median ratio normalization (Akberdin et al., 2018).

2.3 Functional and spatial enrichment analysis (FSEA)

Functional enrichment visualizes comprehensive analysis of differentially expressed genes. FPKM/TPKM/TPM allows one to analyze counts of transcripts or fragments but does not take into consideration the complexity of underlying biological processes which could be influencing the abundance of reads. Abundance normalization methods are also model-independent, meaning they do not consider any kind of relationship between two conditions and do not normalize for within-lane biases, such as GC bias.

For differential expression analysis and functional enrichment analysis, the most commonly reported metric is Log2FoldChange between a reference

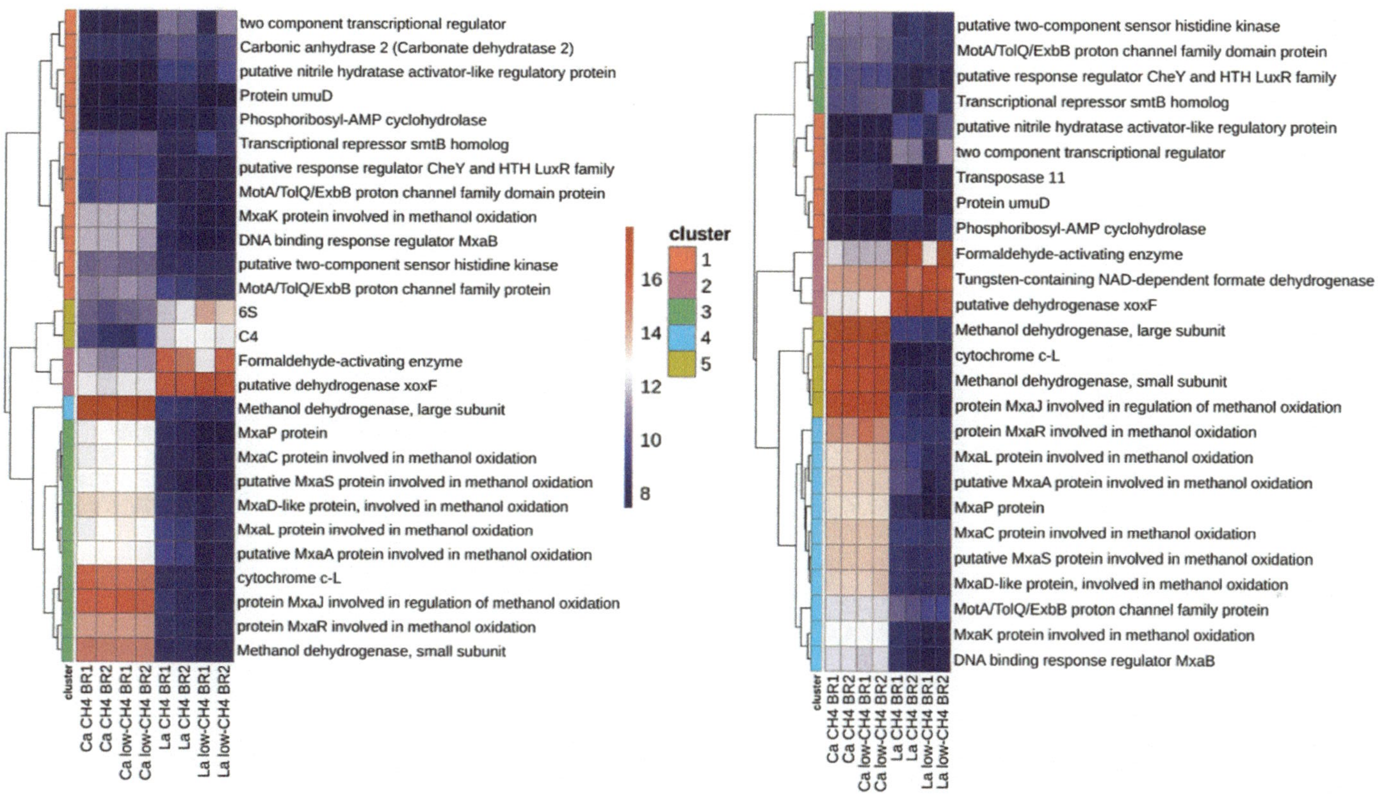

Fig. 3 Clustering by total distance of known genes based on normalization procedure (A) DESeq2 median ratio normalization (B) EDASeq GC normalization. DE genes were filtered by padj <0.01 and $|LFC|>2$. Counts are on a normalized, variance stabilized (vst) transform scale.

condition and an experimental condition. The Log2FoldChange values per gene, between conditions, are used to rank genes from most-highly differentially expressed to most-lowly differentially expressed; DESeq2 was utilized to compare conditions and generate Log2FoldChange values for the entire gene set, modeling the differences between the Ca/La shift.

2.3.1 PROTOCOL 3: Preparation for FSEA

Further details regarding functional enrichment from differentially expressed genes are described in the subsequent sections.

1. Repositories of interest which group genes into functional/integrative units are selected.
2. If the organism being studied has available biological databases, they can be used directly or they are compiled from novel annotations for non-canonical organisms.
3. An input data frame is created from the results of differential gene expression analysis.

2.3.1.1 Gene sets, modules, and pathways from bioinformatic repositories

The primary requirements of functional enrichment are a list of differentially expressed (DE) genes and some form of grouping, organization, or assignment of genes into predefined categories. The most commonly studied organisms (humans, clinical vertebrates, routine viral and bacterial taxa) have a large disposal of repositories which have grouped genes into functional and/or integrative units.

A gene set is an assemblage of genes that are related to each other, typically sharing biological processes, molecular functions, or cellular compartments, which allows them to be grouped into categories such as transporters, DNA-binding proteins, transcription factors, etc. Pathways are also a predefined set of genes but typically do not have functional relationships—rather, a systemic, stepwise enzymatic flow representing cellular or molecular functionalities and are typically visualized with metabolic maps. Modules are the functional units of pathways, representing enzymatic and catalytic conversions.

The most commonly used archives are the Kyoto Encyclopedia of Genes and Genomes (KEGG) and The Gene Ontology (GO) Project, due to their longevity, curation, and broad availability of organismal support. Other widely-used bioinformatic compendiums include wikipathways, The Molecular Signatures Database (MSigDB), The Disease Ontology (DO), and The Disease Gene Network (DisGeNET).

2.3.1.2 Availability/creation of biological databases

Unlike *E. coli* and other commonly studied microbes which have had decades of experimental procedures, high-throughput studies, and procedural/predictive analyses to elucidate cellular and molecular functions, the systems biology of methane-consuming bacteria is less established. The strategy described below can be applied to any non-canonical microbial system. In our pipeline analysis, we chose to focus on the GO and KEGG repositories for functional and systemic enrichment.

Formal GO annotations had not yet been created for the genome of *M. alcaliphilum 20Z*; only a small portion of the strain's genes had been annotated with GO IDs by automated NCBI blastp during its genome submission (Vuilleumier et al., 2012). In order to update the genome to be capable of GO enrichment and for future transcriptomic studies, OmicsBox (2019) was utilized to conduct a whole-genome GO annotation. Blast2GO (Götz et al., 2008) and InterProScan (Jones, Binns, & Chang, 2014) modules were used to retrieve all GO IDs for all genes; the results were merged and the annotation was validated by filtering for IDs associated with bacterial taxa. A Structured Query Language (SQL) organismal database (org.db) was created for *20Z* using AnnotationForge 1.30.1 (Carlson & Pagès, 2020). Transcriptomic analyses which assess gene expression of commonly studied organisms and utilize functional enrichment typically have organisms with available GO organismal database packages through Bioconductor, which allows uniform and organized annotations to be used by multiple groups or different studies.

Similar to how GO organismal databases are available through Bioconductor, KEGG offers species-specific biological databases with an associated organismal code. The KEGG database for *M. alcaliphilum 20Z* also had not been updated since its genome submission; in order to carry out KEGG functional enrichment, the genomic network for *20Z* was re-annotated and re-constructed using KofamKOALA (Aramaki et al., 2020) by assigning KEGG orthology numbers to *20Z's* sequence data through HMMER/HMMSEARCH against the KOfam database. The genome was also cross-referenced with BlastKOALA (Kanehisa, Sato, & Morishima, 2016) and all KEGG ortholog ID's were merged. Multiple locus tags can correspond to a single KEGG ortholog ID because they represent functional units; if this was the case, locus IDs were separated and annotated accordingly.

2.3.1.3 Data extraction from differential gene expression analysis

After performing differential gene expression analysis in DESeq2, the gene identifiers (locus tags), Log2FoldChange values, and adjusted *p*-values are extrapolated to create a new data frame. Differentially expressed genes above a user-defined adjusted *p*-value cutoff are filtered out to reduce false positives during enrichment. The resulting data frame is ordered by descending Log2FoldChange values.

2.3.2 Overrepresented/enrichment analysis (ORA)

Overrepresented/enrichment analysis (Boyle et al., 2004) is an approach to determine if annotated biological functions or processes are overrepresented, or enriched, given a gene list which is subset by user-defined cutoff values, such as a list of differentially expressed genes below a statistical threshold and above an absolute value threshold of Log2FoldChange. Before extracting data above a certain |Log2FoldChange| value, the initial list of differentially expressed genes must be filtered by adjusted *p*-value threshold to account for multiple hypothesis testing.

In order to determine if a group of genes is enriched within a user-defined annotation set, *p*-values are computed by hypergeometric distribution, cross-referencing genes which are considered substantially differentially expressed (i.e., filtered by initial adjusted *p*-value and $|\text{log2foldchange}| > 1$) and occur within the annotation of choice, in contrast to the background distribution which are all of the genes that have annotation. Fisher's exact test is used to determine if the particular differentially expressed genes overlap significantly within a user-defined grouping of genes, considering it enriched.

2.3.2.1 Gene ontology overrepresented analysis

The Gene Ontology is a comprehensive, up-to-date biological compendium which is used as a resource for computational representation of the current scientific availability regarding the functions of genes, coding and non-coding. GO offers the most diverse availability of annotation due to its long-standing contributions and encompassment of a broad array of taxa.

As described above, a GO organismal database was constructed for *M. alcaliphilum 20Z* from the annotation created with OmicsBox. clusterProfiler supports advanced filtering and enrichment options when using an organismal database, such as filtering by specific GO categories—cellular compartment (CC), molecular function (MF), and biological process (BP)—or hierarchical GO-level (1, 2, 3, 4, …). Alternatively, users can create a "universal" list of all GO processes and their nested genes to conduct a generic enrichment; the results will show GO IDs and the terms (genes) enriched under specified statistical cutoffs, but will not cluster GO IDs into

GO categories (biological process) or functions within those categories (such as "cellular response to stimulus" nested under "biological process").

The list of significantly differentially expressed genes in La conditions ($p_{adj} < 0.10$) was subset by $|\text{log2foldchange}| \geq 1$. Using clusterProfiler, users can conduct ORA by specific GO category with no level filter (Fig. 4A, enriched GO BPs with no level restriction) or by GO category and declaring a specific level (Fig. 4B and C, enriched GO MFs restricted to levels two and three).

2.3.3 Gene set enrichment analysis (GSEA)

ORA/enrichment analysis will detect genes where the difference is large, but will not detect small changes in expression that occur in systematic patterns. Gene-set enrichment analysis (GSEA) (Subramanian et al., 2005) overcomes the aforementioned limitation by using all differentially expressed genes, not just those which have been subset above a certain |Log2FoldChange| threshold. Like enrichment analysis, initially derived DE genes should be filtered by adjusted *p*-value to account for multiple hypothesis testing during GSEA and to reduce false positives.

The DE genes are ranked based on their phenotypes; given a defined pre-existing set of genes, such as those sharing the same KEGG pathway/module, or user-defined gene set, such as novel GO annotation or GO-slim annotation, GSEA will determine whether the genes which are shared in the same grouping are randomly distributed through an ordered gene list or primarily occur at the top or bottom.

2.3.3.1 Gene ontology gene-set enrichment analysis

Mentioned previously, the biggest drawback of ORA is the inherent requirement of specifying genes which are substantially differentially expressed, typically determined by a user-defined cutoff of log2foldchange threshold. In the case of DE genes with La shift, of the 4200 computationally derived values, 72.5% (3044) of genes were lost after filtering by p_{adj} to account for multiple hypothesis testing of ORA and GSEA. Furthermore, after subsetting by $|\text{log2foldchange}| > 1$ for ORA, 73.7% of statistically significant DE genes are lost—only 304/1156 remained; those 304 genes were considered during ORA (Fig. 4).

GSEA allows analysis of all (1156) statistically significant differentially expressed genes. Oftentimes in situations pertaining to biological engineering and cell-profiling analyses, small but coordinated changes in gene expression can have substantial and profound biological relevance.

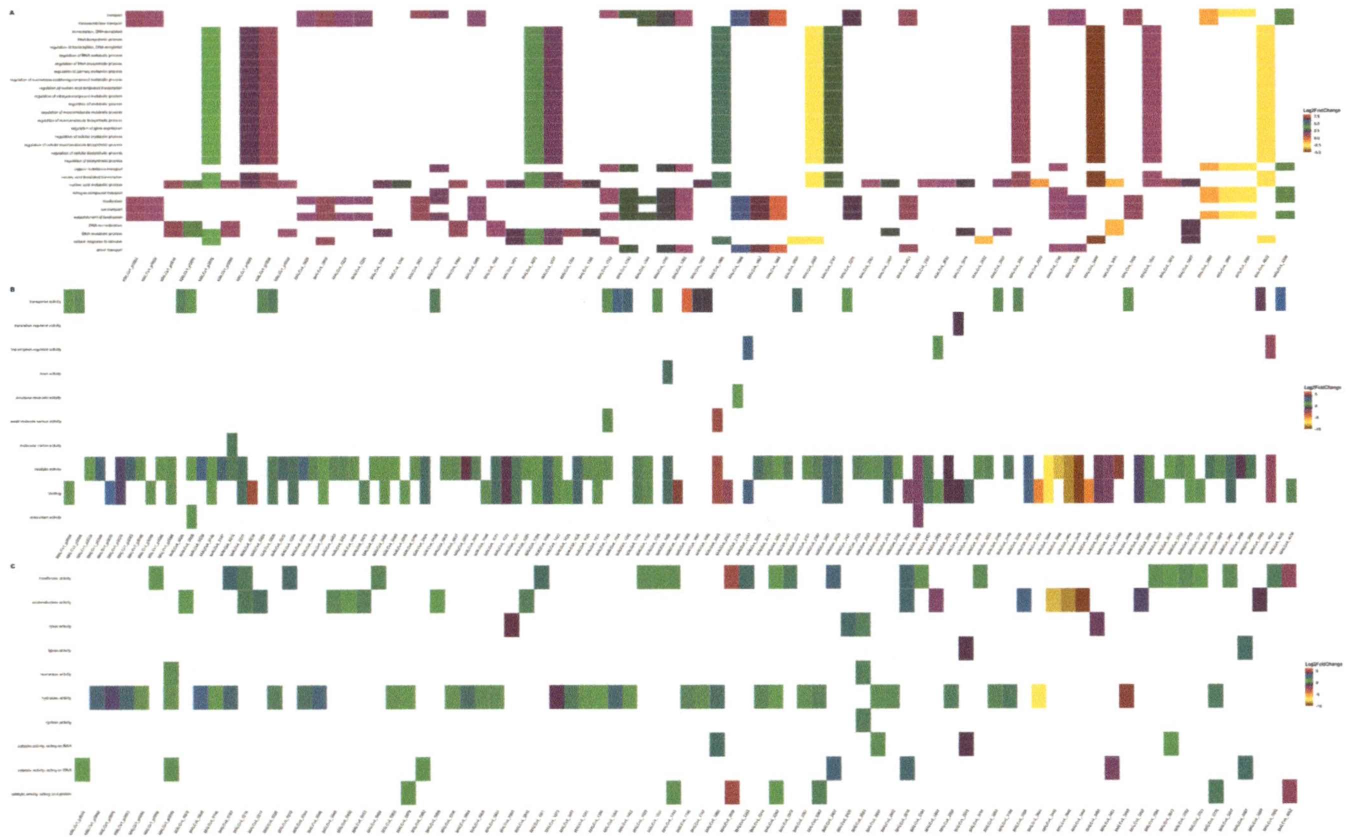

Fig. 4 GO overrepresented (ORA)/enriched heatplots: (A) Enriched GO biological processes with no level restriction; (B) enriched GO molecular functions, restricted to level 2; and (C) enriched GO molecular functions, restricted to level 3.

GSEA can be useful for a top-down perspective of overall expressional shifts; in the case of La shift, on average, the expression of genes is around 70% in comparison to Ca (Log2(Ca vs La) = −0.5); genes involved in oxidation-reduction reactions are particularly negatively affected, upwards of Log2FoldChange = −9.32 (Ca-MDH) (Fig. 5) and Log2FoldChange = −8.80 (Cytochrome *c*).

2.3.4 KEGG functional enrichment and pathway visualization

KEGG organization is structured into three categories: pathways/pathway maps, functional modules, and BRITE database (https://www.genome.jp/kegg/kegg3b.html). Pathway maps are visual representations of cellular and molecular functionalities that have been curated to encompass general biological processes which are usable by different taxa; pathways can be contained within other pathways and operate in a hierarchical structure. Modules are functional units of pathways and represent metabolic processes (such as glycolysis, pyruvate oxidation, and the TCA cycle under "central carbohydrate metabolism"). BRITE organizes proteins into three categories of families: metabolism, genetic information processing, and signaling and cellular processes; the three families categorize proteins according to their functionality (such as "bacterial motility proteins" under "signaling and cellular processes").

ORA and GSEA allow DE genes to be clustered based on their functional relatedness. In some cases, it is more intuitive to spatially visualize expression changes and to infer which metabolic pathways will be affected as a result of up regulation or down regulation of certain genes. Spatial visualization through metabolic pathways is useful as a jumping-off point when considering bioengineering and flux balance analysis for targeted metabolite production. In the case of strain design or determining optimal growth conditions, it is not uncommon to have numerous transcriptomic sampling conditions. If a KEGG organism is available, DE genes between conditions can be visualized on KEGG pathway maps with Pathview (Luo & Brouwer, 2013) to get a broad generalization of which enzymatic steps and metabolic pathways are likely to be affected in comparison to a reference condition (Fig. 6). If a KEGG organism is not available, users can utilize KofamKOALA to conduct a Hidden Markov Model (HMM) search and BlastKOALA for a BLAST-based search to obtain ortholog numbers for genes, then Pathview can be used to visualize transcriptomic data on ortholog maps.

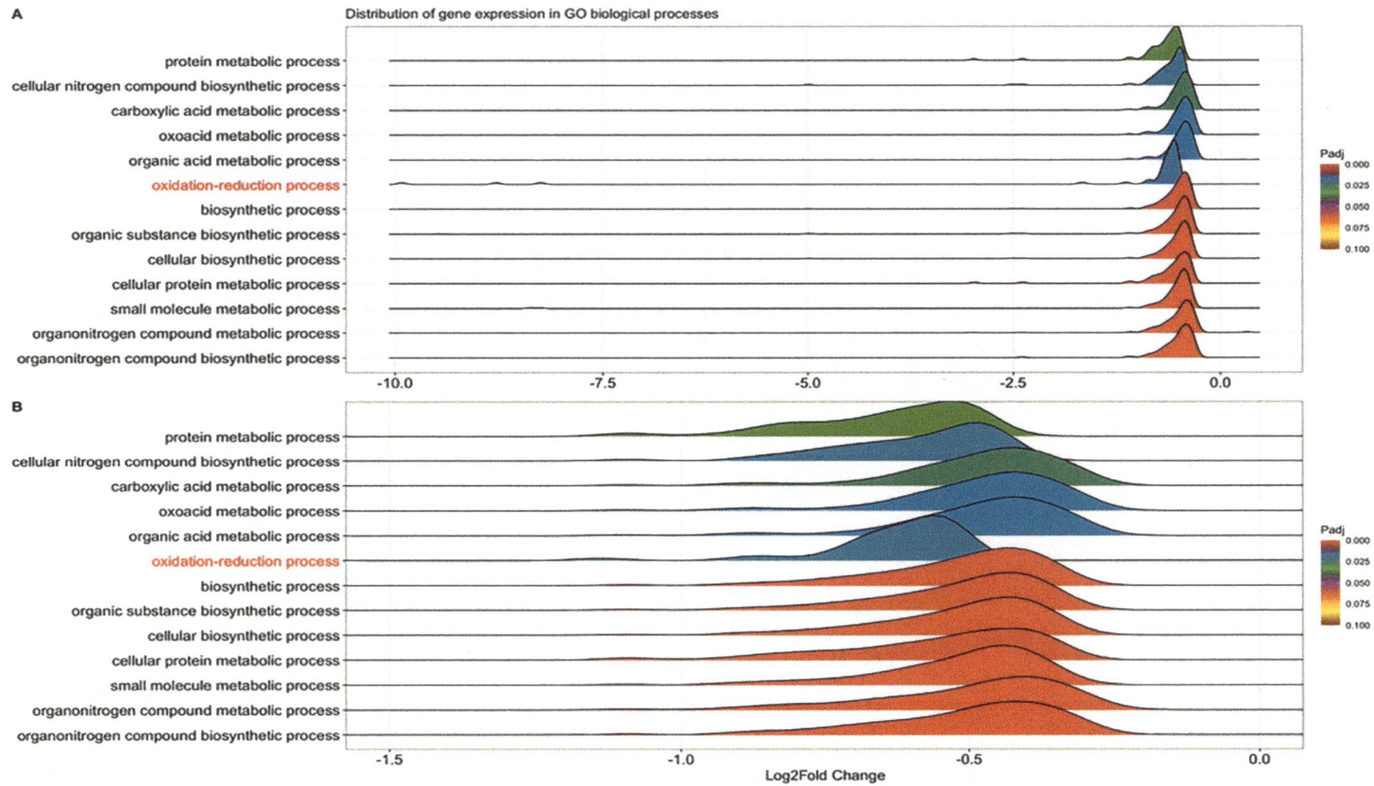

Fig. 5 Ridgeline plots to visualize gene-set enrichment analysis (GSEA) in GO biological processes: (A) Genes involved in the oxidation-reduction process are the most substantially affected during La shift. (B) Concentrated view of expressional change distribution.

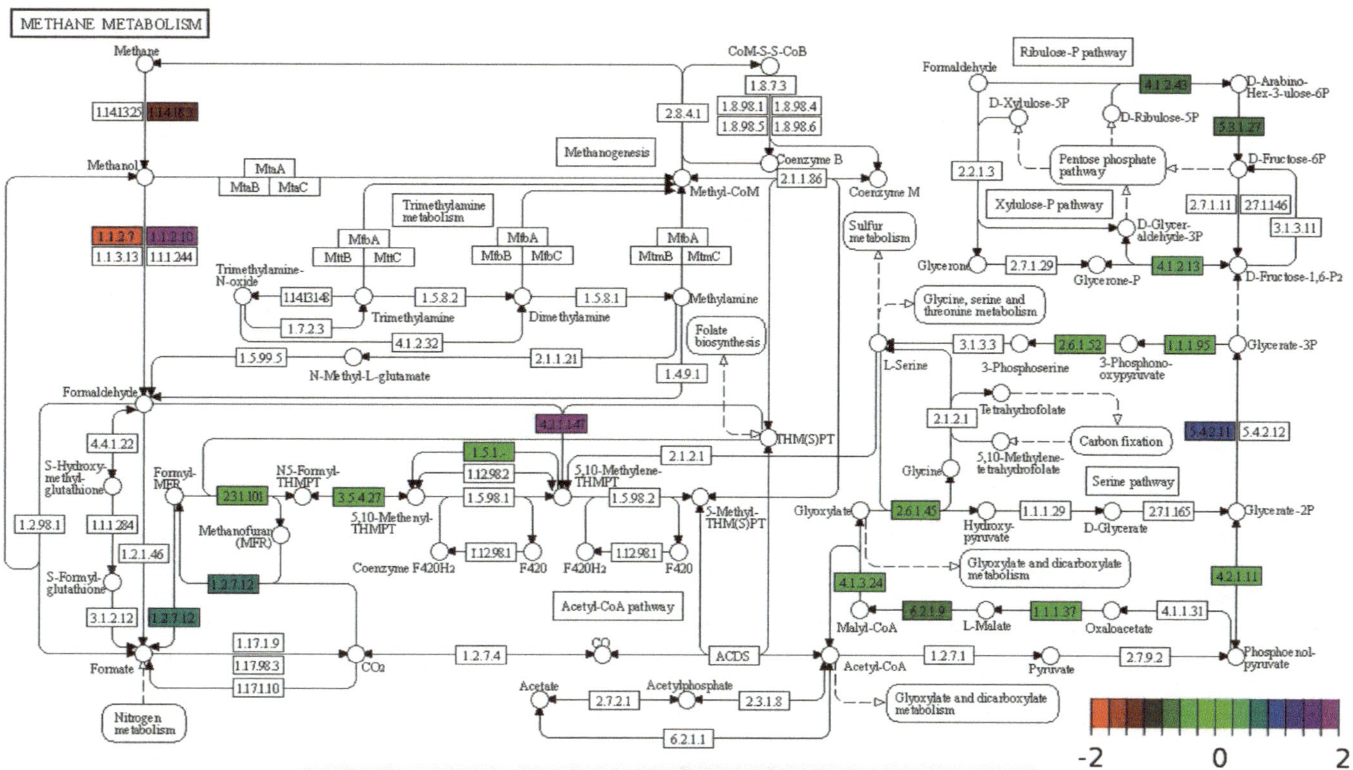

Fig. 6 KEGG Map/Module/BRITE Pathview spatial visualizations. Enzymatic steps affected during La growth conditions in reference to Ca growth conditions (scale of log2fold change) for The Methane Metabolism pathway; the map visualization was abridged to reduce clutter where genes were not significantly changed or if enzymatic steps were absent for the strain *20Z*. Users can specify any map, module, or BRITE hierarchy by their respective organismal code or KEGG ortholog code in Pathview to visualize DE genes. DE genes were filtered by $p_{adj} < 0.05$. Blank KEGG nodes indicate no significant change below statistical threshold. Enzymes are depicted by EC numbers.

Users should have a basic fundamental understanding of which genes belong to their respective pathways, modules, or BRITE hierarchies in order to specify which form of organization is to be visualized. KEGG organizations are not unique for every organism; some pathway maps may contain enzymatic processes which are present in related bacteria, but are not necessarily present for the organism which is being studied. As an example, the Methane Metabolism Pathway map (Fig. 6) contains enzymatic steps for both methanotrophic and methanogenic bacteria, which share interconnected pathways. In reference to Ca-based growth, La growth saw substantially increased expression of La-dependent MDH XoxF (Log2 (Ca vs La) = 3.78, EC: 1.1.2.10) and formaldehyde-activating enzyme (FAE) (Log2(Ca vs La) = 4.22, EC: 4.2.1.147) while having drastically reduced expression of Ca-dependent MDH MxaF (Log2(Ca vs La) = −9.32, EC: 1.1.2.7). If metabolomic data is available which complements experimental conditions, it can also be visualized with Pathview before integrating downstream into stand-alone programs such as The COBRA Toolbox (Heirendt et al., 2019) or Optflux (Rocha et al., 2010) for further analysis.

3. Gene regulatory networks: Overview

The reconstruction of global regulatory networks (GRNs) has never been carried out on any methylotrophic system and very little is known about La-induced gene regulation in methanotrophic bacteria (Groom et al., 2019; Gu & Semrau, 2017; Semrau et al., 2018). However, next-generation sequencing technologies and explosive expansion of available data allow advancement from a gene-level differential gene expression analysis described above to a systems-level analysis of genome-wide co-expression networks and gene regulatory networks (GRNs), and even multi-omics networks. The widely-used weighted gene co-expression network analysis (WGCNA, Zhang & Horvath, 2005) allows unbiased identification and visualization of functional connections between genes, important gene modules, and central players (i.e., hub genes) within modules in specific biological processes or phenotypes (Liu et al., 2018). Indeed, the WGCNA approach can predict functional modules and advance annotation of genes with unknown function through guilt by association even for extremely well-studied model organisms such as *E. coli* (Liu et al., 2018), *S. cerevisiae* (Liu et al., 2017) and *A. thaliana* (Lin et al., 2011). While linear correlation (i.e. Pearson's coefficient) is used in many cases, as a straightforward gene association measure in each pair of genes, it can lead to an overestimated number of not direct or causal (false positive) connections with unknown relationships (Rzhetsky, Gillis, & Pavlidis, 2012) Alternatively, mutual information (MI) method (Meyer, Lafitte, & Bontempi, 2008), Spearman's rank-based correlation, and biweight mid-correlation can be applied to describe the nonlinear relations in pairs of genes. The last two approaches were found to be more robust than Pearson's correlation or simple MI inference methods in the case of nonlinear relations associations or the presence of outliers (Ballouz, Verleyen, & Gillis, 2015; Song, Langfelder, & Horvath, 2012; Zheng, Yuan, Sha, & Sun, 2014). The significant disadvantages of the correlation and MI methods are their inability to distinguish between direct vs indirect interactions and symmetric measures of these interactions excluding the possibility to predict the direction of the relationships between genes. The direction of the relationships in two connected genes has to be derived separately through either heuristic (Opgen-Rhein & Strimmer, 2007) or inductive causation (IC) (Pearl, 1988) algorithms. More sophisticated computational methods have been developed recently to reconstruct GRNs inferring transcriptional factors (TF) directing indirect and causal interaction among genes.

All organisms modulate the expression levels of individual genes in response to changing environmental conditions by regulating the activity of RNA polymerase through a variety of mechanisms including the expression of TFs (Browning & Busby, 2016). Some of the TFs represent genes encoding proteins capable of binding to DNA (individually or as part of a complex) and control the transcription rate of a target gene (TG). The GRNs allow inferring these TFs as well as the regulatory TF-TG relationships, thus revealing the insights of complex intracellular network and guide metabolic engineering efforts. Indeed, it was shown that GRN can be used to predict condition-specific gene expression with 86% accuracy (Fang et al., 2017). The computational approaches for GRN inference include methods based on Boolean networks, ordinary differential equations (ODE) models, modified MI, Bayesian and Neural networks, and ensemble approaches combining advantages of several methods.

While the adoption of artificial neural networks (ANN) for GRN reconstruction remains relatively limited, an increasing number of studies confirms the feasibility of application of ANN with different architectures. Recurrent neural networks (RNNs) inspired by the nervous system structure and function is a potentially excellent tool for modeling GRNs from time-series considering its capacity to learn from temporal, multivariate nonlinear, and noisy data (Kordmahalleh, Sefidmazgi, Harrison, et al., 2017).

A number of RNN-based GRN models have been implemented including neural fuzzy recurrent networks (Maraziotis, Dragomir, & Bezerianos, 2007), RNN coupled with differential evolution and particle swarm optimization (Xu, Venayagamoorthy, & Wunsch, 2007), RNN coupled with generalized extended Kalman filter for weight update in backpropagation (Raza & Alam, 2016), molecular interaction ODE-based RNN model (Ling, Samarasinghe, & Kulasiri, 2013), and hierarchical RNN (Ao & Palade, 2011). Despite this progress, the major problem for the GRN reconstruction using ANN is insufficient sampling (i.e. narrow set of experimental data, a limited number of replicates, etc.) for systems with a large pool of genes. One of the solutions can be clustering the gene expression data as a preprocessing step (Ao & Palade, 2011). Also, similarly to Bayesian networks, other types of biological information could be integrated into the clustering process to improve ANN model predictive power. Moreover, ANN and some other machine learning approaches allow integration of transcriptional data with other omics' datasets as well as structural, spatial, and kinetics knowledge.

4. Final remarks

The described protocols can enable basic gene expression analysis, which often provides new hypotheses for further validation. While DGE is usually sufficient for routing studies, the more in-depth analysis of expression data can provide better resolution, making a more valuable prediction. For example, in the case of La-switch changes, a relatively small set of genes is altered. Cells grown with La showed repression of Ca-MDH and the transcriptional activation of the La-MDH gene, a homolog of the formaldehyde-activating enzyme, a putative transporter, formate dehydrogenase, genes for hemin-transport proteins, and nitrate reductase. Yet, cells exposed to lanthanides display higher growth rates but have lower carbon conversion efficiency than cells supplemented with Ca. The physiological change cannot be captured by DGE analyses with a small set of experimental data. The integration of large datasets across various conditions, including different growth rates, can potentially mitigate the limitations. The Spearman's rank-based correlations between global non-targeted metabolomic data and the flux balance analyses of central metabolic pathways have been used to capture the La-induced redox shift and predict phenotypes (Akberdin et al., 2018).

An ultimate ambition of systems biology is to conceive a comprehensive description of the entire cell by including all complex interactions among various biomolecule classes and link these interactions to cellular level functions or phenotype. While consolidation of multi-omics data is challenging because of its heterogeneity, some of the approaches discussed above can be expanded to infer multi-omics networks. Several recent review studies carefully assessed the challenges of different methods for multi-omics data integration and the inference of integrated networks (Hawe, Theis, & Heinig, 2019; Nicora, Vitali, Dagliati, Geifman, & Bellazzi, 2020; Zhou, Li, & Xia, 2020). WGCNA (correlation and/or MI-based), Bayesian and neural networks are among the most often used models for reconstruction of multi-omics networks including integrated genomics-transcriptomics, genomics-proteomics, transcriptomics-epigenomics, genomics-transcriptomics- proteomics, genomics-transcriptomics-epigenomics networks that can be used for pathway analysis, subgroup identification, and biomarker discovery (Nicora et al., 2020). A more comprehensive multi-scale "Ecomics" model integrating four omics layers (transcriptome, proteome, metabolome, and fluxome with cohesive metadata information) has been developed recently for *E. coli* (Kim, Rai, Zorraquino, & Tagkopoulos, 2016). Importantly, the integration of additional layers and data about the known gene regulatory and protein-protein interactions results in a progressive increase in the prediction performance. The Recurrent Neural Network model with sigmoid activation function and LASSO (least absolute shrinkage and selection operator) constraint regression were used for the transcriptional layer and protein expression values. Another recent study demonstrated the application of neural networks to develop a DCell, a model of the hierarchical structure and function of a eukaryotic cell capable of accurate simulation of cellular growth and subsystem activities that enable a variety of in silico investigations (Ma et al., 2018). Both these studies provide a robust integrative framework and a roadmap for omics-driven predictive modeling allowing in silico guided biological discoveries and advancements for various industrial applications.

Acknowledgments

This work was supported by the U.S. Department of Energy, Office of Science, Office of Biological and Environmental Research, Genomic Science Program under Award Number DE-SC0019181 and the Department of Energy, BioEnergy Technologies Office EE0008501. The Pacific Northwest National Laboratory is operated for DOE by Battelle Memorial Institute under contract DE-AC05-76RL01830.

References

Akberdin, I. R., Collins, D. A., Hamilton, R., Oshchepkov, D. Y., Shukla, A. K., Nicora, C. D., et al. (2018). Rare earth elements alter redox balance in *Methylomicrobium alcaliphilum* 20Z^{R}. *Frontiers in Microbiology*, *9*, 2735. https://doi.org/10.3389/fmicb.2018.02735.

Anaconda Software Distribution. (Nov. 2016). *Computer software. Vers. 2–2.4.0. Anaconda.* https://anaconda.com.

Anders, S., Pyl, P. T., & Huber, W. (2014). HTSeq—A python framework to work with high-throughput sequencing data. *Bioinformatics*, *31*, 166–169. https://doi.org/10.1093/bioinformatics/btu638.

Andrews, S. (2010). *FastQC: A quality control tool for high throughput sequence data*. Available online at: http://www.bioinformatics.babraham.ac.uk/projects/fastqc.

Anthony, C. (1982). *The biochemistry of methylotrophs* (p. 432). New York: Academic Press.

Ao, S. I., & Palade, V. (2011). Ensemble of Elman neural networks and support vector machines for reverse engineering of gene regulatory networks. *Applied Soft Computing*, *11*(2), 1718–1726.

Aramaki, T., Blanc-Mathieu, R., Endo, H., Ohkubo, K., Kanehisa, M., Goto, S., et al. (2020). KofamKOALA: KEGG ortholog assignment based on profile HMM and adaptive score threshold. *Bioinformatics*, *36*(7), 2251–2252. https://doi.org/10.1093/bioinformatics/btz859.

Ballouz, S., Verleyen, W., & Gillis, L. (2015). Guidance for RNA-seq co-expression network construction and analysis: Safety in numbers. *Bioinformatics*, *31*(13), 2123–2130.

Bolger, A. M., Lohse, M., & Usadel, B. (2014). Trimmomatic: A flexible trimmer for Illumina sequence data. *Bioinformatics*, *30*(15), 2114–2120. https://doi.org/10.1093/bioinformatics/btu170.

Boyle, E. I., Weng, S., Gollub, J., Jin, H., Botstein, D., Cherry, J. M., et al. (2004). GO::TermFinder—Open source software for accessing gene ontology information and finding significantly enriched gene ontology terms associated with a list of genes. *Bioinformatics*, *20*(18), 3710–3715.

Browning, D. F., & Busby, S. J. W. (2016). Local and global regulation of transcription initiation in bacteria. *Nature Reviews Microbiology*, *14*(10), 638–650.

Carlson, M., & Pagès, H. (2020). *AnnotationForge: Tools for building SQLite-based annotation data packages. R package version 1.32.0.* https://bioconductor.org/packages/AnnotationForge.

Chistoserdova, L. (2011). Modularity of methylotrophy, revisited. *Environmental Microbiology*, *13*, 2603–2622. https://doi.org/10.1111/j.1462-2920.2011.02464.x.

Chistoserdova, L. (2019). New pieces to the lanthanide puzzle. *Molecular Microbiology*, *111*(5), 1127–1131. https://doi.org/10.1111/mmi.14210.

Chistoserdova, L., Laukel, M., Portais, J. C., Vorholt, J. A., & Lidstrom, M. E. (2004). Multiple formate dehydrogenase enzymes in the facultative methylotroph *Methylobacterium extorquens* AM1 are dispensable for growth on methanol. *Journal of Bacteriology*, *186*(1), 22–28.

Collins, D. A., & Kalyuzhnaya, M. G. (2018). Navigating methane metabolism: Enzymes, compartments, and networks. *Methods in Enzymology*, *613*, 349–383. https://doi.org/10.1016/bs.mie.2018.10.010.

Ewels, P., Magnusson, M., Lundin, S., & Käller, M. (2016). MultiQC: Summarize analysis results for multiple tools and samples in a single report. *Bioinformatics*, *32*(19), 3047–3048. https://doi.org/10.1093/bioinformatics/btw354.

Fang, X., Sastry, A., Mih, N., Kim, D., Tan, J., Yurkovich, J. T., et al. (2017). Global transcriptional regulatory network for Escherichia colirobustly connects gene expression to transcription factor activities. *Proceedings of the National Academy of Sciences*, *114*(38), 10286–10291.

García-Alcalde, F., Okonechnikov, K., Carbonell, J., Cruz, L. M., Götz, S., et al. (2012). Qualimap: Evaluating next-generation sequencing alignment data. *Bioinformatics*, *28*(20), 2678–2679. https://doi.org/10.1093/bioinformatics/bts503.

Götz, S., Garcia-Gomez, J. M., Terol, J., Williams, T. D., Nagaraj, S. H., Nueda, M. J., et al. (2008). High-throughput functional annotation and data mining with the Blast2GO suite. *Nucleic Acids Research*, *36*(10), 3420–3435.

Groom, J. D., Ford, S. M., Pesesky, M. W., & Lidstrom, M. E. (2019). A mutagenic screen identifies a TonB-dependent receptor required for the lanthanide metal switch in the type I methanotroph "*Methylotuvimicrobium buryatense*" 5GB1C. *Journal of Bacteriology*, *201*(15), e00120-19. https://doi.org/10.1128/JB.00120-19.

Grüning, B., Chilton, J., Köster, J., Dale, R., Soranzo, N., Beek, M. V., et al. (2018). Practical computational reproducibility in the life sciences. *Cell Systems*, *6*(6), 631–635. https://doi.org/10.1016/j.cels.2018.03.014.

Gu, W., & Semrau, J. D. (2017). Copper and cerium-regulated gene expression in *Methylosinus trichosporium* OB3b. *Applied Microbiology and Biotechnology*, *101*(23–24), 8499–8516. https://doi.org/10.1007/s00253-017-8572-2.

Hawe, J. S., Theis, F. J., & Heinig, M. (2019). Inferring interaction networks from multi-omics data. *Frontiers in Genetics*, *10*, 535.

Heirendt, L., Arreckx, S., Pfau, T., Mendoza, S. N., Richelle, A., Heinken, A., et al. (2019). Creation and analysis of biochemical constraint-based models using the COBRA toolbox v.3.0. *Nature Protocols*, *14*, 639–702. https://doi.org/10.1038/s41596-018-0098-2.

Jones, P., Binns, D., Chang, H. Y., et al. (2014). InterProScan 5: Genome-scale protein function classification. *Bioinformatics*, *30*(9), 1236–1240. https://doi.org/10.1093/bioinformatics/btu031.

Kalyuzhnaya, M. G., Beck, D. A. C., & Chistoserdova, L. (2011). Functional metagenomics of methylotrophs. *Methods Enzymology*, *495*, 81–98. 1923 https://doi.org/10.1016/B978-0-12-386905-0.00006-1.

Kanehisa, M., Sato, Y., & Morishima, K. (2016). BlastKOALA and GhostKOALA: KEGG tools for functional characterization of genome and metagenome sequences. *Journal of Molecular Biology*, *428*, 726–731. https://doi.org/10.1016/j.jmb.2015.11.006.

Kim, M., Rai, N., Zorraquino, V., & Tagkopoulos, I. (2016). Multi-omics integration accurately predicts cellular state in unexplored conditions for *Escherichia coli*. *Nature Communications*, 7, 13090.

Kordmahalleh, M. M., Sefidmazgi, M. G., Harrison, S. H., et al. (2017). Identifying time-delayed gene regulatory networks via an evolvable hierarchical recurrent neural network. *BioData Mining*, *10*, 29. https://doi.org/10.1186/s13040-017-0146-4.

Langmead, B., & Salzberg, S. L. (2012). Fast gapped-read alignment with bowtie 2. *Nature Methods*, *9*(4), 357–359. https://doi.org/10.1038/nmeth.

Li, H., Handsaker, B., Wysoker, A., Fennell, T., Ruan, J., Homer, N., et al. (2009). The sequence alignment/map format and SAMtools. *Bioinformatics*, *25*(16), 2078–2079. https://doi.org/10.1093/bioinformatics/btp352.

Liao, Y., Smyth, G. K., & Shi, W. (2013). FeatureCounts: An efficient general purpose program for assigning sequence reads to genomic features. *Bioinformatics*, *30*(7), 923–930. https://doi.org/10.1093/bioinformatics/btt656.

Lin, W. D., Liao, Y. Y., Yang, T. J., Pan, C. Y., Buckhout, T. J., & Schmidt, W. (2011). Coexpression-based clustering of Arabidopsis root genes predicts functional modules in early phosphate deficiency signaling. *Plant Physiology*, *155*(3), 1383–1402.

Ling, H., Samarasinghe, S., & Kulasiri, D. (2013). Novel recurrent neural network for modelling biological networks: Oscillatory p53 interaction dynamics. *Biosystems*, *114*(3), 191–205.

Liu, W., Li, L., Long, X., You, W., Zhong, Y., Wang, M., et al. (2018). Construction and analysis of gene co-expression networks in *Escherichia coli*. *Cell*, 7(3), 19. https://doi.org/10.3390/cells7030019.

Liu, W., Li, L., Ye, H., Chen, H., Shen, W., Zhong, Y., et al. (2017). From *Saccharomyces cerevisiae* to human: The important gene co-expression modules. *Biomedical Reports*, 7(2), 153–158.

Love, M. I., Hogenesch, J. B., & Irizarry, R. A. (2016). Modeling of RNA-seq fragment sequence bias reduces systematic errors in transcript abundance estimation. *Nature Biotechnology*, *34*(12), 1287–1291. https://doi.org/10.1038/nbt.3682.

Love, M. I., Huber, W., & Anders, S. (2014). Moderated estimation of fold change and dispersion for RNA-seq data with DESeq2. *Genome Biology*, *15*(12), 550. https://doi.org/10.1186/s13059-014-0550-8.

Luo, W., & Brouwer, C. (2013). Pathview: An R/bioconductor package for pathway-based data integration and visualization. *Bioinformatics*, *29*(14), 1830–1831. https://doi.org/10.1093/bioinformatics/btt285.

Ma, J., Yu, M. K., Fong, S., Ono, K., Sage, E., Demchak, B., et al. (2018). Using deep learning to model the hierarchical structure and function of a cell. *Nature Methods*, *15*(4), 290–298.

Macmanes, M. D. (2014). On the optimal trimming of high-throughput MRNA sequence data. *Frontiers in Genetics*, *5*, 13. https://doi.org/10.3389/fgene.2014.00013.

Maraziotis, I. A., Dragomir, A., & Bezerianos, A. (2007). Gene networks reconstruction and time-series prediction from microarray data using recurrent neural fuzzy networks. *IET Systems Biology*, *1*(1), 41–50.

Meyer, P. E., Lafitte, F., & Bontempi, G. (2008). minet: A R/Bioconductor package for inferring large transcriptional networks using mutual information. *BMC Bioinformatics*, *9*(1), 461.

Mueller, O., Lightfoot, S., & Schroeder, A. (2004). RNA integrity number (RIN)—Standardization of RNA quality control. Agilent Application Note. Publication Number 5989-1165EN.

Nicora, G., Vitali, F., Dagliati, A., Geifman, N., & Bellazzi, R. (2020). Integrated multi-omics analyses in oncology: A review of machine learning methods and tools. *Frontiers in Oncology*, *10*, 1030. https://doi.org/10.3389/fonc.2020.01030.

OmicsBox. (2019). *Bioinformatics made easy*. BioBam Bioinformatics, March 3. https://www.biobam.com/omicsbox.

Opgen-Rhein, R., & Strimmer, K. (2007). From correlation to causation networks: A simple approximate learning algorithm and its application to high-dimensional plant gene expression data. *BMC Systems Biology*, *1*(1), 37.

Oshlack, A., Robinson, M. D., & Young, M. D. (2010). From RNA-seq reads to differential expression results. *Genome Biology*, *11*(12), 220. https://doi.org/10.1186/gb-2010-11-12-220.

Pearl, J. (1988). *Probabilistic reasoning in intelligent systems: Networks of plausible inference*. Morgan Kaufmann Publishers Inc.

Picone, N., & Op den Camp, H. J. M. (2019). Role of rare earth elements in methanol oxidation. *Current Opinion in Chemical Biology*, *49*, 39–44. https://doi.org/10.1016/j.cbpa.2018.09.019.

R Core Team. (2013). *R: A language and environment for statistical computing. R foundation for statistical computing, Vienna, Austria*. http://www.R-project.org/.

Raza, K., & Alam, M. (2016). Recurrent neural network based hybrid model for reconstructing gene regulatory network. *Computational Biology and Chemistry*, *64*, 322–334.

Risso, D. (2018). *EDASeq*. Retrieved November 03, 2020, from https://bioconductor.org/packages/release/bioc/html/EDASeq.html.

Risso, D., Schwartz, K., Sherlock, G., & Dudoit, S. (2011). GC-content normalization for RNA-seq data. *BMC Bioinformatics*, *12*(1), 480. https://doi.org/10.1186/1471-2105-12-480.

Robinson, M. D., Mccarthy, D. J., & Smyth, G. K. (2009). EdgeR: A bioconductor package for differential expression analysis of digital gene expression data. *Bioinformatics*, *26*(1), 139–140. https://doi.org/10.1093/bioinformatics/btp616.

Rocha, I., Maia, P., Evangelista, P., Vilaça, P., Soares, S., Pinto, J. S., et al. (2010). OptFlux: An open-source software platform for in silico metabolic engineering. *BMC Systems Biology*, *4*, 1–12. https://doi.org/10.1186/1752-0509-4-45.

Ross, M. O., & Rosenzweig, A. C. J. (2017). A tale of two methane monooxygenases. *Bioorganic Chemistry*, *22*(2–3), 307–319. https://doi.org/10.1007/s00775-016-1419-y.

Rzhetsky, A., Gillis, J., & Pavlidis, P. (2012). "Guilt by association" is the exception rather than the rule in gene networks. *PLoS Computational Biology*, *8*(3). https://doi.org/10.1371/journal.pcbi.1002444.

Semrau, J. D., DiSpirito, A. A., & Yoon, S. (2010). Methanotrophs and copper. *FEMS Microbiology Reviews*, *34*(4), 496–531. https://doi.org/10.1111/j.1574-6976.2010.00212.x. Epub 2010 Jan 30. PMID: 20236329.

Semrau, J. D., DiSpirito, A. A., Gu, W., & Yoon, S. (2018). Metals and methanotrophy. *Applied Environmental Microbiology*, *84*(6), e02289-17. https://doi.org/10.1128/AEM.02289-17.

Song, L., Langfelder, P., & Horvath, S. (2012). Comparison of co-expression measures: Mutual information, correlation, and model-based indices. *BMC Bioinformatics*, *13*(1), 328.

Stark, R., Grzelak, M., & Hadfield, J. (2019). RNA sequencing: The teenage years. *Nature Reviews Genetics*, *20*(11), 631–656. https://doi.org/10.1038/s41576-019-0150-2.

Subramanian, A., Tamayo, P., Mootha, V. K., Mukherjee, S., Ebert, B. L., Gillette, M. A., et al. (2005). Gene set enrichment analysis: A knowledge-based approach for interpreting genome-wide expression profiles. *Proceedings of the National Academy of Sciences*, *102*(43), 15545–15550.

Vallenet, D. (2006). MaGe: A microbial genome annotation system supported by synteny results. *Nucleic Acids Research*, *34*(1), 53–65. https://doi.org/10.1093/nar/gkj406.

Van Rossum, G., & Drake, F. L. (2009). *Python 3 reference manual*. Scotts Valley, CA: CreateSpace.

Vuilleumier, S., Khmelenina, V. N., Bringel, F., Reshetnikov, A. S., Lajus, A., Mangenot, S., et al. (2012). Genome sequence of the haloalkaliphilic methanotrophic bacterium *Methylomicrobium alcaliphilum* 20Z. *Journal of Bacteriology*, *194*(2), 551–552.

Williams, C. R., Baccarella, A., Parrish, J. Z., & Kim, C. C. (2016). Trimming of sequence reads alters RNA-Seq gene expression estimates. *BMC Bioinformatics*, *17*, 103. https://doi.org/10.1186/s12859-016-0956-2.

Xu, R., Venayagamoorthy, G. K., & Wunsch, D. C. (2007). Modeling of gene regulatory networks with hybrid differential evolution and particle swarm optimization. *Neural Networks*, *20*(8), 917–927.

Yu, G., Wang, L., Han, Y., & He, Q. (2012). clusterProfiler: An R package for comparing biological themes among gene clusters. *OMICS: A Journal of Integrative Biology*, *16*(5), 284–287. https://doi.org/10.1089/omi.2011.0118.

Zhang, B., & Horvath, S. (2005). A general framework for weighted gene co-expression network analysis. *Statistical Applications in Genetics and Molecular Biology*, *4*(17).

Zheng, C.-H., Yuan, L., Sha, W., & Sun, Z. L. (2014). Gene differential co-expression analysis based on biweight correlation and maximum clique. *BMC Bioinformatics*, *15*(3). https://doi.org/10.1186/1471-2105-15-S15-S3.

Zhou, G., Li, S., & Xia, J. (2020). Network-based approaches for multi-omics integration. Computational methods and data analysis for metabolomics. *Methods in Molecular Biology*, *2104*, 469–487.

CHAPTER NINE

Mass spectrometry-based approaches to study lanthanides and lanthanide-dependent proteins in the phyllosphere

Lucas Hemmerle[a], Andrea M. Ochsner[a], Thomas Vonderach[b], Bodo Hattendorf[b], and Julia A. Vorholt[a,*]

[a]Institute of Microbiology, Department of Biology, ETH Zurich, Zurich, Switzerland
[b]Laboratory of Inorganic Chemistry, Department of Chemistry and Applied Biosciences, ETH Zurich, Zurich, Switzerland
*Corresponding author: e-mail address: jvorholt@ethz.ch

Contents

Abstract

Rare-earth elements (REEs) were recently discovered to be biologically significant. The finding was originally made with the methanol dehydrogenase XoxF, which depends on REEs for its activity, and reports of lanthanide-utilizing bacteria have since expanded. Environmental proteomics allows the identification of proteins specifically induced by

Methods in Enzymology, Volume 650
ISSN 0076-6879
https://doi.org/10.1016/bs.mie.2021.01.006

the presence of lanthanides or can provide insights into the preferred use of lanthanide-dependent and -independent isoenzymes, for example. Here we describe protocols for the growth and subsequent mass spectrometry-based proteome analysis of bacteria obtained from controlled artificial media and from the phyllosphere of the model plant *Arabidopsis thaliana*. In addition, the use of inductively coupled plasma mass spectrometry (ICP-MS) is described for the quantification of REEs in biological samples.

1. Introduction

The rare-earth elements (REEs) group contains 17 elements: scandium, yttrium, and the 15 lanthanides (La, Ce, Pr, Nd, Pm, Sm, Eu, Gd, Tb, Dy, Ho, Er, Tm, Yb, Lu). Even though the name "rare" implies a low abundance, the rare-earth elements are common in the upper earth crust. Cerium, in particular, is the 25th most prevalent element in the earth's crust at 63 ppm, but other lanthanides, La, Pr and Nd, are present in the range of 10–40 ppm, which is substantial compared to other metals like Cu, Zn and Mo. To date, more than 200 rare-earth minerals have been identified, most of which occur as phosphate salts or in carbonate minerals such as bastnäsite, monazite, xenotime and loparite (Zhou, Li, & Chen, 2017). The chemical properties of lanthanides and their use in modern technologies such as batteries, magnets and lasers are of great interest to industry and applications have been described in detail over the years (Balaram, 2019). From a biochemical perspective, lanthanides have a higher charge-to-radius ratio than common divalent metals such as Ca^{2+}, Mg^{2+}, which renders them very good Lewis acid catalysts. Nonetheless, in the past lanthanides were considered biologically inactive due to their poor solubility and therefore low bioavailability (Lim & Franklin, 2004). This view changed with the characterization of the pyrroloquinoline quinone (PQQ)- and lanthanide (Ln)-dependent methanol dehydrogenase XoxF in several methylotrophs (Fitriyanto et al., 2011; Hibi et al., 2011; Nakagawa et al., 2012). Methylotrophs are microorganisms that are able to grow on reduced one-carbon (C1) compounds such as methanol or methane as their sole source of carbon and energy. These organisms might be obligate methylotrophs such as *Methylobacillus flagellatus* or facultative methylotrophs such as *Methylobacterium extorquens*. Overall, aerobic methylotrophic bacteria are abundant in natural habitats, including marine environments, soil, and terrestrial plants. In these different environments, they have a major role in the global carbon cycle (Chistoserdova, 2015;

Hanson & Hanson, 1996). Additionally, their ability to use C1 substrates makes them interesting candidates for sugar-independent biotechnology (Chistoserdova, 2018; Schrader et al., 2009). Among the key enzymes in methylotrophic organisms are methanol dehydrogenases (Mdhs), which catalyze the conversion of methanol to formaldehyde and were studied in great detail in different organisms such as *M. extorquens* (Chistoserdova, Chen, Lapidus, & Lidstrom, 2003), *Paracoccus denitrificans* (Harms, de Vries, Maurer, Hoogendijk, & Stouthamer, 1987), *M. flagellatus* (Chistoserdova et al., 2007) and others (Kalyuzhnaya, Hristova, Lidstrom, & Chistoserdova, 2008). One prominent Mdh studied is the PQQ- and Ca^{2+}-dependent Mdh (MxaFI) (Anthony & Williams, 2003), which emerged as a model Mdh and phylogenetic biomarker for methanotrophs (Lau, Fisher, Steudler, & Cavanaugh, 2013), long before XoxF was discovered as an REE-dependent enzyme and with which it shares a phylogenetic relationship. Notably, some methylotrophs only encode XoxF (Jimenez-Infante et al., 2016; Pol et al., 2014; Salcher, Neuenschwander, Posch, & Pernthaler, 2015) and many other methylotrophs (e.g., *M. extorquens* AM1 and PA1) contain both PQQ-dependent Mdhs, namely the Ln-dependent Mdh XoxF and the Ca-dependent Mdh with MxaF as the catalytic subunit (Chistoserdova, 2011). Even though the two enzymes share a significant sequence similarity (Chistoserdova & Lidstrom, 2013) their expression is inversely regulated by Ln^{3+}. While the *mxa* genes are higher expressed in the absence of Ln^{3+}, the *xox* genes are induced in the presence of Ln^{3+} (Chu & Lidstrom, 2016; Gu, Farhan Ul Haque, DiSpirito, & Semrau, 2016; Ochsner et al., 2019). Furthermore, it was shown in *M. extorquens* PA1 that parts of the *xox* gene cluster are fitness-relevant for growth on methanol without REEs (Ochsner et al., 2017) and XoxF is required for expression of MxaF in *M. extorquens* AM1 (Skovran, Palmer, Rountree, Good, & Lidstrom, 2011). The advantages of both isoenzymes and their regulation and other associated REE-binding proteins is not yet fully understood and raises the question of how the Ln-dependent XoxF and other known REE-binding proteins are expressed in natural habitats (Chistoserdova & Kalyuzhnaya, 2018).

Insights into the functionality and prevalence of XoxF *in situ* have been obtained from methylotrophs living in different environments such as the phyllosphere of plants (Delmotte et al., 2009), volcanic mudpots (Pol et al., 2014) and the marine environment (Gifford, Becker, Sosa, Repeta, & DeLong, 2016; Taubert et al., 2015). Particularly, during metagenomics and metaproteomic studies of environmental phyllosphere samples, XoxF was one of the most consistently detected genes as well as one of the most abundant proteins measured (Delmotte et al., 2009; Knief, Delmotte, et al., 2012). The importance of XoxF and the corresponding methylotrophs is further underlined by the consistent presence and rather high abundance of *Methylobacteria* between phylogenetically distinct host plants and geographic regions (Delmotte et al., 2009; Knief, Frances, Cantet, & Vorholt, 2008; Knief, Frances, & Vorholt, 2010; Vorholt, 2012). Some Methylobacteria, also referred to as Pink-Pigmented Facultative Methylotrophs (PPFMs), are abundant and ubiquitous on plants (Vorholt, 2012) and were hypothesized to have plant-beneficial effects leading to increased plant growth (Doronina, Trotsenko, Kuznetsov, Tourova, & Salkinoja-Salonen, 2002; Koenig, Morris, & Polacco, 2002; Madhaiyan et al., 2004). *M. extorquens* PA1 emerged not only as a model system for methylotrophy (Nayak & Marx, 2014; Ochsner et al., 2017) but also as model plant commensal in the phyllosphere where it uses methanol, a by-product of plant pectin synthesis (Knief, Delmotte, et al., 2012; Knief, Dengler, Bodelier, & Vorholt, 2012; Sy, Timmers, Knief, & Vorholt, 2005). The presence of genes for the Ca^{2+} and REE-dependent Mdhs in the genome of *M. extorquens* PA1 and the expression of both in its natural environment, the phyllosphere, make the bacterium a good starting point to address the question of how XoxF and other REE-binding proteins are utilized by bacteria *in situ*.

One approach to study bacteria and their functions is mass spectrometry-based proteomics. By growing the bacteria in their natural environment and comparing to a defined control condition, candidate proteins can be identified that are required specifically in the natural environment. Here we describe a representative workflow depicted in Fig. 1 in which we show how the proteome of a bacterium, here *M. extorquens* PA1, changes upon plant colonization compared to *in vitro* controls. In addition we have shown that REEs are abundant in the *Arabidopsis thaliana* phyllosphere and therefore available to the commensal *in situ* (Ochsner et al., 2019).

2. Cultivation of *A. thaliana* and harvest of the phyllosphere

This section focuses on the plant cultivation and harvesting protocol in order to obtain phyllosphere samples from *A. thaliana* inoculated with a bacterium of interest. While samples may be taken directly from the environment for metaproteome analyses (see Delmotte et al., 2009; Knief, Delmotte, et al., 2012), plants can also be grown under gnotobiotic

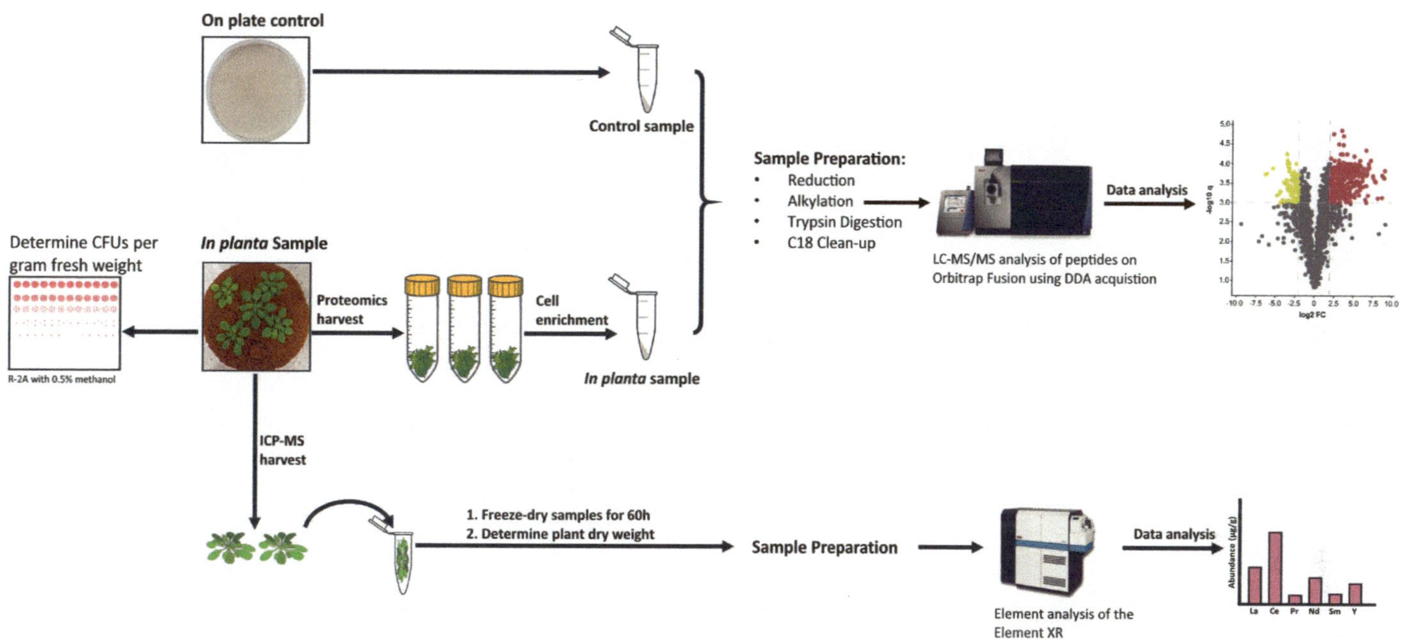

Fig. 1 General overview of the experimental workflow to determine the proteome of bacteria grown in the phyllosphere of *A. thaliana* using mass spectrometry-based proteomics. In parallel, the REE concentration of the calcined clay used for plant cultivation as well as the phyllosphere samples is measured using inductively coupled plasma mass spectrometry (ICP-MS).

conditions, i.e., devoid of bacteria except for bacteria of interest which are specifically inoculated. For plant cultivation under such gnotobiotic conditions different substrates can be used depending on the biological question and frame of the experiment. The MS-agar system is frequently used to study the protective abilities of different bacteria when challenged with the model pathogen *Pseudomonas syringae* DC3000 (Pto) (Innerebner et al., 2011; Vogel et al., 2016). An alternative gnotobiotic system is based on calcined clay (Bai et al., 2015; Carlstrom et al., 2019; Lebeis et al., 2015; Ochsner et al., 2019). Note that calcined clay (Diamond Pro Calcined Clay drying Agent, Cremonini S.R.L., Italy) contains rare-earth elements (REEs), as do the plants grown in this system (Ochsner et al., 2019).

2.1 Seed surface sterilization and plant growth

A. thaliana Col-0 seeds are surface sterilized using 70% ethanol for 2 min. Tubes are spun down, supernatant is removed and seeds are incubated in a 7% NaOCl solution containing 0.2% Triton X-100 for 8 min by constantly inverting the tube. Afterwards, seeds are washed 7 times with ddH_2O by alternating centrifugation and inverting. After the final wash the seeds are stratified at 4 °C in the dark 4–5 days prior to seeding (Innerebner et al., 2011). Before use, the calcined clay is washed 7 times with ddH_2O using a ratio of clay to water of 1:3 (v/v) and heat sterilized for 4 h at 180 °C. In total, for one proteomics experiment we recommend pooling material from about 80–120 plants per condition.

Sterilized seeds are sown in microboxes (O118/80 + OD118 filter: #40, combines, Nazareth, Belgium) containing 175 mL sterile calcined clay and 75 mL ½ Murshige and Skoog medium including vitamins (Murashige & Skoog, 1962) (Sigma-Aldrich, Buchs, Switzerland). For seeding, seeds are distributed in sterile round plates using ddH_2O and pipetted one seed at a time on top of the clay using a P20 pipette adjusted to a volume of 4 μL. A maximum of five plants per pot can be seeded with enough space between one another.

Plants can be grown under a variety of light and temperature conditions. A standard condition is 22 °C, 11 h light and 54% chamber humidity (CU-41L4, Percival Perry, USA) (Carlstrom et al., 2019). Plants can be inoculated with bacteria at chosen time points, e.g., at the seed stage or later such as 10 days of growth. For inoculation, bacteria are pregrown on defined media (to which REE can be added or not, depending on the research question) either in liquid or on 1.5% agar. For growth on solid media, bacteria are suspended and solubilized, e.g., in 10 mM $MgCl_2$. Prior to inoculation of plants the optical density of the suspension should be adjusted; a recommended final OD_{600} is 0.01 (e.g., by adjusting to $OD_{600}=0.1$ and then further diluting 1:10). For inoculation, 10 µL of bacterial solution or 10 mM $MgCl_2$ (mock treatment) is pipetted on each leaf and in the middle of the plant in case of inoculation at the seedling stage. Plants are grown for additional time, e.g., 2 weeks. Plants need to be watered regularly; under the conditions above, it is sufficient to add 1–3 mL ddH_2O per box once a week. *Note*: The water should be added to the side of the boxes to avoid direct contact of liquid and the plants. For harvest of the phyllosphere only, sterilized razor blades can be used to separate the above ground parts from the roots. Make sure to remove the dicotyledons from the rest of the phyllosphere. Further processing of *A. thaliana* phyllosphere samples is described in Sections 2.2–2.4.

Note: The dicotyledons are removed as they are in close contact with the soil (a rich source of nutrients) and often are heavily colonized by bacteria. Furthermore, dicotyledons are in a different growth stage (observed by white, crippled phenotype) from the rest of the phyllosphere and should be therefore not included in the harvest.

2.2 Enumeration of bacterial growth in the phyllosphere

To control for bacterial colonization for the processes described in Sections 2.3–4, it is recommended that bacterial colonization be determined as colony-forming units (CFUs). To do so, a leaf-wash protocol can be used (Innerebner et al., 2011). To this end, 2 mL Eppendorf tubes are filled with 1.3 mL 100 mM phosphate buffer at pH 7. Silwet-L77 at 0.2% (v/v) (Leu + Gygax AG, Birmensdorf, Switzerland) can optionally be added as a non-ionic surfactant in the buffer to increase washing efficiency of bacteria from the phyllosphere. *Note*: Some bacteria might be sensitive to Silwet-L77 and we recommend testing this prior to the plant experiment.

The tubes are weighed before and after the harvest to determine the fresh weight of the plant. For the harvest, single plants are transferred into the buffer-containing tubes and are shaken horizontally for 15 min in a tissue lyzer at 25 Hz, spun down shortly and sonicated for 5 min in a water bath at room temperature. After vortexing, the supernatant containing the bacteria is spotted on R-2A plates (Sigma-Aldrich, Buchs, Switzerland) supplemented with 0.5% methanol (R2A + M) using a tenfold dilution series for the enumeration of CFUs. After 2–3 days of incubation at the appropriate temperature (e.g., 28 °C), depending on the strain(s) used, the plates are counted and the colony-forming units per g plant fresh weight are enumerated.

2.3 Harvest of phyllosphere bacteria for MS-based proteomics

To recover bacteria from the phyllosphere of *A. thaliana*, the phyllosphere of 10 plants are transferred into a 50 mL Falcon tube containing 25 mL ice-cold buffer-containing 10 mM Tris–HCl pH 7.5, 1 mM EDTA and 0.1% (v/v) Silwet L-77 (see above, some bacteria are sensitive). The bacteria are washed off from the phyllosphere through three consecutive cycles of vortexing with sonication in a water bath in-between (first: 1 min vortexing and 30 s sonication; second: 45 s vortexing and 30 s sonication; third: only 10 s sonication). Next, bacteria are separated from leaf material by filtration through a nylon mesh of 200 μm pore size (Spectrum Europe, B.V., Breda, Netherlands). Cells are collected by centrifugation at 3220 × *g* for 10 min at 4 °C, transferred to 2 mL Eppendorf tubes and washed twice with ice-cold 10 mM Tris–HCl pH 7.5 buffer. After the last wash step, cells of each biological replicate are pooled and centrifuged again at 13,000 × *g*. The supernatant is discarded and the cell pellets are snap frozen in liquid N_2 and stored at −80 °C until further processing.

Note: During sampling of aboveground parts of the plant, clay contamination should be avoided. One can carefully scrape off the remaining clay from the lower side of the leaves. Clay particles will lead to shearing of the plant during the harvest increasing the release of plant proteins from lysed plant cells. For MS-based proteomics the contamination of plant proteins will result in a relative decrease of detected bacterial peptides compared to plant peptides, thus reducing the sensitivity toward bacterial peptides.

2.4 Harvest of *A. thaliana* phyllosphere for subsequent analysis by ICP-MS

For element analysis, the phyllosphere of harvested plants are transferred into a 2 mL screw-cap micro tube. *Tip*: One or two plants can be used per tube depending on the size of the plant. We recommend using at least three replicates from plants grown in different microboxes. The plant samples are freeze-dried in a standard freeze drier (Alpha 2-4 LD Plus, Christ, Osterode am Harz, Germany) for 60 h. Lyophilized samples are transferred into new 2 mL tubes and plant dry weight is determined. Samples are stored at −20 °C until further processing.

3. Sample preparation

3.1 Sample preparation for MS-based proteomics of plant-associated and *in vitro* grown bacteria

Below we describe a protocol to analyze bacterial cells using shotgun proteomics.

First prepare 50mM ammonium bicarbonate at pH 8 and the lysis buffer-containing 8M urea, 100mM ammonium bicarbonate and cOmplete EDTA-free protease inhibitor cocktail (1 tablet per 50mL extraction solution) (Sigma-Aldrich, Buchs, Switzerland) with pH 8.

Tip: Before sample preparation let the samples thaw on ice. The lysis buffer without protease inhibitor can be prepared prior to sample processing and stored at −20 °C and be thawed for 30min at RT before use. If the urea is crystalline after thawing of the lysis buffer, it can be solubilized by several cycles of vortexing (5s each).

Reference conditions (e.g. minimal medium with carbon source of choice, +La or −La, etc.) are prepared in liquid artificial media or on plates (see Ochsner et al., 2019). At least five biological replicates per condition are required. It is recommended to plan for cell biomass that results in 2 OD-equivalents (corresponding to 1 mL of $OD_{600}=2$) per sample. Snap freeze cell pellets in liquid N_2 and store at −20 °C until further processing.

Sample preparation for MS-based proteomics of bacterial samples grown *in vitro* and the phyllosphere:

1. Re-suspend samples in 300–600 μL lysis buffer in 2mL tubes.
2. Lyse the cells using a sonicator. For example, the following settings can be used on a VialTweeterHIFU (HIFU, Hielscher, Teltow, Germany):

   A = 100%; C = 80%, 3 × 1 min

   Note: Put the samples on ice between the different cycles to prevent them from heating up.
3. Centrifuge the lysate at 13,000 × *g* for 15min at 4 °C to remove the insoluble fraction.

   Note: If clay particles remain in the supernatant use NucleoSpin columns to remove the particles (proteins are in the flow-through).
4. Transfer lysate into a fresh 2mL tube and take an aliquot for protein concentration determination, e.g., using Pierce™ BCA protein assay kit (Thermo Fisher). *Tip*: A good volume to take from the samples for determination of protein concentration by BCA assay is 20 μL.

   Pause point: After this step the samples can be frozen at −20 °C until further processing.

5. Determine the protein concentration for each sample. *Tip*: Normalize the protein concentration of each sample to approximately the same concentration. A suitable starting amount of total protein is 50–100 μg.
6. Put samples on ice to thaw them if they were stored at −20 °C.
7. Reduce protein disulfide bonds using tris(2-carboxyethyl)phosphine (TCEP) with a final concentration of 5 mM on a tabletop shaker for 30 min at 37 °C with shaking at 300 rpm.
8. Alkylate cysteine residues with iodoacetamide (IAA) for 30 min at room temperature in the dark. *Note*: Iodoacetamide is light sensitive.
9. Fill the samples up with 50 mM ammonium bicarbonate to dilute the 8 M urea to a final concentration of less than 1.5 M. *Note*: Proteases such as trypsin tend to work better at lower urea concentrations.
10. Add sequencing-grade modified trypsin (Promega AG, Dübendorf, Switzerland) at a ratio of 1:20 to 1:100 (trypsin to protein, w/w) per sample to digest the proteins into peptides. *Note*: Adding too little trypsin to the protein solution may result in incomplete digest. This leads to a reduced number of tryptic peptides and thus to a lower protein coverage of the LC-MS/MS assay.
11. Incubate overnight at 37 °C with shaking at 300 rpm. *Note*: In addition, LysC (Promega AG, Dübendorf, Switzerland), or RapiGest (Waters Corporation, Baden-Dättswil, Switzerland) can be added to increase trypsin cleavage leading to more peptides and to reduce the number of missed cleavages.
12. Inactivate trypsin by heating the samples to 95 °C for 5 min and afterwards acidifying samples with formic acid to a final pH of 3.
13. Centrifuge samples at 20,000 × *g* for 10 min to remove insoluble material and transfer supernatant into new 2 mL Eppendorf tubes.

Next, the supernatant is desalted using Sep-Pak Vac C18 reversed phase columns (Waters Corporation, Baden-Dättswil, Switzerland). Here we present a short protocol for the 1 cc/100 mg 55–105 μm cartridge. For the C18 clean-up protocol prepare the following solvent solutions and ensure that the acetonitrile and water used are of ultra-trace purity (e.g., Optima™, FisherScientific, Ultrapure, Merck) and that the purity of the formic acid is ≥98:

Solvent A: contains 80% acetonitrile in 20% water and 0.1% formic acid (v/v/v)

Solvent B: contains 3% acetonitrile in 97% water and 0.1% formic acid (v/v/v)

Solvent C: contains 50% acetonitrile in 50% water and 0.1% formic acid (v/v/v)

Prior to sample loading, pre-wash the column with 1 mL 100% methanol.

14. Precondition the column using 2 × 1 mL solvent A containing 80% acetonitrile (ACN) and 0.1% formic acid (FA).
15. Equilibrate column using 2 × 1 mL solvent B containing 3% acetonitrile (ACN) and 0.1% FA.
16. Load samples on column. *Note*: Carefully load samples in smaller volumes 2–3 × 300 μL and make sure that the column never runs dry.
17. Wash column using solvent B 4 × 1 mL.
18. Slowly (drop by drop) elute samples with 2 × 300 μL solvent C containing 50% ACN and 0.1% FA. *Tip*: You can use a 1 mL pipette tip that is cut off partially to push the liquid slowly through the cartridge into a fresh 1.5 mL tube.
19. Dry samples under vacuum until all of the liquid is evaporated.

    Pause point: After this step the samples can be stored at −80 °C until further processing.

Prior to LC-MS/MS measurement, re-solubilize samples in solvent B containing 3% ACN with 0.1% FA to a final concentration of 0.1–1 $mg\,mL^{-1}$. The peptide samples are now ready for LC-MS/MS analysis.

Note: The sample preparation described in 3.1 is also applicable with different cell lysis approaches. We recommend using an approach with minimal sample loss, as the amount of sample that can be retrieved from the phyllosphere is rather low. Therefore, it is important to work with small volumes and minimize tube switching to minimize the risk of contamination. Another sample preparation approach that has worked in our hands is the Filter-Aided Sample Preparation (Wisniewski, Zougman, & Mann, 2009).

3.2 Sample preparation for element analysis using ICP-MS

To determine its bulk element composition, plant material can be readily dissolved by oxidative digestion.

Various methods can be applied (Korn et al., 2008), while the most suitable for ICP-MS analyses is microwave-assisted digestion using nitric acid (HNO_3). A small amount of hydrogen peroxide (H_2O_2) may be included in the reaction mixture if a black residue remains after digestion. Traces of hydrofluoric acid (HF) can be added if silica (SiO_2) precipitated during the digestion of samples with high silicon content. In such a case, the excess fluoride ions need to be evaporated together with the volatile SiF_4 to avoid precipitation of the barely soluble REE-fluorides. Instead of evaporation, silica can also be removed by filtration if necessary but it needs to be kept in mind that REEs may co-precipitate and will be filtered off too. All reagents used should be of ultra-trace purity (HNO_3: e.g., Optima™, FisherScientific, Ultrapure, Merck, Ultrex,® JT Baker) or purified by sub-boiling distillation before use. Also, all digestion vessels, sample containers and pipette tips should be cleaned by immersion in diluted HNO_3 for 24 h, followed by rinsing 2 times with ultra-pure water and drying before use. *Note*: While working with HNO_3, HF and H_2O_2, we strongly recommend wearing chemical safety goggles, gloves and working under a fume hood or a workplace with proper ventilation.

Digestion of plant material for bulk element analyses of REEs by ICP-MS:

1. Weigh a suitable amount of the sample into a clean digestion vessel. Higher sample amounts allow for lower REE concentrations to be detected. Still, an upper limit exists, which is determined by the capacity of the digestion system used. Consult the manufacturer's recommendation of the maximum sample load.
2. Per 100 mg of sample add 2 mL concentrated HNO_3 (60–69%, ultra-trace purity), and 0.2–0.5 mL high purity water.
3. To minimize the amount of CO_2 gas forming inside the closed digestion vessel, allow the mixture to sit loosely capped for 15–30 min at room temperature.
4. Close the digestion vessel and place it into the microwave digestion system. Then carry out the digestion according to the manufacturers' recommendations of temperature and pressure. Usually a protocol, suitable for food (leaves, vegetables), is provided from the instrument manufacturer. The following protocol (Carbohydrates) was successfully used with a Turbowave™ (MLS GmbH, Germany) high pressure digestion system:

Time	MW power	Temperature	Pressure
2:30	800 W	25–80 °C	100 bar
8:00	1000 W	160 °C	120 bar
3:00	1200 W	160–220 °C	120 bar
25:00	1200 W	220 °C	120 bar

5. Remove the digestion vials, allow to cool to room temperature if necessary and inspect the digest:
   5.1. If a clear solution was obtained and total volume of the digest is below 2 mL proceed with *Step 6*.

5.2. If the solution remained colored, the organic matrix may have not been decomposed completely, which can cause measurement artifacts. Repeat the digestion (*Step 4*) after adding 0.2 mL of H_2O_2 (30%, ultra-trace purity) per 100 mg of sample. If a clear solution was obtained, proceed with *Step 5.4*.

5.3. If precipitation of silica has occurred, add 0.1 mL of concentrated HF (48%, ultra-trace purity) and repeat the digestion (*Step 4*).

5.3.1. Afterwards place the open digestion vessel on a hotplate at 110–150 °C and evaporate to incipient dryness. Always keep the vessel below boiling temperature of the solution to avoid splashing.

5.3.2. Add 0.5 mL of concentrated HNO_3 and evaporate again. Repeat 3 times.

5.3.3. Add 2 mL of diluted HNO_3 (2%, ultra-trace purity in ultra-pure water) and verify that a clear solution is obtained. Proceed with *Step 6*.

5.4. Place the open digestion vessel on a hotplate at 110–150 °C and evaporate to incipient dryness. Add 2 mL of diluted HNO_3 (2%, ultra-trace purity in ultra-pure water) and verify that a clear solution is obtained and no precipitation occurred. Proceed with *Step 6*. Otherwise continue with *Step 5.2* or *Step 5.3*.

6. Transfer the digest into a clean, suitable vial (e.g., 15 mL poly-propylene centrifuge tubes). Wash the walls of the digestion vessel 3–5 times with 1 mL of diluted HNO_3 (2%, ultra-trace purity diluted in ultra-pure water) and combine all wash solution into the vial.
7. Add 0.1 mL of a solution containing 200 ng/g Indium as internal standard for quantification by ICP-MS.
8. Make up to a total of 10 g of solution with the diluted HNO_3. Seal the vial and mix thoroughly.

Pause point: Store the solution in a refrigerator (4 °C) until analysis. It may be advantageous to add a small volume (e.g., 0.1 mL) of a solution with known amount (e.g., 200 ng/L) of an element with negligible abundance in the sample (e.g., Rh) as a recovery standard in *Step 2* to monitor possible sample loss during preparation.

4. Mass spectrometry acquisition parameters

In this section, we will describe the acquisition parameters for both mass spectrometry-based approaches (LC-MS/MS and ICP-MS) used in

Ochsner et al. (2019). These are example guidelines; depending on the LC setup and MS instrument, parameters will have to be adjusted.

Note: Different commercial software tools for MS-based proteomics analysis exist. In the methods description below Progenesis QI is used; an alternative free tool for label-free quantification is MaxQuant (www.maxquant.org) (Cox & Mann, 2008).

4.1 LC-MS/MS-based proteomics

The mass spectrometry analysis of the peptide samples can be performed on an EASY-nLC 1000 system (Thermo Fisher Scientific) coupled to an Orbitrap Fusion (Thermo Fisher Scientific). Chromatographic separation is performed using an acetonitrile/water solvent system containing two channels with 0.1% (v/v) formic acid for channel A and 0.1% (v/v) formic acid, 99.9% (v/v) acetonitrile for channel B. Two microliters of peptides at a concentration of ca. 0.5 μg μL^{-1} are loaded on an EASY-Spray C18 LC column (75 μm × 500 mm, Thermo Fisher Scientific) heated to 50 °C and eluted at a flow rate of 300 nL min^{-1} by a gradient from 2% to 5% B in 2 min, 25% B in 93 min, 35% B in 10 min and 95% B in 10 min. The mass spectrometer is configured to data-dependent acquisition and operating in top speed mode with a cycle time of 3 s. Full-scan MS spectra are acquired in the Orbitrap analyzer with a mass range of 300–1500 *m/z* and a resolution of 120 k with an automated gain control (AGC) targeted value of 4×10^5. HCD peptide fragments (isolation window 1.6 *m/z*) are obtained using a normalized collision energy of 30 with an AGC target value of 2000 and acquired in the ion trap with rapid scan rate. To avoid multiple scans of dominant ions, dynamic exclusion is set to 25 s. The internal lock mass calibration is set on *m/z* 371.10124 and 445.12003. *Tip*: In addition, injection of a pooled sample containing equal amounts of all individual samples can be performed and later used as an artificial alignment reference for Progenesis QI enabling better alignment results.

The raw MS files are then analyzed. Commercial software packages can be used such as Progenesis QI (Nonlinear Dynamics) using the High-Mass Accuracy Instrument option. Automatic alignment is performed using the artificial alignment reference. *Tip*: In the alignment step, 3–5 vectors along the retention time gradient can be manually seeded to aid the automatic alignment. From each Progenesis peptide ion (default sensitivity in peak picking), a maximum of the top five tandem mass spectra are exported using the charge deconvolution, the deisotoping option and a maximum number of 200 peaks per MS/MS. The Mascot generic file (.mgf) is then searched using the Mascot server (Matrix Science) against a forward and reversed protein sequence database (in FASTA format) containing the annotated proteins of the bacterium of interest (e.g., *M. extorquens* PA1 (NC_010172.1)). In addition, it is recommended to concatenate the database with a larger database such as the reversed decoyed database of the 6721 yeast proteins for better false discovery rate assessment during protein identification. Furthermore, we recommend adding known mass spectrometry contaminants, such as keratins, the protease used for digestion (e.g., trypsin) and, if used, the 11 iRT peptides (Escher et al., 2012). *Note*: The iRT peptides are commercial synthetic peptides that can be used for retention time normalization and as internal LC-MS quality control (www.biognosys.org). In addition, when bacterial samples obtained from plants are included, the respective database should be added such as the *A. thaliana* TAIR database to identify potential plant protein contaminations. Parameters for precursor ion tolerance and fragment ion tolerance are set to ± 10 ppm and ± 0.5 Da, respectively. The corresponding protease is set as the protein cleaving enzyme (e.g., trypsin) and two missed cleavages are allowed. Carbamidomethylation of cysteine is set as a fixed modification, and oxidation of methionine, and carbamylation of the N-terminus and lysine are set as variable modifications. The Mascot database search result is loaded into Scaffold (Proteome Software) using 5% peptide and 10% protein false discovery rate (FDR). The Scaffold Spectrum Report is exported and imported back into Progenesis QI. Normalization is performed using all proteins. For quantification, it is recommended that only proteins that have been identified with at least three unique peptide ions be used. Proteins are grouped with Progenesis and the normalized abundance of the three most abundant peptide ions (relative quantification using Hi-3) from the same protein group are averaged together individually for each sample. For statistical testing, one-way ANOVA is applied on the normalized protein abundance. The resulting p-values are corrected using the Benjamini-Hochberg correction directly in Progenesis QI.

Note: There are some basic but important guidelines in how to handle MS-based proteomics-derived data. Most of the guidelines can be found in the journals specialized in proteomics, such as *Molecular & Cellular Proteomics* (https://www.mcponline.org/page/content/mass-spec-guidelines), *Journal of Proteome Research* and the *Journal of Proteomics*. Furthermore, most journals request that authors deposit the original MS raw files, the database used for protein identification, and the processed data in specific repositories such as the ProteomeXchange consortium or its members, respectively (http://www.proteomexchange.org/) (Deutsch et al., 2017).

4.2 ICP-MS

The determination of REEs using ICP-MS is routine in many laboratories, especially in the earth sciences. Due to the comparably small amount of dissolved salts in the plant digests, non-spectral interferences are typically not very pronounced. Spectral interferences can occur however, in particular when REE heavier than Sm shall be quantified as well. Depending on the instrument used and the operating conditions, an ICP instrument produces varying amounts of molecular ions, which can cause spectral overlap. In the case of REE analyses, the most critical of these occur through O- and OH-containing molecular ions of Ba and the lighter REE. Even so called "high-resolution ICP-MS" can usually not resolve these molecular ions from the atomic ions of the heavier REEs by mass spectrometric means. Ar-containing molecular ions can also occur, but typically at much lower abundances. While, e.g., BaO^+ can occur in the mass spectrum at % values relative to the atomic Ba^+ ions, $BaAr^+$ was only observed at sub-ppm levels. Table 1 lists the isotopes recommended for the analysis of the REEs together with the most likely origin of spectral overlap. Spectral interferences on the recommended isotopes of Y, La, Ce, Pr, Nd and Sm are not likely in plant samples, but the analysis of heavier REEs can be substantially affected.

Different methods can be applied to mitigate spectral overlap from the low-m/z REE-$O(H)^+$ ions. The simplest way includes mathematical correction of the measured ion signals via experimentally determined intensity ratios of the interfering species (Zhao et al., 2019). If available, collision cell ICP-MS in so called "energy discrimination" mode (Hattendorf & Gunther, 2004) or ICP-MS/MS methods (Santoro et al., 2016) can be used. It is further recommended to include Ba and the lighter REEs in the analysis of heavy REEs to monitor their abundance in the sample.

The quantitative analysis is performed using external calibration standards of the target REE with Indium as internal standard. If a recovery standard was used, it must be included in the calibration as well. REE calibration standards can be made from multi- or single element standards available in sufficient purity from various manufacturers. REE where spectral overlaps are likely to occur (e.g., Gd), should, however, not be mixed with those causing the interference (e.g., Ce, Pr, Nd). Thus, if the entire suite of REEs needs to be quantified, at least three sets of calibration standards should be used:

1. a multi-element set containing Y, La, Ce, Nd, Sm, Eu, Yb and Lu
2. a multi-element set containing Gd, Tb, Dy, Ho, Er and Tm

Table 1 Isotope selection for REE analyses and potential O-, OH- and Ar-based spectral overlaps.

Isotope	Spectral interference from[a]	
	With O, OH	**With Ar**
^{89}Y		Ti
^{139}La	Sb, Te	Ru
^{140}Ce	Sn, Te	Mo, Ru
^{141}Pr	Te	Ru
^{143}Nd, **^{146}Nd**	I, Te	Rh, Pd, Sn
^{147}Sm, ^{149}Sm	Cs	Ag
^{151}Eu, ^{153}Eu	***Ba***	Cd, In
^{157}Gd, **^{158}Gd**[b]	***Ce, Pr, Nd***	Sn
^{159}Tb	***Nd***	Sn
^{161}Dy, **^{163}Dy**	***Nd, Sm***	Sb, Te
^{165}Ho	Nd, ***Sm***	Te
^{166}Er, ^{167}Er	***Nd, Sm***	Te, I
^{169}Tm	Sm, ***Eu***	
^{172}Yb, ^{173}Yb	***Gd***, Dy	Cs
^{175}Lu	Gd, ***Tb***	Ba

[a]Elements that may form molecular ions at the same *m/z* of the preferred isotope.
[b]The spectral overlaps from $^{158}Dy^+$, $^{142}Ce^{16}O^+$ and $^{142}Nd^{16}O^+$ are often less intense than those from $^{141}Pr^{16}O^+$ and $^{140}Ce^{16}OH^+$, making ^{158}Gd the preferred isotope for analysis.
Isotopes in bold letters are typically least affected by spectral overlap. Elements in bold italic indicate the dominant source of a spectral interference.

3. a single element set for Nd

Thereby the direct spectral overlap of the dominant O- and OH-containing spectral interferences can be avoided and it is possible to determine the formation rate of these interferences across the REEs. The concentration range to be covered by the calibration standards depends also on the sample mass used in the digestion. Modern ICP-MS instruments should allow for linear calibration from 0.01 to 10 s of ng/g in solution, which is recommended for samples with unknown REE content. If the concentration range can be estimated beforehand, the range may be adjusted individually. All calibration standards are prepared in diluted HNO_3 and contain 2 ng/g of In as internal standard. If a recovery standard was used in the digestion, an additional set of calibration standards ranging from 0.5 to 1.5 times its final concentration must be included in the calibration as well.

The ICP-MS should be operated according to the manufacturer's recommendations. During instrument setup it is advised to adjust the ICP operating conditions to minimize the abundance of O- and OH-containing molecular ions, if heavy REEs shall be determined. The analysis method includes all recommended isotopes of the target REE, ^{115}In, and the recovery standard (e.g., ^{103}Rh). Measurements are performed in peak-hopping mode using integration times of at least 100 ms per isotope and at least 10 replicates per analysis. The digests can be run directly after having warmed up to ambient temperature and the REE concentration calculation is done via the instrument's data reduction software. For quality control purposes it is recommended to monitor the sensitivity for the internal standard In, and the concentration for the recovery standard if used. ^{115}In sensitivity in a digest should not differ by more than 25% from that of the calibration standards and the concentration of the recovery standard should be within 15% of the expected value. Further recommended controls are analysis of certified reference materials, spike recovery testing, and duplicate analyses of several samples in a random sequence.

If quality controls fail, a matrix effect must be considered and the analysis repeated after further diluting the digest (5–10 times) in 2% HNO_3.

5. General experimental considerations

5.1 Experimental considerations while working with REEs

Glassware can contain variable amounts of REEs originating from the manufacturing process, which leads to experimental variability (Pol et al., 2014). Furthermore, despite harsh washing protocols REEs were found to remain in shake flasks from previous cultivations (Vu et al., 2016). It is therefore preferable to work with single-use plastic containers as much as possible, especially for the *in vitro* growth of microorganisms for comparisons.

Another challenge is posed by the low solubility of certain REE salts in water. Of special importance for biological experiments is the extremely low solubility of REE phosphate salts (Firsching & Brune, 1991). REE stock solutions can be prepared from the commercially available chloride salts, which are readily soluble in water, but when preparing growth media thorough mixing before aliquoting or pouring is crucial to achieve homogenous distribution of REEs.

6. Summary

Here, we give an example of the application of an MS-based approach to study bacteria grown both *in vitro* as well as *in situ*. This approach can help to effectively identify adaptation strategies and mechanisms bacteria apply to survive and thrive in this environment (Muller, Schubert, Rost, Aebersold, & Vorholt, 2016; Ochsner et al., 2019). In particular, the approach has demonstrated so far that REE-dependent enzymes are indeed produced by a model methylotrophic bacterium in its natural habitat (Ochsner et al., 2019). This further illustrated that REEs are of biological significance for organisms that thrive in this environment. In summary, the methods described here can be used to study the adaptation of other bacteria to the plant environment and other natural habitats in order to further illuminate the importance of rare-earth elements in biology.

Acknowledgment

This work was supported by a grant from the Swiss National Science Foundation (grant no. 31003A_173094).

References

Anthony, C., & Williams, P. (2003). The structure and mechanism of methanol dehydrogenase. *Biochimica et Biophysica Acta*, *1647*(1–2), 18–23.

Bai, Y., Muller, D. B., Srinivas, G., Garrido-Oter, R., Potthoff, E., Rott, M., et al. (2015). Functional overlap of the Arabidopsis leaf and root microbiota. *Nature*, *528*(7582), 364–369.

Balaram, V. (2019). Rare earth elements: A review of applications, occurrence, exploration, analysis, recycling, and environmental impact. *Geoscience Frontiers*, *10*(4), 1285–1303.

Carlstrom, C. I., Field, C. M., Bortfeld-Miller, M., Muller, B., Sunagawa, S., & Vorholt, J. A. (2019). Synthetic microbiota reveal priority effects and keystone strains in the Arabidopsis phyllosphere. *Nature Ecology and Evolution*, *3*(10), 1445–1454.

Chistoserdova, L. (2011). Modularity of methylotrophy, revisited. *Environmental Microbiology*, *13*(10), 2603–2622.

Chistoserdova, L. (2015). Methylotrophs in natural habitats: Current insights through metagenomics. *Applied Microbiology and Biotechnology*, *99*(14), 5763–5779.

Chistoserdova, L. (2018). Applications of methylotrophs: Can single carbon be harnessed for biotechnology? *Current Opinion in Biotechnology*, *50*, 189–194.

Chistoserdova, L., Chen, S.-W., Lapidus, A., Lidstrom, M. E., et al. (2003). Methylotrophy in *Methylobacterium extorquens* AM1 from a genomic point of view. *Journal of Bacteriology*, *185*(10), 2980–2987.

Chistoserdova, L., & Kalyuzhnaya, M. G. (2018). Current trends in methylotrophy. *Trends in Microbiology*, *26*(8), 703–714.

Chistoserdova, L., Lapidus, A., Han, C., Goodwin, L., Saunders, L., Brettin, T., et al. (2007). Genome of *Methylobacillus flagellatus*, molecular basis for obligate methylotrophy, and polyphyletic origin of methylotrophy. *Journal of Bacteriology*, *189*(11), 4020–4027.

Chistoserdova, L., & Lidstrom, M. E. (2013). Aerobic methylotrophic prokaryotes. In E. Rosenberg, E. F. DeLong, S. Lory, E. Stackebrandt, & F. Thompson (Eds.), *The prokaryotes: Prokaryotic physiology and biochemistry* (pp. 267–285). Berlin, Heidelberg: Springer Berlin Heidelberg.

Chu, F., & Lidstrom, M. E. (2016). XoxF acts as the predominant methanol dehydrogenase in the type I methanotroph *Methylomicrobium buryatense*. *Journal of Bacteriology*, *198*(8), 1317–1325.

Cox, J., & Mann, M. (2008). MaxQuant enables high peptide identification rates, individualized p.p.b.-range mass accuracies and proteome-wide protein quantification. *Nature Biotechnology*, *26*(12), 1367–1372.

Delmotte, N., Knief, C., Chaffron, S., Innerebner, G., Roschitzki, B., Schlapbach, R., et al. (2009). Community proteogenomics reveals insights into the physiology of phyllosphere bacteria. *Proceedings of the National Academy of Sciences of the United States of America*, *106*(38), 16428–16433.

Deutsch, E. W., Csordas, A., Sun, Z., Jarnuczak, A., Perez-Riverol, Y., Ternent, T., et al. (2017). The ProteomeXchange consortium in 2017: Supporting the cultural change in proteomics public data deposition. *Nucleic Acids Research*, *45*(D1), D1100–D1106.

Doronina, N. V., Trotsenko, Y. A., Kuznetsov, B. B., Tourova, T. P., & Salkinoja-Salonen, M. S. (2002). *Methylobacterium suomiense* sp. nov. and *Methylobacterium lusitanum* sp. nov., aerobic, pink-pigmented, facultatively methylotrophic bacteria. *International Journal of Systematic and Evolutionary Microbiology*, *52*(Pt. 3), 773–776.

Escher, C., Reiter, L., MacLean, B., Ossola, R., Herzog, F., Chilton, J., et al. (2012). Using iRT, a normalized retention time for more targeted measurement of peptides. *Proteomics*, *12*(8), 1111–1121.

Firsching, F. H., & Brune, S. N. (1991). Solubility products of the trivalent rare-earth phosphates. *Journal of Chemical and Engineering Data*, *36*(1), 93–95.

Fitriyanto, N. A., Fushimi, M., Matsunaga, M., Pertiwiningrum, A., Iwama, T., & Kawai, K. (2011). Molecular structure and gene analysis of Ce3+-induced methanol dehydrogenase of *Bradyrhizobium* sp. MAFF211645. *Journal of Bioscience and Bioengineering*, *111*(6), 613–617.

Gifford, S. M., Becker, J. W., Sosa, O. A., Repeta, D. J., & DeLong, E. F. (2016). Quantitative transcriptomics reveals the growth- and nutrient-dependent response of a streamlined marine methylotroph to methanol and naturally occurring dissolved organic matter. *mBio*, 7(6), 01279-16.

Gu, W., Farhan Ul Haque, M., DiSpirito, A. A., & Semrau, J. D. (2016). Uptake and effect of rare earth elements on gene expression in *Methylosinus trichosporium* OB3b. *FEMS Microbiology Letters*, *363*(13), fnw129.

Hanson, R. S., & Hanson, T. E. (1996). Methanotrophic bacteria. *Microbiological Reviews*, *60*(2), 439–471.

Harms, N., de Vries, G. E., Maurer, K., Hoogendijk, J., & Stouthamer, A. H. (1987). Isolation and nucleotide sequence of the methanol dehydrogenase structural gene from *Paracoccus denitrificans*. *Journal of Bacteriology*, *169*(9), 3969–3975.

Hattendorf, B., & Gunther, D. (2004). Suppression of in-cell generated interferences in a reaction cell ICP-MS by bandpass tuning and kinetic energy discrimination. *Journal of Analytical Atomic Spectrometry*, *19*(5), 600–606.

Hibi, Y., Asai, K., Arafuka, H., Hamajima, M., Iwama, T., & Kawai, K. (2011). Molecular structure of La3+-induced methanol dehydrogenase-like protein in *Methylobacterium radiotolerans*. *Journal of Bioscience and Bioengineering*, *111*(5), 547–549.

Innerebner, G., Knief, C., & Vorholt, J. A. (2011). Protection of Arabidopsis thaliana against leaf-pathogenic *Pseudomonas syringae* by *Sphingomonas* strains in a controlled model system. *Applied and Environmental Microbiology*, 77(10), 3202–3210.

Jimenez-Infante, F., Ngugi, D. K., Vinu, M., Alam, I., Kamau, A. A., Blom, J., et al. (2016). Comprehensive genomic analyses of the OM43 clade, including a novel species from the Red Sea, indicate ecotype differentiation among marine methylotrophs. *Applied and Environmental Microbiology*, *82*(4), 1215–1226.

Kalyuzhnaya, M. G., Hristova, K. R., Lidstrom, M. E., & Chistoserdova, L. (2008). Characterization of a novel methanol dehydrogenase in representatives of *Burkholderiales*: Implications for environmental detection of methylotrophy and evidence for convergent evolution. *Journal of Bacteriology*, *190*(11), 3817–3823.

Knief, C., Delmotte, N., Chaffron, S., Stark, M., Innerebner, G., Wassmann, R., et al. (2012). Metaproteogenomic analysis of microbial communities in the phyllosphere and rhizosphere of rice. *The ISME Journal*, *6*(7), 1378–1390.

Knief, C., Dengler, V., Bodelier, P. L., & Vorholt, J. A. (2012). Characterization of *Methylobacterium* strains isolated from the phyllosphere and description of *Methylobacterium longum* sp. nov. *Antonie Van Leeuwenhoek*, *101*(1), 169–183.

Knief, C., Frances, L., Cantet, F., & Vorholt, J. A. (2008). Cultivation-independent characterization of methylobacterium populations in the plant phyllosphere by automated ribosomal intergenic spacer analysis. *Applied and Environmental Microbiology*, *74*(7), 2218–2228.

Knief, C., Frances, L., & Vorholt, J. A. (2010). Competitiveness of diverse *Methylobacterium* strains in the phyllosphere of Arabidopsis thaliana and identification of representative models, including *M. extorquens* PA1. *Microbial Ecology*, *60*(2), 440–452.

Koenig, R. L., Morris, R. O., & Polacco, J. C. (2002). tRNA is the source of low-level trans-zeatin production in *Methylobacterium* spp. *Journal of Bacteriology*, *184*(7), 1832–1842.

Korn, M. D. A., Morte, E. S. D., dos Santos, D. C. M. B., Castro, J. T., Barbosa, J. T. P., Teixeira, A. P., et al. (2008). Sample preparation for the determination of metals in food samples using spectroanalytical methods—A review. *Applied Spectroscopy Reviews*, *43*(2), 67–92.

Lau, E., Fisher, M. C., Steudler, P. A., & Cavanaugh, C. M. (2013). The methanol dehydrogenase gene, mxaF, as a functional and phylogenetic marker for proteobacterial methanotrophs in natural environments. *PLoS One*, *8*(2), e56993.

Lebeis, S. L., Paredes, S. H., Lundberg, D. S., Breakfield, N., Gehring, J., McDonald, M., et al. (2015). Salicylic acid modulates colonization of the root microbiome by specific bacterial taxa. *Science*, *349*, 860–864.

Lim, S., & Franklin, S. J. (2004). Lanthanide-binding peptides and the enzymes that might have been. *Cellular and Molecular Life Sciences*, *61*(17), 2184–2188.

Madhaiyan, M., Saravanan, V. S., Jovi, D. B., Lee, H., Thenmozhi, R., Hari, K., et al. (2004). Occurrence of *Gluconacetobacter diazotrophicus* in tropical and subtropical plants of Western Ghats, India. *Microbiological Research*, *159*(3), 233–243.

Muller, D. B., Schubert, O. T., Rost, H., Aebersold, R., & Vorholt, J. A. (2016). Systems-level proteomics of two ubiquitous leaf commensals reveals complementary adaptive traits for phyllosphere colonization. *Molecular & Cellular Proteomics*, *15*(10), 3256–3269.

Murashige, T., & Skoog, F. (1962). A revised medium for rapid growth and bio assays with tobacco tissue cultures. *Physiologia Plantarum*, *15*(3), 473–497.

Nakagawa, T., Mitsui, R., Tani, A., Sasa, K., Tashiro, S., Iwama, T., et al. (2012). A catalytic role of XoxF1 as La3+-dependent methanol dehydrogenase in *Methylobacterium extorquens* strain AM1. *PLoS One*, 7(11), e50480.

Nayak, D. D., & Marx, C. J. (2014). Genetic and phenotypic comparison of facultative methylotrophy between *Methylobacterium extorquens* strains PA1 and AM1. *PLoS One*, *9*(9), e107887.

Ochsner, A. M., Christen, M., Hemmerle, L., Peyraud, R., Christen, B., & Vorholt, J. A. (2017). Transposon sequencing uncovers an essential regulatory function of phosphoribulokinase for methylotrophy. *Current Biology*, *27*(17), 2579–2588. e2576.

Ochsner, A. M., Hemmerle, L., Vonderach, T., Nussli, R., Bortfeld-Miller, M., Hattendorf, B., et al. (2019). Use of rare-earth elements in the phyllosphere colonizer *Methylobacterium extorquens* PA1. *Molecular Microbiology*, *111*(5), 1152–1166.

Pol, A., Barends, T. R., Dietl, A., Khadem, A. F., Eygensteyn, J., Jetten, M. S., et al. (2014). Rare earth metals are essential for methanotrophic life in volcanic mudpots. *Environmental Microbiology*, *16*(1), 255–264.

Salcher, M. M., Neuenschwander, S. M., Posch, T., & Pernthaler, J. (2015). The ecology of pelagic freshwater methylotrophs assessed by a high-resolution monitoring and isolation campaign. *The ISME Journal*, *9*(11), 2442–2453.

Santoro, A., Thoss, V., Guevara, S. R., Urgast, D., Raab, A., Mastrolitti, S., et al. (2016). Assessing rare earth elements in quartz rich geological samples. *Applied Radiation and Isotopes*, *107*, 323–329.

Schrader, J., Schilling, M., Holtmann, D., Sell, D., Filho, M. V., Marx, A., et al. (2009). Methanol-based industrial biotechnology: Current status and future perspectives of methylotrophic bacteria. *Trends in Biotechnology*, *27*(2), 107–115.

Skovran, E., Palmer, A. D., Rountree, A. M., Good, N. M., & Lidstrom, M. E. (2011). XoxF is required for expression of methanol dehydrogenase in *Methylobacterium extorquens* AM1. *Journal of Bacteriology*, *193*(21), 6032–6038.

Sy, A., Timmers, A. C., Knief, C., & Vorholt, J. A. (2005). Methylotrophic metabolism is advantageous for *Methylobacterium extorquens* during colonization of Medicago truncatula under competitive conditions. *Applied and Environmental Microbiology*, *71*(11), 7245–7252.

Taubert, M., Grob, C., Howat, A. M., Burns, O. J., Dixon, J. L., Chen, Y., et al. (2015). XoxF encoding an alternative methanol dehydrogenase is widespread in coastal marine environments. *Environmental Microbiology*, *17*(10), 3937–3948.

Vogel, C., Bodenhausen, N., Gruissem, W., & Vorholt, J. A. (2016). The Arabidopsis leaf transcriptome reveals distinct but also overlapping responses to colonization by phyllosphere commensals and pathogen infection with impact on plant health. *The New Phytologist*, *212*(1), 192–207.

Vorholt, J. A. (2012). Microbial life in the phyllosphere. *Nature Reviews. Microbiology*, *10*(12), 828–840.

Vu, H. N., Subuyuj, G. A., Vijayakumar, S., Good, N. M., Martinez-Gomez, N. C., & Skovran, E. (2016). Lanthanide-dependent regulation of methanol oxidation systems in *Methylobacterium extorquens* AM1 and their contribution to methanol growth. *Journal of Bacteriology*, *198*(8), 1250–1259.

Wisniewski, J. R., Zougman, A., & Mann, M. (2009). Combination of FASP and StageTip-based fractionation allows in-depth analysis of the hippocampal membrane proteome. *Journal of Proteome Research*, *8*(12), 5674–5678.

Zhao, W., Zong, K. Q., Liu, Y. S., Hu, Z. C., Chen, H. H., & Li, M. (2019). An effective oxide interference correction on Sc and REE for routine analyses of geological samples by inductively coupled plasma-mass spectrometry. *Journal of Earth Science*, *30*(6), 1302–1310.

Zhou, B. L., Li, Z. X., & Chen, C. C. (2017). Global potential of rare earth resources and rare earth demand from clean technologies. *Minerals*, 7(11), 203.

CHAPTER TEN

Cultivation techniques to study lanthanide metal interactions in the haloalkaliphilic Type I methanotroph *"Methylotuvimicrobium buryatense"* 5GB1C

Joseph D. Groom[a,*] and Mary E. Lidstrom[a,b]

[a]Department of Chemical Engineering, University of Washington, Seattle, WA, United States
[b]Department of Microbiology, University of Washington, Seattle, WA, United States
[*]Corresponding author: e-mail address: groomjd@uw.edu

Contents

Abstract

Lanthanide metals are commonly used in technological devices including batteries, computers, catalysts and magnets. Despite their important properties, mining difficulties and pollution concerns limit the number of mines worldwide. Because of these concerns, biometallurgy is an attractive possibility for lanthanide extraction from recycled materials or from contaminated sites. Methylotrophs, bacteria that grow on reduced carbon substrates like methane and methanol, utilize lanthanides for a central reaction in their metabolisms. They must have some mechanism for uptake or trafficking, and are therefore excellent candidates for applying small molecules or proteins for selective lanthanide metal recycling. The haloalkaliphilic methanotroph *"Methylotuvimicrobium buryatense"* 5GB1C is the fastest growing methanotroph isolated to date, and thus has great industrial potential. The MxaFI enzyme complex uses calcium as a Lewis acid in conjunction with the pyroquinoline quinone cofactor to oxidize methanol, while the alternative enzyme XoxF uses lanthanide metals (e.g. lanthanum and cerium) for the same function. Lanthanide metals, abundant in the earth's crust, strongly repress

Methods in Enzymology, Volume 650
ISSN 0076-6879
https://doi.org/10.1016/bs.mie.2021.01.042

the transcription of *mxaF* yet activate the transcription of *xoxF*, implying that XoxF may be the predominant methanol dehydrogenase in the bacterium's native environment. It may be that lanthanum interaction mechanisms are different from those in other microorganisms. In addition, the facile genetics in this strain and existing background information make it a good study organism for biological lanthanum uptake. The interesting physiology of this organism required empirical work to develop cultivation methods that allow robust assays of gene expression and measurement of lanthanum associated with cell biomass. In this chapter, we show that altering the metal chelator increased the availability of lanthanum to the cell as measured by the specific gene expression response. We also made further alterations to prevent lanthanum precipitation in medium for the growth of haloalkaliphiles.

1. Introduction

Rare earth elements were once believed to be biologically inert, but a new field of research emerged when lanthanide-dependent activities were demonstrated in methylotrophs, organisms that utilize reduced single-carbon compounds (Pol et al., 2014; Skovran, Raghuraman, & Martinez-Gomez, 2019). There remains much to be learned about the interactions between organisms and metals found in the earth's crust. Since numerous studies have shown that lanthanides are essential metal cofactors for alcohol dehydrogenases (Good et al., 2016; Huang et al., 2019; Wehrmann, Billard, Martin-Meriadec, Zegeye, & Klebensberger, 2017), there is much to be explored; there are very likely additional enzymes that require lanthanides. For instance, *Bradyrhizobium* and *Streptomyces* species exhibit relatively uncharacterized lanthanide-induced microbiological processes, including exopolysaccharide and antibiotic production (Fitriyanto, Fushimi, et al., 2011; Fitriyanto, Nakamura, 2011; Kawai, Wang, Okamoto, & Ochi, 2007).

It has been shown in various bacteria that lanthanides can associate with and bind at the cell surface. Microscopy has demonstrated in *E. coli* lanthanide metals seem to accumulate in the periplasmic space between the outer membrane and inner membrane (Bayer & Bayer, 1991), a region specific to Gram-negative bacteria. Lanthanide metals associate with phosphate and carboxylate groups at the outer cell surface in the Gram-positive bacterium *Bacillus subtilis* (Takahashi, Châtellier, Hattori, Kato, & Fortin, 2005). In the Gram-negative bacterium *Pseudomonas fluorescens*, lanthanides were similarly shown to adsorb to carboxylate groups at the cell surface, but also to extracellular organic ligands like cellulose and chitin, and even more interestingly to the iron-binding siderophore molecule desferrioxamine B (Ozaki et al., 2006). However, in the methylotroph *Methylorubrum extorquens* AM1, lanthanides enter the cytoplasm (Mattocks, Ho, & Cotruvo, 2019; Roszczenko-Jasińska et al., 2020). More recent work has begun to unravel the diverse mechanisms of lanthanide uptake machinery and the means of conveying regulatory signals in bacteria like *M. extorquens* AM1 that use lanthanides for key biological reactions (Chu, Beck, & Lidstrom, 2016; Chu & Lidstrom, 2016; Farhan Ul Haque et al., 2015; Gu, Haque, DiSpirito, & Semrau, 2016; Vu et al., 2016; Wehrmann, Berthelot, Billard, & Klebensberger, 2018, 2019; Wehrmann et al., 2017).

Methylotrophic bacteria provide an excellent system to study the interactions of lanthanide metals with the bacterial periplasm, as many are Gram-negative proteobacteria on which lanthanide metals have clear in vivo effects (Nakagawa et al., 2012; Pol et al., 2014; Skovran, Palmer, Rountree, Good, & Lidstrom, 2011; Vu et al., 2016). The methanol dehydrogenase enzyme XoxF uses lanthanide metal ions as its metal cofactor to produce formaldehyde from methanol in the periplasm (Keltjens, Pol, Reimann, & Op den Camp, 2014). The most detailed studies of lanthanide uptake have taken place in the non-methanotrophic methylotroph *Methylorubrum extorquens* AM1 and its close relative *Methylorubrum extorquens* PA1. Uptake machinery includes a TonB-dependent transporter, periplasmic binding proteins and an ATP-binding cassette transport system, in addition to uncharacterized proteins (Ochsner et al., 2019; Roszczenko-Jasińska et al., 2020). Another protein, lanmodulin, tightly binds lanthanide metals but is not required for lanthanide utilization in tested growth conditions (Cook, Featherston, Showalter, & Cotruvo, 2019; Cotruvo, Featherston, Mattocks, Ho, & Laremore, 2018; Ochsner et al., 2019). Small molecules for binding lanthanides analogous to iron siderophores have been hypothesized but not yet identified.

Methanotrophs like "*Methylotuvimicrobium buryatense*" 5GB1C, unlike non-methanotrophic methylotrophs, can grow on methane in addition to methanol. Such organisms have potential for direct application to greenhouse gas mitigation and to industrial bioproduction using methane. Gammaproteobacterial (Type I) methanotrophs are also interesting for industrial applications because of their robust growth and efficient central metabolism using the ribulose monophosphate pathway, characteristics that facilitate the production of valuable chemicals (Kalyuzhnaya, Puri, & Lidstrom, 2015). Studies on lanthanide interactions in haloalkaliphilic gammaproteobacterial methanotrophs will complement those carried out in other methanotrophs and non-methanotrophic methylotrophs. "*M. buryatense*" 5GB1C is phylogenetically distinct (the others studied so far are alphaproteobacteria), it grows at high pH (optimum = 8.5), and requires salt (optimum = 0.75% weight/volume) (Kaluzhnaya et al., 2001). It is likely that the mechanisms involved are different from those in the alphaproteobacterial strains, and this will serve as a good comparison for the diversity of lanthanide interaction mechanisms.

"*M. buryatense*" 5GB1C is a particularly good candidate for methanotrophic lanthanide studies because of the rapid genetic tools available (Puri et al., 2015; Yan, Chu, Puri, Fu, & Lidstrom, 2016). It is also an excellent model for metal regulation studies, as it has only one copy of *xoxF* encoding the lanthanide-dependent methanol dehydrogenase (Khmelenina et al., 2013). The most obvious phenotype in response to lanthanides is the methanol dehydrogenase switch: the canonical *mxaF* methanol dehydrogenase gene is drastically down-regulated, while *xoxF* is up-regulated (Chu et al., 2016; Chu & Lidstrom, 2016). To orchestrate this gene regulation, lanthanides must somehow associate with the cell—either by ligand binding or by transport into the cell—but the actual mechanism of this interaction is unknown. An outer membrane TonB-dependent receptor, named LanA, is absolutely required for normal lanthanide gene regulation, and is necessary to supply the methanol dehydrogenase enzyme XoxF with its required lanthanide metal cofactor to allow growth in an Δ*mxaF* deletion strain (Groom, Ford, Pesesky, & Lidstrom, 2019). The discovery of this novel component of the lanthanide regulatory system highlights the complexity of this circuit and suggests further components are likely involved.

Although the known *M. extorquens* PA1 lanthanide transporter is not a close homolog to "*M. buryatense*" 5GB1C LanA, our experiments show clearly that LanA is involved in lanthanide signal transduction (Groom et al., 2019). With the physiological differences between haloalkaliphiles and neutrophiles like *M. extorquens* strains, different growth media are required (Delaney et al., 2013; Puri et al., 2015). These differences include salt content, pH, phosphate amount, and chelators for metals that are traditionally used. New methods are required to prevent lanthanum precipitation in haloalkaliphilic medium, so that studies with lanthanides can be performed. Combining the powerful genetic system of "*M. buryatense*" 5GB1C with various microscopic, biophysical, and high-throughput sequencing techniques would comprise a cohesive yet diverse plan aimed towards a better understanding of these fast-growing microorganisms and metals. Learning more about the regulatory and uptake mechanisms of lanthanide metals in methanotrophs is critical to understand the microbiological impact on atmospheric methane levels, and it is necessary to apply these organisms to industrial bioproduction and biometallurgy. By changing the metal chelator to a citrate chelator instead of the previously used EDTA chelator, and altering various medium components, we provide new growth conditions for more accurate measurements of lanthanum uptake in haloalkaliphilic organisms like *"M. buryatense"* 5GB1C.

2. Methods

2.1 Measuring the effect of different metal chelators in a lanthanum-dependent gene expression assay

2.1.1 Before you begin

Timing: Acid wash tubes 2 days before cell culture begins, prepare medium immediately before cell culture begins

1. Prepare 1 M hydrochloric acid-washed glassware.
   - **(a)** Fill clean glassware with acid and allow to sit overnight.
   - **(b)** Decant acid the following day for storage and re-use.
   - **(c)** Fill glassware with MilliQ-purified water for 1 h.
   - **(d)** Dry several hours or overnight.
   - **(e)** Sterilize dried glassware before use.
2. Prepare NMS2 medium with filter-sterilized components (see Table 1 for medium component recipes).
   - **(a)** Dilute an appropriate volume of Component A (nitrate mineral salts) to 1 × in MilliQ-purified water.
   - **(b)** Add appropriate amount for 1 X concentration of Component B (carbonate buffer).
   - **(c)** Add appropriate amount for 1 X concentration of Component D (phosphate buffer).
   - **(d)** Add appropriate amount for 1 X concentration of Component E (Trace Elements with the EDTA chelator) or Component F (Trace Elements with the citrate chelator).
   - **(e)** Distribute medium among acid-washed Bellco tubes or standard culture tubes for cell growth.
   - **(f)** Add appropriate amount of Component G (lanthanum chloride stock solution) for desired concentration (generally from 0 to 30 μM).
3. Prepare 50 mM Tris–HCl, pH 7.5 for catechol-2,3-dioxygenase reporter gene assay.

Table 1 Recipes for solutions of concentrated media components.

NMS2 media components	**Concentration (g/L)**
Component A: Nitrate mineral salts (10 X)	
NaCl	75
$MgSO_4 \cdot 7H_2O$	2
$CaCl_2 \cdot 2H_2O$	0.14
KNO_3	1
Component B: 1M Carbonate buffer (20 X, prepared fresh)	
$NaHCO_3$	58.8
Na_2CO_3	31.4
Adjust to pH 9.5	
Component C: 1M Trizma buffer	
Trizma base	121.14
Adjust to pH 7.0, 7.5, 8.2, or 9.0	
Component D: Phosphate buffer (50 X in standard NMS2)	
KH_2PO_4	5.44
$NaHPO_4 \cdot H_2O$	10.73
Adjust to pH 6.8	
Component E: Trace Elements with EDTA chelator (500 X)	
Na_2-EDTA	1
$FeSO_4 \cdot 7H_2O$	2
$ZnSO_4 \cdot 7H_2O$	0.8
$MnCl_2 \cdot 4H_2O$	0.03
H_3BO_3	0.03
$CoCl_2 \cdot 6H_2O$	0.2
$CuCl_2 \cdot 2H_2O$	0.6
$NiCl_2 \cdot 6H_2O$	0.02
$Na_2MoO_4 \cdot 2H_2O$	0.05
Component F: Trace Elements with citrate chelator (500 X)	
$Na_3C_6H_5O_7$	6.7

Table 1 Recipes for solutions of concentrated media components.—cont'd

NMS2 media components	**Concentration (g/L)**
$FeSO_4 \cdot 7H_2O$	2
$ZnSO_4 \cdot 7H_2O$	0.8
$MnCl_2 \cdot 4H_2O$	0.03
H_3BO_3	0.03
$CoCl_2 \cdot 6H_2O$	0.2
$CuCl_2 \cdot 2H_2O$	0.6
$NiCl_2 \cdot 6H_2O$	0.02
$Na_2MoO_4 \cdot 2H_2O$	0.05
Component G: Lanthanum (2.83 mM stock, prepared fresh)	
Lanthanum (III) chloride hydrate ($LaCl_3 \cdot xH_2O$, $x = 6–7$)	1

Units are in g/L. Sources of reagents are listed separately in the Key resources table.

2.1.2 Key resources table

Reagent or resource	**Source**	**Identifier**
Bacterial strains		
"*Methylotuvimicrobium buryatense*" 5GB1C	Mary Lidstrom, University of Washington	
Chemicals		
Sodium chloride (NaCl)	Fisher Scientific	7647-14-5
Magnesium sulfate heptahydrate ($MgSO_4 \cdot 7H_2O$)	Sigma-Aldrich	10034-99-8
Calcium chloride dihydrate ($CaCl_2 \cdot 2H_2O$)	Sigma-Aldrich	10035-04-8
Potassium nitrate (KNO_3)	Sigma-Aldrich	7757-79-1
Potassium phosphate monobasic (KH_2PO_4)	Fisher Scientific	7778-77-0
Sodium phosphate dibasic heptahydrate ($NaHPO_4 \cdot 7H_2O$)	Sigma-Aldrich	7782-85-6
Sodium bicarbonate ($NaHCO_3$)	Fisher Scientific	144-55-8
Sodium carbonate (Na_2CO_3)	Fisher Scientific	497-19-8
Ethylenediaminetetraacetic acid disodium salt (Na_2-EDTA)	Sigma-Aldrich	6381-92-6

Continued

—cont'd

Reagent or resource	**Source**	**Identifier**
Sodium citrate ($Na_3C_6H_5O_7$)	Van Waters & Rogers Ltd.	6132-04-3
Iron (II) sulfate heptahydrate ($FeSO_4 \bullet 7H_2O$)	Sigma-Aldrich	7782-63-0
Zinc sulfate heptahydrate ($ZnSO_4 \bullet 7H_2O$)	Sigma-Aldrich	7746-20-0
Manganese chloride tetrahydrate ($MnCl_2 \bullet 4H_2O$)	Sigma-Aldrich	13446-34-9
Boric acid (H_3BO_3)	Sigma-Aldrich	10043-35-3
Cobalt chloride hexahydrate ($CoCl_2 \bullet 6H_2O$)	J.T. Baker	7791-13-1
Cupric chloride dihydrate ($CuCl_2 \bullet 2H_2O$)	J.T. Baker	10125-13-0
Nickel chloride hexahydrate ($NiCl_2 \bullet 6H_2O$)	Sigma-Aldrich	7718-54-9
Sodium molybdate dihydrate ($Na_2MoO_4 \bullet 2H_2O$)	Sigma-Aldrich	10102-40-6
Catechol	Sigma-Aldrich	120-80-9
Lanthanum (III) chloride hydrate ($LaCl_3 \bullet xH_2O$)	Sigma-Aldrich	20211-76-1
Trizma hydrochloride (Tris–HCl)	Sigma-Aldrich	1185-53-1
Hydrochloric acid 32–38%	Fisher Scientific	7647-01-0
Methane gas	Airgas, Inc.	74-82-8
Methanol	Fisher Scientific	67-56-1
Experimental models: Organisms/strains		
"*Methylotuvimicrobium buryatense*" 5GB1C strain FC31, harboring P_{mxaF}-*xylE* reporter gene construct	Mary Lidstrom, University of Washington	
"*Methylotuvimicrobium buryatense*" 5GB1C strain JG15, Δ*lanA* harboring P_{mxaF}-*xylE* reporter gene construct	Mary Lidstrom, University of Washington	
Software and algorithms		
SoftMax Pro plate reader software	Molecular Devices	
Other		
Anaerobic Tubes, 18 × 150 mm	Bellco Glass, Inc.	2048-00150

2.1.3 Materials and equipment

- *"Methylomicrobium buryatense"* 5GB1C strains of interest
- Methane gas or methanol
- Anaerobic tubes for cell culture with methane
- Prepared NMS2 growth medium
- 10 mM catechol solution, prepared immediately before reporter gene assay
- Molecular Devices SpectraMax 190 Plate Reader

2.1.4 Step-by-step method details

2.1.4.1 Cell growth and harvesting

Timing: 3 + days for strain revival from frozen stock, 1 overnight incubation for starter culture, 1 overnight incubation for cell growth for the reporter gene assay, 10–30 min for sample collection.

1. Revive strain of "*M. buryatense*" 5GB1C strain FC31 and/or mutant strains 3–5 days prior to experiment on NMS2 agar plates (1.5% w/v). Incubate under and atmosphere of 25% methane/75% air at 30 °C, with a trace element solution chelated either with EDTA or citrate depending on the experiment.
2. Inoculate plate-grown biomass into liquid NMS2 medium under an atmosphere of 25% methane/75% air, or with 0.4% (v/v) methanol.
3. Transfer starter culture to new medium at new OD_{600} of ~0.03, and grow cells to late exponential phase (usually greater than $OD_{600} = 0.6$) or early stationary phase.
   - **(a)** Choosing a growth phase depends on the mutant strain and the gene of interest.
   - **(b)** Reporter gene results are similar for P_{mxaF} and P_{xoxF} at both exponential and stationary phases.
4. Harvest cells at 4000 x *g* for 10 min, preferably in a desktop centrifuge.
5. Resuspend the pellet with 50 mM Tris-HCl, pH7.5, to final $OD_{600} = 0.5$ to normalize the samples.

Note: As an example of harvesting and resuspension, for a sample $OD_{600} = 0.8$, harvest 625 μL of cells and resuspend with 1000 μL of 50 mM Tris–HCl, pH 7.5, to achieve final $OD_{600} = 0.5$.

2.1.4.2 XylE reporter assay

Timing: 10–30 min plate setup, 10 min kinetic assay.

6. Make fresh 100 mM catechol solution in 50 mM Tris–HCl, pH 7.5, and dilute to a 10 mM catechol stock.

7. Warm SpectraMax 190 Plate Reader or other suitable plate reader to 30 °C.
8. Select a kinetic measurement, and set absorbance to 375 nm.
9. Set the timing to 10 min with absorbance measurements at intervals of 20 s.
10. Set plate reader to shake for 5 s before the first read.
11. Add 10 μL 10 mM catechol to each well that will contain a reaction.
12. Add >100 μL of well-mixed cell suspension into a corresponding well.
13. Use a multi-channel pipetter to transfer 90 μL of cell suspension to the corresponding catechol well.
14. Immediately begin assay once cell suspensions are added to catechol wells.

Critical: Mixing the cell suspensions is very important for assay precision.

Pause Point: For accurate and precise results, it is best not to pause the procedure after cells are harvested.

2.1.5 Expected outcomes

The experiment should provide results of gene expression in response to the presence of lanthanides. This could use the *mxaF* gene promoter (P_{mxaF}) as the sensor, in this case, or other lanthanide-responsive gene regulatory elements. An example of the result of using two different metal chelators is shown in Fig. 1. Citrate makes lanthanum more available to wild-type "*M. buryatense*" 5GB1C, leading to full P_{mxaF} repression at 1 μM, whereas 3 μM is required for full repression in EDTA-containing cultures. At high lanthanum concentrations (30 μM), an appreciable level of P_{mxaF} repression can be observed in the Δ*lanA* transporter mutant strain, suggesting that citrate may facilitate La uptake by the cell. It may be that lanthanum can be allowed into the periplasm through alternative, non-specific transporters if an appropriate chelator is present. This would be consistent with results in a transporter mutant of *Methylorubrum extorquens* AM1 (Roszczenko-Jasińska et al., 2020). We chose the specific concentration of citrate so that it was comparable to other media (Delaney et al., 2013). However, it cannot be ruled out that the higher concentration of citrate medium plays a role in the availability, as the molar amount of citrate was higher than the molar amount of EDTA.

2.1.6 Quantification and statistical analysis

The XylE reporter gene assay measures the production of 2-hydroxymuconate semialdehyde from catechol, catalyzed by the XylE enzyme. The rate of

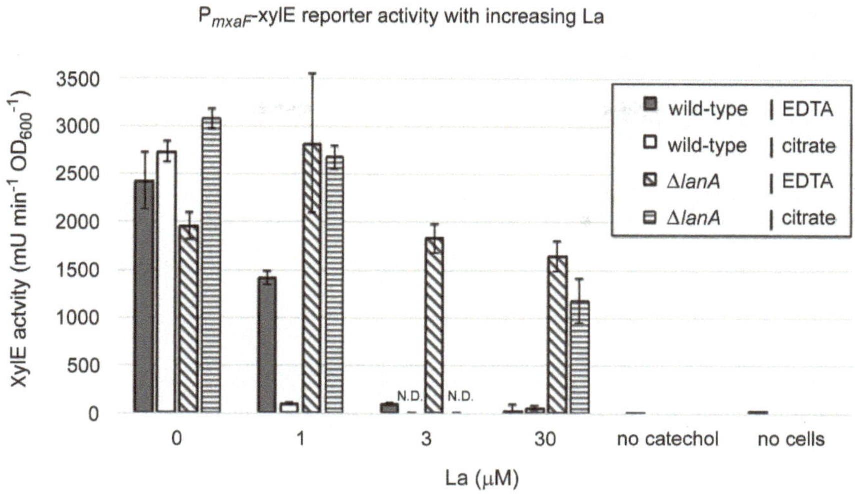

Fig. 1 Citrate makes lanthanum more bio-available to cells. Whole cell catechol-2, 3-dioxygenase reporter gene assays were performed for wild-type and mutant strains. The wild-type was FC31 (P_{mxaF}-*xylE*) and the Δ*lanA* mutant was created in this genetic background. Cells were either grown without La or in the presence of 1 μM, 3 μM, or 30 μM La. The final medium concentration of EDTA was 6.8 μM and the final concentration of citrate was 51.9 μM. Results are the average ± standard deviation of two biological replicates ($n = 2$).

product formation is measured by absorbance at 375 nm (A375) over specified time points. The more XylE enzyme present, the more 2-hydroxymuconate semialdehyde is formed over a specified time. The amount of XylE is, in turn, a function of the activity of the promoter. To accurately interpret this kinetic reading, one must measure only the linear portion of the slope of A375 plotted over time. A reliably linear slope resulted from a time course of 2 min 20 s, with readings taken every 20 s. To convert the value of this slope to the units seen in Fig. 1 (mU min^{-1} OD_{600}^{-1}), the slope is multiplied by 1000 to convert to milli-Units and divided by the OD_{600} value (0.5).

2.1.7 Advantages

The reporter gene assay is a rapid, high-throughput measurement of gene expression. Many conditions can be quantified simultaneously for multiple mutant strains. This approach was able to quickly show that the chelator is important for the bioavailability of lanthanum to this organism—that is, citrate makes lanthanum more available to the cell than does EDTA.

2.1.8 Limitations

The reporter gene assay has a lower dynamic range than reverse transcriptase polymerase chain reaction (RT-PCR), and it is likely that more subtle phenotypes will need to be detected by RT-PCR or even RNA-seq.

2.1.9 Optimization and troubleshooting

2.1.9.1 Problem: Kinetic curve not linear or too steep

Some promoters may return kinetic values at an adjusted $OD_{600} = 0.5$ that are too great, or non-linear.

2.1.9.2 Potential solution to optimize the procedure

Adjust the resuspension volume until a reliable slope of A375 over a given time can be achieved. Make sure to adjust the calculation using the actual value that was the final adjusted OD_{600}.

2.1.9.3 Safety considerations and standards

Proper storage and usage of compressed methane gas must be followed.

2.1.10 Alternative methods/procedures

Examples of methods and results using RT-PCR and RNA-seq for similar lanthanum assays using EDTA as the chelator can be found in other manuscripts (Chu et al., 2016; Chu & Lidstrom, 2016; Groom et al., 2019).

2.2 Cell-associated lanthanum assay

2.2.1 Before you begin

Timing: Prepare medium immediately before cell culture begins.

1. Prepare *modified* NMS2 medium for lanthanum uptake using filter-sterilized medium components (See Table 1 for medium component recipes)
   - **(a)** Dilute an appropriate volume of Component A (nitrate mineral salts) to 1 × in MilliQ-purified water.
   - **(b)** Add appropriate amount for 1 X concentration of Component C (Trizma buffer, pH 8.2).
   - **(c)** Add appropriate amount for 0.05 X concentration of Component D (phosphate buffer).
   - **(d)** Add appropriate amount for 1 X concentration of Component F (Trace Elements with the citrate chelator).
   - **(e)** Distribute medium among acid-washed Bellco tubes or standard culture tubes for cell growth.

(f) Add appropriate amount of Component G (lanthanum chloride stock solution) for desired concentration (2 μM in the experiments shown).

(g) Add 0.4% methanol (volume/volume) for growth in plastic shake flasks.

2. Choose and revive strain of *M. buryatense* or other haloalkaliphile of interest. Strains FC31 and JG15 were used for experiments shown.

2.2.2 Key resources table

Reagent or resource	Source	Identifier
Bacterial strains		
"*Methylotuvimicrobium buryatense*" 5GB1C	Mary Lidstrom, University of Washington	
Chemicals		
Sodium chloride (NaCl)	Fisher Scientific	7647-14-5
Magnesium sulfate heptahydrate ($MgSO_4 \cdot 7H_2O$)	Sigma-Aldrich	10034-99-8
Calcium chloride dihydrate ($CaCl_2 \cdot 2H_2O$)	Sigma-Aldrich	10035-04-8
Potassium nitrate (KNO_3)	Sigma-Aldrich	7757-79-1
Potassium phosphate monobasic (KH_2PO_4)	Fisher Scientific	7778-77-0
Sodium phosphate dibasic heptahydrate ($NaHPO_4 \cdot 7H_2O$)	Sigma-Aldrich	7782-85-6
Sodium bicarbonate ($NaHCO_3$)	Fisher Scientific	144-55-8
Sodium carbonate (Na_2CO_3)	Fisher Scientific	497-19-8
Sodium citrate ($Na_3C_6H_5O_7$)	Van Waters & Rogers Ltd.	6132-04-3
Iron sulfate heptahydrate ($FeSO_4 \cdot 7H_2O$)	Sigma-Aldrich	7782-63-0
Zinc sulfate heptahydrate ($ZnSO_4 \cdot 7H_2O$)	Sigma-Aldrich	7746-20-0
Manganese chloride tetrahydrate ($MnCl_2 \cdot 4H_2O$)	Sigma-Aldrich	13446-34-9
Boric acid (H_3BO_3)	Sigma-Aldrich	10043-35-3
Cobalt chloride hexahydrate ($CoCl_2 \cdot 6H_2O$)	J.T. Baker	7791-13-1
Cupric chloride dihydrate ($CuCl_2 \cdot 2H_2O$)	J.T. Baker	10125-13-0
Nickel chloride hexahydrate ($NiCl_2 \cdot 6H_2O$)	Sigma-Aldrich	7718-54-9

Continued

—cont'd

Reagent or resource	Source	Identifier
Sodium molybdate dihydrate ($Na_2MoO_4 \cdot 2H_2O$)	Sigma-Aldrich	10102-40-6
Lanthanum chloride hydrate ($LaCl_3 \cdot xH_2O$)	20211-76-1	20211-76-1
Trizma hydrochloride	Sigma-Aldrich	1185-53-1
Trizma base	Sigma-Aldrich	77-86-1
Yttrium (Y) ICP Standard	GFS Chemicals, Inc.	13494-98-9
Nitric acid, 12N (HNO_3)	Fisher Scientific	7697-37-2
Methanol	Fisher Scientific	67-56-1
Experimental models: organisms/strains		
"*Methylotuvimicrobium buryatense*" 5GB1C strain FC31, harboring P_{mxaF}-*xylE* reporter gene construct	Mary Lidstrom, University of Washington	
"*Methylotuvimicrobium buryatense*" 5GB1C strain JG15, Δ*lanA* harboring P_{mxaF}-*xylE* reporter gene construct	Mary Lidstrom, University of Washington	
Other		
Polypropylene Erlenmeyer Flask, 250mL	Kartell	241345

2.2.3 Materials and equipment

- *"Methylomicrobium buryatense"* 5GB1C strains of interest
- Plastic flask or other suitable cell culture vessel
- PerkinElmer ELAN DRC-e Inductively Coupled Plasma Mass Spectrometer
- Modified medium, as detailed in Table 1 and Section 2.1.1
- Nitric acid, 12N (70% v/v)
- Methanol
- Centrifuge tubes and small volume disposable tubes for sample harvesting and cell lysis
- Yttrium ICP Standard

2.2.4 Step-by-step method details

2.2.4.1 Cell growth and harvesting

Timing: 3 + days for strain revival from frozen stock, 1 overnight incubation for starter culture, variable length incubation for cell growth for the lanthanum uptake assay, 10–30 min for sample collection

1. Revive strain of "*M. buryatense*" 5GB1C strain FC31 and/or mutant strains 3–5 days prior to experiment on NMS2 agar plates.
2. Inoculate plate-grown biomass into liquid modified NMS2 medium with 0.4% (v/v) methanol.
3. Transfer 1 mL of starter culture to 50 mL modified NMS2 medium in polypropylene flask, and grow cells desired OD_{600}.
   (a) Choosing a growth phase for lanthanide measurement depends on the mutant strain. In the data shown, 5 mL samples were taken at OD_{600} ~0.3 and ~1.3 to assess uptake at early exponential phase and early stationary phase.
   (b) Taking samples at multiple time points allows an observation of cell-associated lanthanum over time.
   (c) Prepare a blank sample of medium, and replicate the procedure for it, to account for any precipitation that may still occur. This can be subtracted from the final calculation later.
4. Remove 5 mL of culture, and harvest cells at 4000 x *g* for 10 min at room temperature.
   (a) Save supernatants and add 142 μL 70% (v/v) nitric acid for a final solution of 2% (v/v) nitric acid.

2.2.4.2 Cell lysis

Timing: 2 h for cell lysis, 5–10 min for sample preparation.

5. Resuspend pellets in 358 μL of modified NMS2 medium in small-volume disposable tubes.
6. Add 142 μL 70% (v/v) nitric acid to achieve 20% (v/v) nitric acid for cell lysis.
7. Lyse cells at 95 °C for 2 h.
   (a) Seal tubes with parafilm so they do not open during heat treatment.
8. Add 4.5 mL modified NMS2 medium to dilute to 2% (v/v) nitric acid final concentration.
9. Add 10 μL 1:100 diluted yttrium standard to every 5 mL sample, including saved supernatants. Mix well.

Critical: An internal yttrium standard is necessary to be able to reliably compare spectral counts.

2.2.4.3 Metal measurement

Timing: Allow ~3 h for measurement, including machine initialization.

10. Measure the spectral counts of lanthanum or other lanthanide metal in each sample with the inductively – coupled plasma mass spectrometer (ICP-MS).

   (a) Record spectral counts for the lanthanide of interest and yttrium, which is the internal standard.

11. Follow standard operating procedures for the ICP-MS.

   (a) These are facility-specific.

   (b) An example of a detailed standard operating protocol (SOP) can be found at https://cse.wwu.edu/files/AMSEC/Instrumentation/ICPMS_SOPs_Liquids.pdf.

Pause Point: Samples can be stored at room temperature after nitric acid addition, until ICP-MS machine is available for use. Measurements were precise before and after storage for several weeks.

2.2.5 Expected outcomes

For lanthanide uptake experiments, the medium was altered to prevent the precipitation of lanthanum, which has been previously observed in nitrate mineral salt (NMS) medium used to grow the alphaproteobacterial methanotroph *Methylosinus trichosporium* OB3b (Gu & Semrau, 2017). To avoid lanthanum precipitation, several alterations were required:

(1) The phosphate concentration was decreased 20-fold from the published NMS2 medium (Figs. 2 and 4).

(2) Polypropylene flasks rather than glassware were used to avoid adsorption of lanthanum to glass, which interferes with cell-associated lanthanum measurements (Fig. 3). Much of the lanthanum in the glass vials was lost to the growth vessel and was not quantified in either the supernatant or the pellet after centrifugation. Cells were grown on 0.4% (v/v) methanol because methane containment in the atmosphere above a culture requires sealed glass tubes or vials.

(3) The 500 × concentrated trace metal stock was chelated with 300 mM sodium citrate rather than EDTA to make lanthanum more bioavailable (Fig. 1) and to potentially help in avoiding lanthanum adsorption to glass (Figs. 3 and 4).

(4) The medium was buffered at pH 8.2 with 50 mM Trizma to maintain cell growth at high pH, yet prevent lanthanum precipitation (Figs. 4 and 5).

These alterations decreased the lanthanum precipitation to $<$100 nM of added lanthanum (Fig. 5). Previous studies with "*M. buryatense*" 5GB1C used NMS2 medium with 0.75% (w/v) NaCl, relatively high phosphate concentrations, and EDTA as the metal chelator (Chu et al., 2016; Chu & Lidstrom, 2016; Puri et al., 2015).

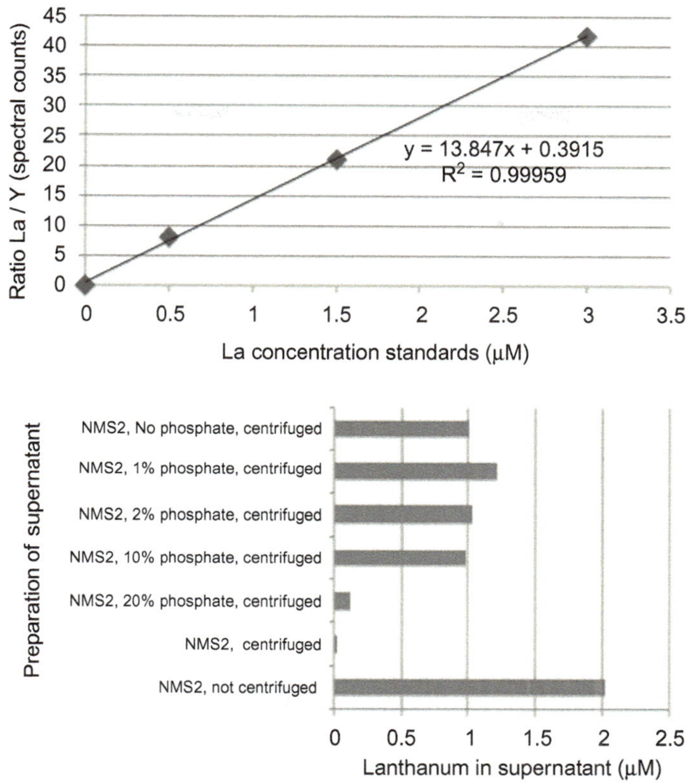

Fig. 2 Decreasing phosphate reduces lanthanum precipitation in standard NMS2 medium. Blank medium with 2 μM La was varied for phosphate concentration. Centrifuged samples were analyzed by ICP-MS to determine La concentration, and the La concentration in the supernatants is reported.

2.2.6 Quantification and statistical analysis

Measure the spectral counts of the lanthanide of interest (e.g. La) and the spectral counts of yttrium (Y) that was spiked into the sample as an internal standard. Create a standard curve of the La/Y ratio for each of the standards used (Fig. 2). In these experiments, at least four standards between 0 and 3 μM La were measured, and for each sample, La/Y ratios were interpolated using the standard curve to calculate La concentration. The La/Y ratios from the blank sample with no cells can be subtracted from the final calculations of La concentrations in the experimental samples. This will account for any La precipitation that occurs in the medium alone.

2.2.7 Advantages

Precipitation of lanthanum has been minimized (Fig. 5), especially compared to precipitation that occurs in standard NMS2 medium (Fig. 2).

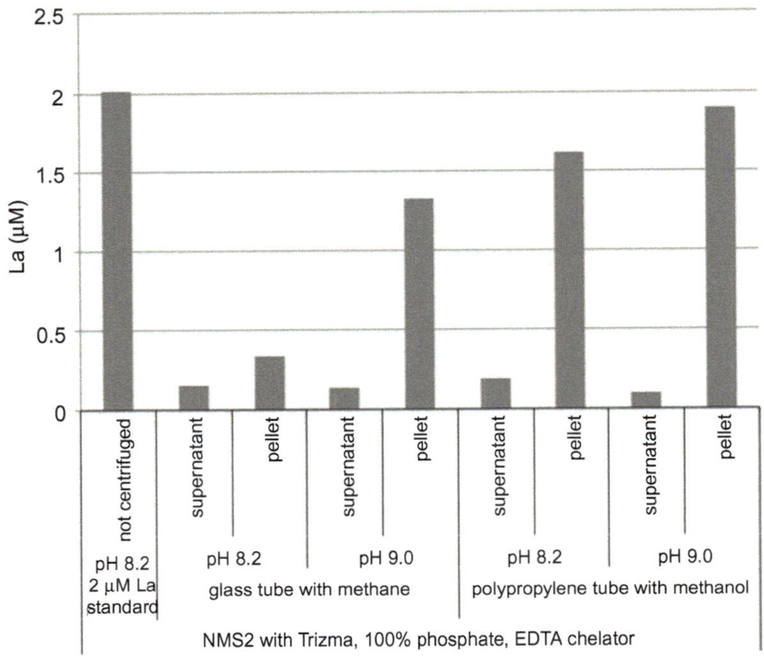

Fig. 3 Growth vessel material and pH affect La that remains in solution. Blank medium with 2 µM La was varied for buffer pH and vessel to contain medium. Centrifuged samples were analyzed by ICP-MS to determine La concentration, and the La concentration in the supernatants and acid-lysed pellets is reported. When the La concentrations in the supernatant and pellet do not total 2 µM in columns 2–3 and columns 4–5, it is assumed that much of the La adheres to the glass in these conditions.

2.2.8 Limitations

It is likely that lanthanide metals adhere to ligands on the cell wall as observed in other organisms (Ozaki et al., 2006; Takahashi et al., 2005; Takahashi, Yamamoto, Yamamoto, & Tanaka, 2010). Therefore, the amount that is actually taken up and utilized by the bacterium cannot be measured accurately with the cell-association assay. This could be the reason that the mutant lacking the LanA transporter, which is responsible for lanthanum signaling and very likely lanthanum import, exhibits the same amount of lanthanum in the cell pellet as the wild-type. The assay could be further developed to selectively wash away lanthanides attached to the outside of the cell before lysis, and in doing so differentiate the quantity of lanthanides taken up into the periplasm and/or cytoplasm from the total cell-associated lanthanides.

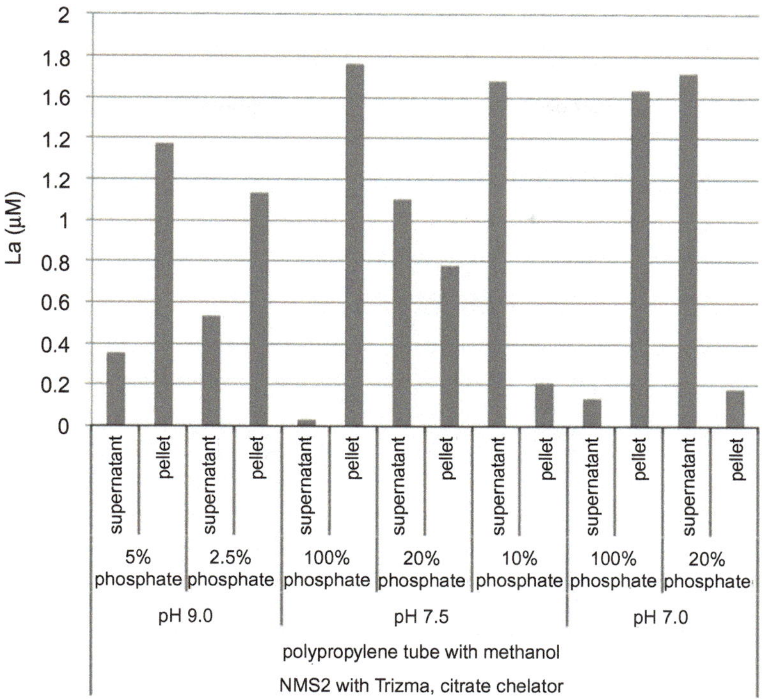

Fig. 4 Decreasing phosphate at lower pH prevents precipitation. Blank medium with 2 µM La in polypropylene tubes was varied for phosphate concentration and buffer pH. Centrifuged samples were analyzed by ICP-MS to determine La concentration, and the La concentration in the supernatants is reported.

2.2.9 Optimization and troubleshooting

2.2.9.1 Problem: Precipitation of La occurs (level 2 heading)

As detailed in "Expected Outcomes," determination of the medium composition is an empirical process. It is clearly important for lanthanide metals to have a carefully prepared medium to prevent precipitation in solution.

2.2.9.2 Potential solution to optimize the procedure (level 2 heading)

Medium components must be tested and combined in different ways, while still allowing cell growth. In the case of this haloalkaliphilic organism, lanthanides were kept in solution by decreasing the phosphate concentration (Figs. 2 and 4), using polypropylene growth vessels (Fig. 3), using a different chelator (Figs. 1 and 4), and gently acidifying the growth pH (Fig. 5).

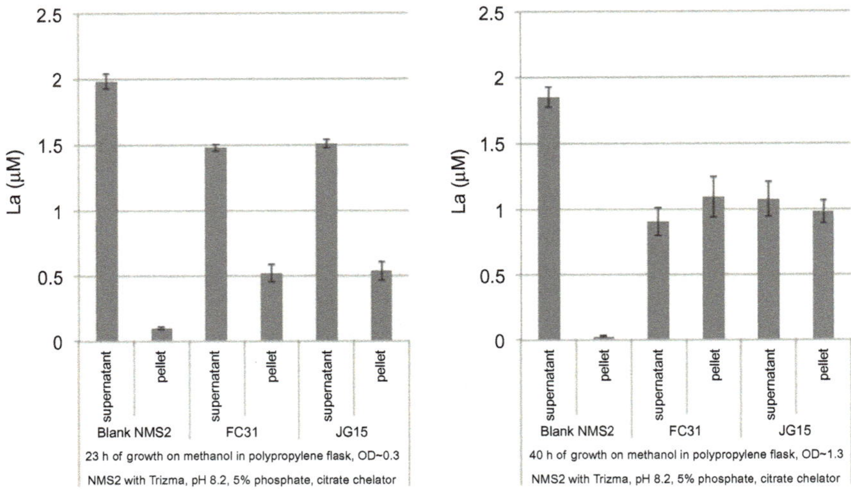

Fig. 5 Modified NMS2 for lanthanum uptake studies minimized precipitation and allowed quantification of cell-associated lanthanum. Cells were inoculated into medium containing 2 µM La. Cells were harvested at the indicated optical densities, corresponding to early exponential phase ($OD_{600} \sim 0.3$) and early stationary phase ($OD_{600} \sim 1.3$). Lysed cell pellets and supernatants were analyzed by ICP-MS to determine La concentration, and the concentration associated with the pellets adjusted to the original sample volume is reported. Results are the average ± standard deviation of two biological replicates, including the results from medium not inoculated with cells ($n = 2$).

2.2.9.3 Safety considerations and standards

Nitric acid should be handled in a fume hood with lab goggles, gloves and a lab coat. Site-specific training for the ICP-MS must be obtained.

2.2.10 Alternative methods/procedures

An Arsenazo assay for lanthanide quantification has been developed for the methylotroph *Methylorubrum extorquens* AM1 and the acidophilic verrucomicrobial methanotroph *Methylacidiphilum fumariolicum* SolV (Hogendoorn et al., 2018). This has not been tested with "*M. buryatense*" 5GB1C, and difficulties may occur due to the medium pH and other medium components that differ from the media for these strains.

3. Conclusion

While haloalkaliphilic methanotrophic bacteria pose many more challenges than neutrophilic bacteria that grow on non-gaseous substrates, the fast growth of some strains on a natural gaseous waste stream offers many industrial advantages. In this chapter, we have detailed our efforts to accurately measure lanthanide metal interaction and uptake. These include altering the chelator used in previously published medium, and altering the concentrations and pH levels of medium components that affect lanthanum precipitation.

References

Bayer, M. E., & Bayer, M. H. (1991). Lanthanide accumulation in the periplasmic space of Escherichia coli B. *Journal of Bacteriology*, *173*(1), 141–149.

Chu, F., Beck, D., & Lidstrom, M. E. (2016). MxaY regulates the lanthanide-mediated methanol dehydrogenase switch in Methylomicrobium buryatense. *PeerJ*, *4*, e2435. https://doi.org/10.7717/peerj.2435.

Chu, F., & Lidstrom, M. E. (2016). XoxF acts as the predominant methanol dehydrogenase in the type I Methanotroph *Methylomicrobium buryatense*. *Journal of Bacteriology*, *198*(8), 1317–1325. https://doi.org/10.1128/JB.00959-15.

Cook, E. C., Featherston, E. R., Showalter, S. A., & Cotruvo, J. A., Jr. (2019). Structural basis for rare earth element recognition by *Methylobacterium extorquens* Lanmodulin. *Biochemistry*, *58*(2), 120–125. https://doi.org/10.1021/acs.biochem.8b01019.

Cotruvo, J. A., Jr., Featherston, E. R., Mattocks, J. A., Ho, J. V., & Laremore, T. N. (2018). Lanmodulin: A highly selective lanthanide-binding protein from a lanthanide-utilizing bacterium. *Journal of the American Chemical Society*, *140*(44), 15056–15061. https://doi.org/10.1021/jacs.8b09842.

Delaney, N. F., Kaczmarek, M. E., Ward, L. M., Swanson, P. K., Lee, M. C., & Marx, C. J. (2013). Development of an optimized medium, strain and high-throughput culturing methods for Methylobacterium extorquens. *PLoS One*, *8*(4), e62957. https://doi.org/10.1371/journal.pone.0062957.

Farhan Ul Haque, M., Kalidass, B., Bandow, N., Turpin, E. A., DiSpirito, A. A., & Semrau, J. D. (2015). Cerium regulates expression of alternative methanol dehydrogenases in Methylosinus trichosporium OB3b. *Applied and Environmental Microbiology*, *81*(21), 7546–7552. https://doi.org/10.1128/AEM.02542-15.

Fitriyanto, N. A., Fushimi, M., Matsunaga, M., Pertiwiningrum, A., Iwama, T., & Kawai, K. (2011). Molecular structure and gene analysis of Ce^{3+}-induced methanol dehydrogenase of *Bradyrhizobium* sp. MAFF211645. *Journal of Bioscience and Bioengineering*, *111*(6), 613–617. https://doi.org/10.1016/j.jbiosc.2011.01.015.

Fitriyanto, N. A., Nakamura, M., Muto, S., Kato, K., Yabe, T., Iwama, T., et al. (2011). Ce^3+-induced exopolysaccharide production by Bradyrhizobium sp. MAFF211645. *Journal of Bioscience and Bioengineering*, *111*(2), 146–152. https://doi.org/10.1016/j.jbiosc.2010.09.008.

Good, N. M., Vu, H. N., Suriano, C. J., Subuyuj, G. A., Skovran, E., & Martinez-Gomez, C. N. (2016). Pyrroloquinoline quinone-containing ethanol dehydrogenase in Methylobacterium extorquens AM1 extends lanthanide-dependent metabolism to multi-carbon substrates. *Journal of Bacteriology*, *198*(22), 3109–3118. https://doi.org/10.1128/JB.00478-16.

Groom, J. D., Ford, S. M., Pesesky, M. W., & Lidstrom, M. E. (2019). A mutagenic screen identifies a TonB-dependent receptor required for the lanthanide metal switch in the type I Methanotroph "Methylotuvimicrobium buryatense" 5GB1C. *Journal of Bacteriology*, *201*(15), 1–17. https://doi.org/10.1128/JB.00120-19.

Gu, W., Haque, M., DiSpirito, A. A., & Semrau, J. D. (2016). Uptake and effect of rare earth elements on gene expression in *Methylosinus trichosporium* OB3b. *FEMS Microbiology Letters*, *363*(13), 1–6. https://doi.org/10.1093/femsle/fnw129.

Gu, W., & Semrau, J. D. (2017). Copper and cerium-regulated gene expression in *Methylosinus trichosporium* OB3b. *Applied Microbiology and Biotechnology*, *101*(23–24), 8499–8516. https://doi.org/10.1007/s00253-017-8572-2.

Hogendoorn, C., Roszczenko-Jasińska, P., Martinez-Gomez, C. N., de Graaff, J., Grassl, P., Pol, A., et al. (2018). Facile Arsenazo III-based assay for monitoring rare earth element depletion from cultivation media for methanotrophic and methylotrophic bacteria. *Applied and Environmental Microbiology*, *84*(8), 17. https://doi.org/10.1128/AEM.02887-17.

Huang, J., Yu, Z., Groom, J., Cheng, J. F., Tarver, A., Yoshikuni, Y., et al. (2019). Rare earth element alcohol dehydrogenases widely occur among globally distributed, numerically abundant and environmentally important microbes. *The ISME Journal*, *13*(8), 2005–2017. https://doi.org/10.1038/s41396-019-0414-z.

Kaluzhnaya, M., Khmelenina, V., Eshinimaev, B., Suzina, N., Nikitin, D., Solonin, A., et al. (2001). Taxonomic characterization of new alkaliphilic and alkalitolerant methanotrophs from soda lakes of the Southeastern Transbaikal region and description of *Methylomicrobium buryatense* sp. nov. *Systematic and Applied Microbiology*, *24*(2), 166–176. https://doi.org/10.1078/0723-2020-00028.

Kalyuzhnaya, M. G., Puri, A. W., & Lidstrom, M. E. (2015). Metabolic engineering in methanotrophic bacteria. *Metabolic Engineering*, *29*, 142–152. https://doi.org/10.1016/j.ymben.2015.03.010.

Kawai, K., Wang, G., Okamoto, S., & Ochi, K. (2007). The rare earth, scandium, causes antibiotic overproduction in Streptomyces spp. *FEMS Microbiology Letters*, *274*(2), 311–315. https://doi.org/10.1111/j.1574-6968.2007.00846.x.

Keltjens, J. T., Pol, A., Reimann, J., & Op den Camp, H. J. (2014). PQQ-dependent methanol dehydrogenases: Rare-earth elements make a difference. *Applied Microbiology and Biotechnology*, *98*(14), 6163–6183. https://doi.org/10.1007/s00253-014-5766-8.

Khmelenina, V. N., Beck, D. A. C., Munk, C., Davenport, K., Daligault, H., Erkkila, T., et al. (2013). Draft genome sequence of *Methylomicrobium buryatense* strain 5G, a haloalkaline-tolerant methanotrophic bacterium. *Genome Announcements*, *1*(4), e00053-00013. https://doi.org/10.1128/genomeA.00053-13.

Mattocks, J. A., Ho, J. V., & Cotruvo, J. A., Jr. (2019). A selective, protein-based fluorescent sensor with picomolar affinity for rare earth elements. *Journal of the American Chemical Society*, *141*(7), 2857–2861. https://doi.org/10.1021/jacs.8b12155.

Nakagawa, T., Mitsui, R., Tani, A., Sasa, K., Tashiro, S., Iwama, T., et al. (2012). A catalytic role of XoxF1 as La3+−dependent methanol dehydrogenase in Methylobacterium extorquens strain AM1. *PLoS One*, 7(11), e50480. https://doi.org/10.1371/journal.pone.0050480.

Ochsner, A. M., Hemmerle, L., Vonderach, T., Nüssli, R., Bortfeld-Miller, M., Hattendorf, B., et al. (2019). Use of rare-earth elements in the phyllosphere colonizer *Methylobacterium extorquens PA*1. *Molecular Microbiology*, *111*(5), 1152–1166.

Ozaki, T., Suzuki, Y., Nankawa, T., Yoshida, T., Ohnuki, T., Kimura, T., et al. (2006). Interactions of rare earth elements with bacteria and organic ligands. *Journal of Alloys and Compounds*, *408–412*, 1334–1338.

Pol, A., Barends, T. R., Dietl, A., Khadem, A. F., Eygensteyn, J., Jetten, M. S., et al. (2014). Rare earth metals are essential for methanotrophic life in volcanic mudpots. *Environmental Microbiology*, *16*(1), 255–264. https://doi.org/10.1111/1462-2920.12249.

Puri, A. W., Owen, S., Chu, F., Chavkin, T., Beck, D. A. C., Kalyuzhnaya, M. G., et al. (2015). Genetic tools for the industrially promising Methanotroph *Methylomicrobium buryatense*. *Applied and Environmental Microbiology*, *81*(5), 1775–1781. https://doi.org/10.1128/AEM.03795-14.

Roszczenko-Jasińska, P., Vu, H. N., Subuyuj, G. A., Crisostomo, R. V., Cai, J., Lien, N. F., et al. (2020). Gene products and processes contributing to lanthanide homeostasis and methanol metabolism in *Methylorubrum extorquens* AM1. *Scientific Reports, 10*(1), 12663. https://doi.org/10.1038/s41598-020-69401-4.

Skovran, E., Palmer, A. D., Rountree, A. M., Good, N. M., & Lidstrom, M. E. (2011). XoxF is required for expression of methanol dehydrogenase in *Methylobacterium extorquens* AM1. *Journal of Bacteriology, 193*(2), 6032–6038. https://doi.org/10.1128/JB.05367-11.

Skovran, E., Raghuraman, C., & Martinez-Gomez, N. C. (2019). Lanthanides in methylotrophy. *Current Issues in Molecular Biology, 33*, 101–116. https://doi.org/10.21775/cimb.033.101.

Takahashi, Y., Châtellier, X., Hattori, K. H., Kato, K., & Fortin, D. (2005). Adsorption of rare earth elements onto bacterial cell walls and its implication for REE sorption onto natural microbial mats. *Chemical Geology, 219*(1–4), 53–67. https://doi.org/10.1016/j.chemgeo.2005.02.009.

Takahashi, Y., Yamamoto, M., Yamamoto, Y., & Tanaka, K. (2010). EXAFS study on the cause of enrichment of heavy REEs on bacterial cell surfaces. *Geochimica et Cosmochimica Acta, 74*(19), 5443–5462. https://doi.org/10.1016/j.gca.2010.07.001.

Vu, H. N., Subuyuj, G. A., Vijayakumar, S., Good, N. M., Martinez-Gomez, C. N., & Skovran, E. (2016). Lanthanide-dependent regulation of methanol oxidation systems in *Methylobacterium extorquens* AM1 and their contribution to methanol growth. *Journal of Bacteriology, 198*(8), 1250–1259. https://doi.org/10.1128/JB.00937-15.

Wehrmann, M., Berthelot, C., Billard, P., & Klebensberger, J. (2018). The PedS2/PedR2 two-component system is crucial for the rare earth element switch in Pseudomonas putida KT2440. *mSphere, 3*(4), 1–12. https://doi.org/10.1128/mSphere.00376-18.

Wehrmann, M., Berthelot, C., Billard, P., & Klebensberger, J. (2019). Rare earth element (REE)-dependent growth of *Pseudomonas putida* KT2440 relies on the ABC-transporter PedA1A2BC and is influenced by iron availability. *Frontiers in Microbiology, 10*, 2494. https://doi.org/10.3389/fmicb.2019.02494.

Wehrmann, M., Billard, P., Martin-Meriadec, A., Zegeye, A., & Klebensberger, J. (2017). Functional role of lanthanides in enzymatic activity and transcriptional regulation of pyrroloquinoline quinone-dependent alcohol dehydrogenases in *Pseudomonas putida* KT2440. *MBio, 8*(3), e00570-00517. https://doi.org/10.1128/mBio.00570-17.

Yan, X., Chu, F., Puri, A. W., Fu, Y., & Lidstrom, M. E. (2016). Electroporation-based genetic manipulation in type I methanotrophs. *Applied and Environmental Microbiology, 82*(7), 2062–2069. https://doi.org/10.1128/AEM.03724-15.

FSC
www.fsc.org
MIX
Papier | Fördert
gute Waldnutzung
FSC® C083411

Kolibri 360 GmbH
Ferdinand-Jühlke-Straße 7
99095 Erfurt, Deutschland
produktsicherheit@kolibri360.de